Seventh International Workshop: Group Analysis of Differential Equations and Integrable Systems 2014

(GADEIS VII)

Journal of Physics: Conference Series Volume 621

Larnaca, Cyprus
15 – 19 June 2014

ISBN: 978-1-5108-0756-3
ISSN: 1742-6588

Printed from e-media with permission by:

Curran Associates, Inc.
57 Morehouse Lane
Red Hook, NY 12571

Some format issues inherent in the e-media version may also appear in this print version.

Printed by Curran Associates, Inc. (2015)

For permission requests, please contact the Institute of Physics
at the address below.

Institute of Physics
Dirac House, Temple Back
Bristol BS1 6BE UK

Phone: 44 1 17 929 7481
Fax: 44 1 17 920 0979

techtracking@iop.org

Additional copies of this publication are available from:

Curran Associates, Inc.
57 Morehouse Lane
Red Hook, NY 12571 USA
Phone: 845-758-0400
Fax: 845-758-2634
Email: curran@proceedings.com
Web: www.proceedings.com

TABLE OF CONTENTS

Journal of Physics: Conference Series **621** (2015) 011001 doi:10.1088/1742-6596/621/1/011001

PREFACE

The Seventh International Workshop "Group Analysis of Differential Equations and Integrable Systems" (GADEIS-VII) took place at Flamingo Beach Hotel, Larnaca, Cyprus during the period June 15–19, 2014. Fifty nine scientists from nineteen countries participated in the Workshop, and forty one lectures were presented. The Workshop topics ranged from theoretical developments of group analysis of differential equations, hypersymplectic structures, theory of Lie algebras, integrability and superintegrability to their applications in various fields.

The Series of Workshops is a joint initiative by the Department of Mathematics and Statistics, University of Cyprus, and the Department of Applied Research of the Institute of Mathematics, National Academy of Sciences, Ukraine. The Workshops evolved from close collaboration among Cypriot and Ukrainian scientists. The first three meetings were held at the Athalassa campus of the University of Cyprus (October 27, 2005, September 25–28, 2006, and October 4–5, 2007). The fourth (October 26–30, 2008), the fifth (June 6–10, 2010) and the sixth (June 17–21, 2012) meetings were held at the coastal resort of Protaras.

We would like to thank all the authors who have published papers in the Proceedings. All of the papers have been reviewed by at least two independent referees. We express our appreciation of the care taken by the referees. Their constructive suggestions have improved most of the papers. The importance of peer review in the maintenance of high standards of scientific research can never be overstated.

Olena Vaneeva, Christodoulos Sophocleous,
Roman Popovych, Vyacheslav Boyko, Pantelis Damianou

Journal of Physics: Conference Series **621** (2015) 011002 doi:10.1088/1742-6596/621/1/011002

Peer review statement

All papers published in this volume of *Journal of Physics: Conference Series* have been peer reviewed through processes administered by the proceedings Editors. Reviews were conducted by expert referees to the professional and scientific standards expected of a proceedings journal published by IOP Publishing.

Algebraic method for finding equivalence groups

Alexander Bihlo[1,2], **Elsa Dos Santos Cardoso-Bihlo**[3] **and Roman O Popovych**[3,4]

[1] Department of Mathematics and Statistics, Memorial University of Newfoundland, St. John's (NL) A1C 5S7, Canada

[2] Department of Mathematics, University of British Columbia, Vancouver (BC) V6T 1Z2, Canada

[3] Wolfgang Pauli Institut, Universität Wien, Oskar-Morgenstern-Platz 1, 1090 Wien, Austria

[4] Institute of Mathematics of NAS of Ukraine, 3 Tereshchenkivska Str., 01601 Kyiv, Ukraine

E-mail: `abihlo@mun.ca`, `elsa.cardoso@univie.ac.at`, `rop@imath.kiev.ua`

Abstract. The algebraic method for computing the complete point symmetry group of a system of differential equations is extended to finding the complete equivalence group of a class of such systems. The extended method uses the knowledge of the corresponding equivalence algebra. Two versions of the method are presented, where the first involves the automorphism group of this algebra and the second is based on a list of its megaideals. We illustrate the megaideal-based version of the method with the computation of the complete equivalence group of a class of nonlinear wave equations with applications in nonlinear elasticity.

1. Introduction

Classes of (systems of) differential equations are (systems of) differential equations that contain parameters, which may be constants or functions, and are collectively referred to as arbitrary elements. Most systems of differential equations of science and engineering involve such arbitrary elements and hence are classes rather than single systems. The wide occurrence of classes of differential equations is the reason why their study has a prominent place in group analysis of differential equations [1–4]. The classification of Lie symmetry properties of systems from a class depending on the values of the arbitrary elements is called the group classification problem for this class.

Equivalence transformations of a class of differential equations are point (resp. contact) transformations in the underlying space of independent and dependent variables, involved derivatives and arbitrary elements that map each system from the class to another system from the same class [1, 3]. Such transformations play a central role in group classification of differential equations since they can be used to select the simplest representative among similar systems. Continuous equivalence transformations form a connected Lie group, which can thus be found using the infinitesimal method by computing the corresponding Lie algebra, the so-called equivalence algebra, i.e., the set of vector fields which generate one-parameter subgroups of equivalence transformations. Due to its algorithmic nature, this is a standard way for computing equivalence transformations in papers on group classification of differential equations [1, 5].

The main shortcoming of the infinitesimal method is that it misses discrete equivalence transformations. That is, not the complete equivalence group is obtained but only its continuous component. The situation is completely analogous to the construction of the maximal Lie invariance group for a single system of differential equations, which is carried out within the framework of the infinitesimal method; but in order to present the corresponding complete group of point symmetries, also discrete symmetry transformations have to be computed.

Finding complete point symmetry groups (and, analogously, complete equivalence groups) is a nontrivial task if the *direct method* is applied [3, 6–9]. In the direct method one computes point symmetries without the detour of infinitesimal generators. The direct method hence also allows one to find discrete point symmetries and thus to construct the complete point symmetry group.[1] This method does not use the linearization of the determining equations, which enables the algorithmic computation of Lie symmetries, and thus usually one has to solve a system of nonlinear partial differential equations; which is a challenging endeavor. For systems with a nontrivial continuous component of the complete point symmetry group, a refinement for the pure direct method can be used, the so-called *algebraic method*. The algebraic method is simpler in that it allows one to restrict the principal form of point transformations before invoking the direct method. In particular, the solution of nonlinear systems of partial differential equations can often be completely avoided.

The central idea of the algebraic method for the computation of discrete symmetries of differential equations is the following: Given a system of differential equations $\mathcal{L}$, let $\mathfrak{g}_\mathcal{L}$ denote its maximal Lie invariance algebra, which is known. Push-forwards of vector fields defined on the corresponding space of independent and dependent variables by point symmetries of $\mathcal{L}$ induce automorphisms of $\mathfrak{g}_\mathcal{L}$. This property can be used in order to restrict the form of those point transformations that can be symmetries of $\mathcal{L}$. The restricted form can then be substituted into the system $\mathcal{L}$ from which point one still proceeds with the direct method. In fact, the algebraic method just simplifies the application of the direct method.

The original version of the algebraic method was proposed by Hydon [10–12]; see also [13]. It is based on the immediate use of the above property of point symmetries of $\mathcal{L}$ and requires the explicit computation of the entire automorphism group of $\mathfrak{g}_\mathcal{L}$; this is why we call it the automorphism-based version of the algebraic method or, shortly, the *automorphism-based method*. The computation of the entire automorphism group is a comprehensive or even impossible task for infinite-dimensional Lie algebras. Therefore, for a proper application of the automorphism-based method the algebra $\mathfrak{g}_\mathcal{L}$ has to be finite dimensional, which is a severe limitation of the applicability of this version of the algebraic method.

There are countless examples of systems of differential equations with infinite-dimensional maximal Lie invariance algebras, especially in hydrodynamics. This is why in [9] an essential refinement of the algebraic method was proposed. It is built around the notion of fully characteristic ideals (shortly called *megaideals*); these are vector subspaces of a Lie algebra $\mathfrak{g}$ that are invariant under any transformation from the automorphism group of $\mathfrak{g}$ [9, 14, 15], see also [16, Exercise 14.1.1]. Usually, the requirement for a linear nondegenerate mapping of $\mathfrak{g}$ to leave invariant each of the megaideals of $\mathfrak{g}$ produces enough constraints to understand the rough structure of automorphisms. Moreover, in [9, 14, 15] several propositions were proved that enable the systematic computation of new megaideals using known ones. Given a system of differential equations $\mathcal{L}$, let $\mathfrak{g}_\mathcal{L}$ be its maximal Lie invariance algebra. If a sufficient number of megaideals of the maximal Lie invariance algebra $\mathfrak{g}_\mathcal{L}$ of the system $\mathcal{L}$ is known, the condition of their invariance with respect to automorphisms of $\mathfrak{g}_\mathcal{L}$ can be used in order to derive maximal restrictions on the form of a general point symmetry transformation of $\mathcal{L}$ that are obtainable within the framework of the algebraic method. This is the essence of the megaideal-based version of the algebraic method or, shortly, the *megaideal-based method*.

[1] Discrete symmetries like alternating signs of independent and/or dependent variables can easily be found ad hoc, by empiric methods. Even if such trivial symmetries exhaust a complete set of discrete symmetries that are independent up to composing with continuous symmetries, which is often the case, the problem is to prove that this is really the case. Such a proof cannot be realized within an empirical approach.

The effectiveness of this version has been demonstrated by computing the complete point symmetry group of the barotropic vorticity equation on the beta-plane [9], the vorticity equation on the sphere [14], the quasi-geostrophic two-layer model [17] and the system of primitive equations [18]. For each of these (systems of) differential equations, the corresponding maximal Lie invariance algebra is infinite dimensional.

The same techniques as applied to compute the complete point symmetry group of a system of differential equations can also be used to compute the complete equivalence group of a class of systems of differential equations. Moreover, the megaideal-based method is even more effective here than for symmetry groups as equivalence groups are often infinite dimensional. Showing how to realize the computation of equivalence groups using the algebraic method is the purpose of the present paper. Both versions of the algebraic method are discussed.

We illustrate the megaideal-based method by the computation of the equivalence group for the class of nonlinear wave equations of the general form

$$u_{tt} = f(x, u_x)u_{xx} + g(x, u_x), \tag{1a}$$

with the nonvanishing condition

$$(f_{u_x}, g_{u_x u_x}) \neq (0, 0). \tag{1b}$$

This condition is essential as it guarantees that all equations from the class (1) are really nonlinear, cf. [4]. Physically, this class arises, in particular, in the study of nonlinear elasticity.

The study of the class (1) plays a prominent role in the field of group classification of differential equations. It was initially considered in [22], where one-dimensional symmetry extensions of the kernel algebra with respect to a fixed finite-dimensional subalgebra of the infinite-dimensional equivalence algebra of this class were found. In other words, a partial preliminary group classification problem was solved. A similar direction was taken in [23], where the partial preliminary group classification of the class (1) with respect to one-dimensional subalgebras of an infinite-dimensional subalgebra of the equivalence algebra was considered. The group classification problem for systems of the form $v_t = a(x, v)w_x$ and $w_t = b(x, v)v_x$ related to equations from the original class (1) was first investigated in [24] and then exhaustively solved using the algebraic method of group classification in [25]. A study related to the group classification problem of the class (1) was carried out in [26], resulting in the construction of second-order differential invariants of the equivalence algebra. Finally, the complete group classification problem for the class (1) was solved in [4] by the enhanced algebraic method of group classification, thus completing the over twenty years long investigation of the point symmetry properties of this class.

The further structure of the present paper is as follows. In Section 2, we recall some definitions related to classes of systems of differential equations, their equivalence algebras and their equivalence groups. Section 3 contains the necessary material explaining the algebraic method for the computation of (complete point) equivalence groups. Both versions of the method—automorphism- and megaideal-based—are introduced and presented in the form of step-by-step procedures. In Section 4, we illustrate the megaideal-based method with the computation of the equivalence group of the class (1). In the final Section 5, we summarize our results and give some thoughts about future directions of investigation.

2. The equivalence group and equivalence algebra

In this section we recall the central definition of the equivalence group and the equivalence algebra of a class of (systems of) differential equations. These are two of the fundamental concepts that play a key role in group classification of differential equations. As such, they are discussed extensively in the papers and books on this subject, see, e.g., [1–5, 27, 28]. It is natural to first rigorously define the notion of a class of differential equations.

Let $\mathcal{L}_\theta$: $L(x, u_{(p)}, \theta_{(q)}(x, u_{(p)})) = 0$ denote a system of differential equations parameterized by the tuple of arbitrary elements $\theta(x, u_{(p)}) = (\theta^1(x, u_{(p)}), \dots, \theta^k(x, u_{(p)}))$. Here and in the following, the tuple $x = (x_1, \dots, x_n)$ consists of the independent variables and $u_{(p)}$ includes both the tuple of dependent variables $u = (u^1, \dots, u^m)$ as well as all the derivatives of u with respect to x of order up to p. By $\theta_{(q)}$ we denote the partial derivatives of the arbitrary elements θ of order not exceeding q for which both x and $u_{(p)}$ act as the independent variables.

In the definition of the class (1) we explicitly included the inequality $(f_{u_x}, g_{u_x u_x}) \neq (0, 0)$, which restricts the values that the two arbitrary elements f and g of the class (1) can take. Such a set of so-called auxiliary conditions on the arbitrary elements θ is a necessary and important component of the precise definition of the class of differential equations but it is often posed in an implicit way. The procedure can be formalized as follows.

Consider the system of auxiliary differential equations $S(x, u_{(p)}, \theta_{(q')}(x, u_{(p)})) = 0$ and inequalities $\Sigma(x, u_{(p)}, \theta_{(q')}(x, u_{(p)})) \neq 0$,[2] in which again both x and $u_{(p)}$ act as independent variables. Components of the tuples S and Σ are smooth functions of x, $u_{(p)}$ and $\theta_{(q')}$. It is thus required that the tuple of arbitrary elements θ runs through the solution set, denoted by $\mathcal{S}$, of both the auxiliary equations $S = 0$ and inequalities $\Sigma \neq 0$.

The above discussion can be summarized in the following definition:

Definition 1. The set $\{\mathcal{L}_\theta \mid \theta \in \mathcal{S}\}$ denoted by $\mathcal{L}|_\mathcal{S}$ is called a *class of differential equations* defined by parameterized systems $\mathcal{L}_\theta$ and the set $\mathcal{S}$ of arbitrary elements θ.

In order to define the equivalence group of the class $\mathcal{L}|_\mathcal{S}$ it is convenient to first introduce the notion of admissible transformations and the notion of equivalence groupoid [2–4,28]. Given the arbitrary elements $\theta, \tilde\theta \in \mathcal{S}$ with the associated systems $\mathcal{L}_\theta$ and $\mathcal{L}_{\tilde\theta}$ from the class $\mathcal{L}|_\mathcal{S}$, the set $\mathrm{T}(\theta, \tilde\theta)$ of point transformations that map the system $\mathcal{L}_\theta$ to the system $\mathcal{L}_{\tilde\theta}$ is called the set of admissible (point) transformations from $\mathcal{L}_\theta$ to $\mathcal{L}_{\tilde\theta}$. In particular, the maximal point symmetry (pseudo)group G_θ of the system $\mathcal{L}_\theta$ coincides, by definition, with the set $\mathrm{T}(\theta, \theta)$.

A triplet $(\theta, \tilde\theta, \phi)$, where $\theta, \tilde\theta \in \mathcal{S}$ with $\mathrm{T}(\theta, \tilde\theta) \neq \varnothing$ and $\phi \in \mathrm{T}(\theta, \tilde\theta)$, is called an *admissible transformation* in the class $\mathcal{L}|_\mathcal{S}$. In other words, an admissible transformation is a triplet consisting of a pair of similar systems (a source and a target system) and a point transformation connecting these two systems.

The partial binary operation of composition is naturally defined for pairs of admissible transformations for which the target system of the first admissible transformation and the source equation of the second admissible transformation coincides, $(\theta, \tilde\theta, \phi) \circ (\tilde\theta, \bar\theta, \tilde\phi) = (\theta, \bar\theta, \tilde\phi \circ \phi)$. Each admissible transformation is invertible, $(\theta, \tilde\theta, \phi)^{-1} = (\tilde\theta, \theta, \phi^{-1})$, i.e., the inversion of admissible transformations is a unitary operation defined everywhere. The set of admissible transformations of the class $\mathcal{L}|_\mathcal{S}$ that is endowed with the operations of composition and taking the inverse is called the *equivalence groupoid* of this class and denoted by $\mathcal{G}^\sim = \mathcal{G}^\sim(\mathcal{L}|_\mathcal{S})$. All the groupoid axioms are obviously satisfied for $\mathcal{G}^\sim$.

For a general class of differential equations, it is not possible to relate its equivalence groupoid with a group of point transformations that act in the extended space of $(x, u_{(p)}, \theta)$, respect the contact structure on the space of $(x, u_{(p)})$ and preserve the class. However, it makes sense to single out a maximal part of the equivalence groupoid that admits such a relation; then the associated group is referred to as the equivalence group.

Definition 2. The *(usual) equivalence group*, denoted by $G^\sim = G^\sim(\mathcal{L}|_\mathcal{S})$, of the class $\mathcal{L}|_\mathcal{S}$ is the (pseudo)group of point transformations in the space of $(x, u_{(p)}, \theta)$, each element Φ of which satisfies the following properties: It is projectable to the space of $(x, u_{(p')})$ for any p' with $0 \leqslant p' \leqslant p$. The projection $\Phi|_{(x,u_{(p')})}$ is the p'th order prolongation of $\Phi|_{(x,u)}$. For any θ from $\mathcal{S}$ its image $\Phi\theta$ also belongs to $\mathcal{S}$. Finally, $\Phi|_{(x,u)} \in \mathrm{T}(\theta, \Phi\theta)$.

[2] Other kinds of inequalities ($>$, $<$, etc.) as well as collections of them are also relevant.

Here we say that a point transformation φ: $\tilde{z} = \varphi(z)$ in the space of variables $z = (z_1, \dots, z_k)$ is projectable on the space of variables $z' = (z_{i_1}, \dots, z_{i_{k'}})$ with $1 \leqslant i_1 < \dots < i_{k'} \leqslant k$ if the expressions for the transformed variables $\tilde{z}'$ depend only on z'. The projection of φ to the z'-space is denoted by $\varphi|_{z'}$, $\tilde{z}' = \varphi|_{z'}(z')$.

Elements of $G^\sim$ are called *equivalence transformations* of the class $\mathcal{L}|_\mathcal{S}$. Each equivalence transformation Φ induces a family of admissible transformations parameterized by the arbitrary elements, $\{(\theta, \Phi\theta, \Phi|_{(x,u)}) \mid \theta \in \mathcal{S}\}$. If the entire equivalence groupoid $\mathcal{G}^\sim$ is induced by the equivalence group $G^\sim$ in the above way, then transformational properties of the class $\mathcal{L}|_\mathcal{S}$ are particularly nice and the class is called *normalized* [2–4, 28]. The notion of normalized classes of differential equations serves as basis for developing the algebraic method of group classification of differential equations.

If the arbitrary elements θ do not depend on derivatives of u of order greater than p', $p' < p$, then the group $G^\sim$ can be assumed to act in the space of $(x, u_{(p')}, \theta)$.

In some cases, e.g., if the arbitrary elements θ depend on x and u only, we can neglect the condition that the transformation components for (x, u) of equivalence transformations do not involve θ, which gives the *generalized equivalence group* $G^\sim_{\rm gen} = G^\sim_{\rm gen}(\mathcal{L}|_\mathcal{S})$ of the class $\mathcal{L}|_\mathcal{S}$ [2, 3, 29]. Each element Φ of $G^\sim_{\rm gen}$ is a point transformation in the (x, u, θ)-space such that for any θ from $\mathcal{S}$ its image $\Phi\theta$ also belongs to $\mathcal{S}$, and $\Phi(\cdot, \cdot, \theta(\cdot, \cdot))|_{(x,u)} \in \mathrm{T}(\theta, \Phi\theta)$.

It was mentioned in the introduction that the computation of point transformations of systems of differential equations using the direct method usually involves the solution of nonlinear systems of differential equations. Thus, rather than studying finite equivalence transformations, it is common in the framework of group analysis of differential equations to only consider their infinitesimal counterparts. This restriction linearizes the nonlinear determining equations for finding such transformations, making their computation essentially algorithmic. Thus, to the continuous component of the equivalence group $G^\sim$ one may associate the Lie algebra $\mathfrak{g}^\sim$ of vector fields in the space of $(x, u_{(p)}, \theta)$, which for any $0 \leqslant p' \leqslant p$ are projectable to the space of $(x, u_{(p')})$ with the property that their projections to the space of $(x, u_{(p')})$ are the p'th order prolongations of their projections to the space of (x, u). Moreover, these vector fields are the generators of one-parameter groups of equivalence transformations of the class $\mathcal{L}|_\mathcal{S}$. The Lie algebra $\mathfrak{g}^\sim$ is called the *(usual) equivalence algebra* of the class $\mathcal{L}|_\mathcal{S}$.

In a similar way, by considering the generalized equivalence group of the class $\mathcal{L}|_\mathcal{S}$ instead of the usual one and by allowing the components of vector fields that correspond to x and u to also depend on θ, one can define the *generalized equivalence algebra* of the class $\mathcal{L}|_\mathcal{S}$.

The (usual or generalized) equivalence algebra $\mathfrak{g}^\sim$ of a class $\mathcal{L}|_\mathcal{S}$ can be found by invoking an infinitesimal invariance criterion completely analogous to that used for the computation of the infinitesimal generators of one-parameter point symmetry groups of a single system of differential equations [1, 5].

In order to use the infinitesimal invariance criterion properly, one has to consider a vector field Q on the extended space of independent variables x, derivatives $u_{(p)}$ and arbitrary elements θ. The components of Q corresponding to derivatives of u with respect to x should respect the contact structure of the space of independent and dependent variables. In other words, these coefficients should be obtained using the general prolongation formula [1, 30]. The vector field Q should be prolonged to the derivatives of the arbitrary elements assuming both x and $u_{(p)}$ as the independent variables; again, the general prolongation formula should be used for this purpose. The prolonged vector field is then applied to the joint system Δ of the general form $L(x, u_{(p)}, \theta_{(q)}) = 0$ of systems from the class $\mathcal{L}|_\mathcal{S}$ and of the auxiliary conditions defining the set $\mathcal{S}$ of values of the arbitrary elements θ. If necessary, differential consequences of equations from the system Δ have to be considered. The remaining computational procedure then parallels the computation of infinitesimal generators of Lie point symmetries and involves the solution of an overdetermined linear system of partial differential equations.

3. Algebraic method

As was mentioned in the introduction, any point symmetry transformation $\mathcal{T}$ of a system of differential equations $\mathcal{L}$ generates an automorphism of the maximal Lie invariance algebra of $\mathcal{L}$ via push-forwarding of vector fields in the space of system variables. This condition implies constraints for the transformation $\mathcal{T}$ which are then taken into account in further calculations using the direct method [9–14, 18]. The set of transformations found in the way described constitute the complete point symmetry group of the system $\mathcal{L}$ including both continuous and discrete point transformations.

The above algebraic method can be easily extended to the framework of equivalence transformations. A basis for this is given by the following simple theorem, which is proved in the same way as the similar assertion for point symmetries.

Theorem 1. *Let $\mathcal{L}|_\mathcal{S}$ be a class of (systems of) differential equations, $G^\sim$ and $\mathfrak{g}^\sim$ the equivalence group and the equivalence algebra of this class (of the same type, namely, either usual or generalized ones).*[3] *Any transformation $\mathcal{T}$ from $G^\sim$ induces an automorphism of $\mathfrak{g}^\sim$ via push-forwarding of vector fields in the relevant space of independent variables, derivatives of unknown functions and arbitrary elements of the class.*

Proof. Consider an arbitrary vector field $Q \in \mathfrak{g}^\sim$. The local one-parameter transformation group $G_Q = \{\exp(\varepsilon Q)\}$ associated with Q is contained in $G^\sim$. Then the one-parameter transformation group $\tilde G_Q = \{\mathcal{T}\exp(\varepsilon Q)\mathcal{T}^{-1}\}$, which is similar to G_Q with respect to $\mathcal{T}$ and whose generator is $\mathcal{T}_*Q$, is also contained in $G^\sim$. This means that the vector field $\mathcal{T}_*Q$ belongs to $\mathfrak{g}^\sim$. An arbitrary push-forward respects the Lie bracket of vector fields, $[\mathcal{T}_*Q, \mathcal{T}_*Q'] = \mathcal{T}_*[Q, Q']$ for any $Q, Q' \in \mathfrak{g}^\sim$. Therefore, $\mathcal{T}_*$ is an automorphisms of $\mathfrak{g}^\sim$. □

In other words, for each element $\mathcal{T}$ of $G^\sim$ there exists an automorphism $\mathfrak{T}$ of $\mathfrak{g}^\sim$ such that the condition $\mathcal{T}_*Q = \mathfrak{T}Q$ is satisfied for any Q from $\mathfrak{g}^\sim$.

It should be noted that while the correspondence $\mathcal{T} \to \mathcal{T}_*$ defines a representation of $G^\sim$ in $\mathrm{Aut}(\mathfrak{g}^\sim)$, this representation is often unfaithful. This is always the case if the equivalence group $G^\sim$ (resp. the equivalence algebra $\mathfrak{g}^\sim$) has a nontrivial center. Moreover, the representation image $G^\sim_*$ is a subgroup of the automorphism group $\mathrm{Aut}(\mathfrak{g}^\sim)$ that may be essentially smaller than the entire automorphism group $\mathrm{Aut}(\mathfrak{g}^\sim)$. In other words, there may exist many automorphisms of $\mathfrak{g}^\sim$ that are not induced by elements of $G^\sim$.

The discussion on continuous point symmetries of a single system of differential equations $\mathcal{L}$ given in [14] can be adopted to equivalence transformations as well. In particular, continuous equivalence transformations of the class $\mathcal{L}|_\mathcal{S}$ can be composed and thus generate a connected normal subgroup, denoted by $U^\sim$, of the equivalence group $G^\sim$. The transformations from $U^\sim$ induce mappings on the equivalence algebra $\mathfrak{g}^\sim$ that are internal automorphisms of $\mathfrak{g}^\sim$ thus generating the normal subgroup $\mathrm{Int}(\mathfrak{g}^\sim)$ of the automorphism group $\mathrm{Aut}(\mathfrak{g}^\sim)$. So, $\mathrm{Int}(\mathfrak{g}^\sim) = U^\sim_*$. The representatives of the elements of the factor group $G^\sim/U^\sim$ in $G^\sim$ will be interpreted as *discrete equivalence transformations* of the class $\mathcal{L}|_\mathcal{S}$. It is clear from the construction that it makes sense to consider only discrete equivalence transformations that are independent up to composing them with continuous equivalence transformations of the class $\mathcal{L}|_\mathcal{S}$. When $G^\sim$ is infinite dimensional, the factorization is usually difficult but it is sometimes still possible to realize it and thus to identify the discrete equivalence transformations. An example will be given in Section 4.

[3] The further consideration does not depend on what kind of equivalence groups and equivalence algebras (usual or generalized) are involved. For consistency, we should just handle the usual (resp. generalized) equivalence algebra when looking for the usual (resp. generalized) equivalence group and impose relevant a priori restrictions on the admitted structure of equivalence transformations.

Remark 1. Classes of systems of differential equations with a single dependent variable (i.e., with $m = 1$) may possess contact equivalence transformations that are not prolongations of point transformations. In this case the entire theory can be extended to contact equivalence transformations.

3.1. Version based on automorphisms

Theorem 1 allows us to extend the automorphism-based method for finding point symmetry groups of systems of differential equations [10–12] to equivalence groups of classes of such systems. The extension properly works if the corresponding equivalence algebra $\mathfrak{g}^{\sim}$ is of finite nonzero (and, moreover, low) dimension since the knowledge of the entire group $\mathrm{Aut}(\mathfrak{g}^{\sim})$ is needed. A new feature is that one should consider *appropriate* point transformations and vector fields in the relevant extended space of independent variables, derivatives of unknown functions and arbitrary elements, i.e., of $(x, u_{(p)}, \theta)$. If the arbitrary elements θ do not depend on derivatives of u of order greater than p', $p' < p$, then these appropriate objects can be restricted to the space of $(x, u_{(p')}, \theta)$. In the case $p' > 0$ they should be consistent with the contact structure of the space of $(x, u_{(p')})$. In the course of computing usual equivalence groups, they should also be projectable to the space of the system variables (x, u).

We present the extension of the automorphism-based method in the form of a step-by-step procedure. In what follows the indices i, j and k run from 1 to $n = \dim \mathfrak{g}^{\sim}$, and we assume summation over repeated indices.

(i) Let $\mathcal{L}|_{\mathcal{S}}$ be a class of (systems of) differential equations, find the equivalence algebra $\mathfrak{g}^{\sim}$ of this class by the infinitesimal invariance criterion. Suppose that $n = \dim g^{\sim} < \infty$.
(ii) Fix a basis $\{Q_1, \dots, Q_n\}$ of $\mathfrak{g}^{\sim}$, and compute the structure constants c_{ij}^k of $\mathfrak{g}^{\sim}$ in this basis, $[Q_i, Q_j] = c_{ij}^k Q_k$.
(iii) The general form (a_j^i) of automorphism matrices in the basis $\{Q_1, \dots, Q_n\}$ is obtained via solving the system of algebraic equations $c_{i'j'}^{k'} a_i^{i'} a_j^{j'} = c_{ij}^k a_k^{k'}$.[4]
(iv) Take the general form of an appropriate point transformation $\mathcal{T}$ in the space of x, $u_{(p)}$ and θ and push-forward the vector fields $Q_1, \dots, Q_n$. Then we set $\mathcal{T}_* Q_i = a_i^j Q_j$, $i = 1, \dots, n$, where (a_i^j) is the general form of automorphism matrices. Consequently equating the corresponding vector-field components of right- and left-hand sides of these equations produces a system of differential equations for components of the transformation $\mathcal{T}$.[5]
(v) Integrate the system and use the obtained intermediate form of $\mathcal{T}$ within the framework of the direct method in order to complete the system of constraints for $\mathcal{T}$ and to produce the final form for $\mathcal{T}$.

Recall once more that the difference in the application of the above procedure to the computation of the usual (resp. generalized) equivalence group is that $\mathfrak{g}^{\sim}$ is the usual (resp. generalized) equivalence algebra; additionally, for the usual equivalence group one restricts oneself to transformations that are projectable to the space of x and u while for the generalized equivalence group one takes $\mathcal{T}$ to be a general point transformation in the space of (x, u, θ) without any additional a priori restrictions.

[4] Automorphism groups have been computed for many finite-dimensional Lie algebras, in particular for all semi-simple Lie algebras [19] and Lie algebras of dimension not greater than six, see [13,15,20,21] and references therein. The automorphism group of a decomposable Lie algebra is easily constructed from the automorphism groups of the decomposition components [20]. Finding the automorphism group of a Lie algebra $\mathfrak{g}$ can be simplified if the chosen basis of $\mathfrak{g}$ is consistent with a known megaideal hierarchy of $\mathfrak{g}$.

[5] The system derived involves parameters of the automorphism group $\mathrm{Aut}(\mathfrak{g}^{\sim})$. Therefore, in the course of integrating the system one in fact needs to solve a compatibility problem in order to find values of the parameters for which the system is consistent with respect to components of $\mathcal{T}$. If the group $G_*^{\sim}$ does not coincide with $\mathrm{Aut}(\mathfrak{g}^{\sim})$, which is often the case, then there exist values of the parameters for which the system is inconsistent.

Journal of Physics: Conference Series **621** (2015) 012001 doi:10.1088/1742-6596/621/1/012001

Remark 2. The procedure described above can be modified by factoring out the group $U^{\sim}$ of continuous equivalence transformations of the class $\mathcal{L}|_{\mathcal{S}}$, which can easily be computed once the equivalence algebra $\mathfrak{g}^{\sim}$ is known. As $U^{\sim}_{*} = \mathrm{Int}(\mathfrak{g}^{\sim})$, in order to do the modification, instead of considering the general form of automorphism matrices in Step 3, one can factorize general automorphisms by inner automorphisms. More precisely, suppose that for each equivalence class of automorphisms with respect to $\mathrm{Int}(\mathfrak{g}^{\sim})$ one can take a representative in such a way that these representatives constitute a matrix group. Then, this group is isomorphic to the factor group $\mathrm{Aut}(\mathfrak{g}^{\sim})/\mathrm{Int}(\mathfrak{g}^{\sim})$. In general, the form of representatives is more restrictive than the general form of automorphism matrices, which may simplify the equations for $\mathcal{T}$ derived in Step 3 of the above procedure. The realization of the modified procedure results in the discrete equivalence transformations of the class $\mathcal{L}|_{\mathcal{S}}$ composed with elements of the center of $U^{\sim}$.

For an equivalence algebra of large finite dimension or, especially, for an infinite-dimensional equivalence algebra the computation of its automorphism groups may be a difficult problem, which also needs additional efforts for its rigorous formulation in the case of infinite dimension. Moreover, a significant part of automorphisms may not be realized via push-forwarding of vector fields by appropriate point transformations in the relevant space of independent and dependent variables, involved derivatives and arbitrary elements. In such a case, the exhaustive description of automorphisms is not too principal as then the equivalence transformations induce only a (small) proper subgroup of the automorphism group.

3.2. Version based on megaideals

It has been pointed out in Section 3.1 that for the automorphism-based method, it is crucial that the equivalence algebra $\mathfrak{g}^{\sim}$ is low dimensional. If this is not the case, the computation of the automorphism group $\mathrm{Aut}(\mathfrak{g}^{\sim})$ becomes an intricate problem. It is a quite common situation for a class of differential equations to be studied that its equivalence algebra $\mathfrak{g}^{\sim}$ is infinite dimensional and thus a more appropriate version of the algebraic method is needed. Mathematically, the main problem with the version reviewed in Section 3.1 is that in computing the automorphism group $\mathrm{Aut}(\mathfrak{g}^{\sim})$ one also has to find the elements $\mathrm{Aut}(\mathfrak{g}^{\sim}) \setminus G^{\sim}_{*}$, which cannot be used for the computation of $G^{\sim}_{*}$.

This deficiency is remedied by working with megaideals of $\mathfrak{g}^{\sim}$ rather than with $\mathrm{Aut}(\mathfrak{g}^{\sim})$ itself [9, 14, 18]. That is, we employ the fact that $G^{\sim}_{*}\mathfrak{i} \subseteq \mathfrak{i}$ for any megaideal $\mathfrak{i}$ of $\mathfrak{g}^{\sim}$, rather than invoking the condition $G^{\sim}_{*} \subseteq \mathrm{Aut}(\mathfrak{g}^{\sim})$.[6]

We recall that a *fully characteristic ideal* (or, shortly, *megaideal*) $\mathfrak{i}$ of a Lie algebra $\mathfrak{g}$ is a vector subspace of $\mathfrak{g}$ that is invariant under any transformation from the automorphism group $\mathrm{Aut}(\mathfrak{g})$ of $\mathfrak{g}$ [9,15], cf. [16, Exercise 14.1.1]. We thus have that $\mathfrak{T}\mathfrak{i} = \mathfrak{i}$ for each megaideal $\mathfrak{i}$ of $\mathfrak{g}$, whenever $\mathfrak{T}$ is a transformation from $\mathrm{Aut}(\mathfrak{g})$.[7] Megaideals of $\mathfrak{g}$ are ideals and, moreover, characteristic ideals of $\mathfrak{g}$. In the present context, the Lie algebra in this definition is the equivalence algebra $\mathfrak{g}^{\sim}$ of the class of differential equations $\mathcal{L}|_{\mathcal{S}}$.

In order to make the version of the megaideal-based method most effective, it is essential to be able to construct wide sets of megaideals. The more megaideals are known, the better restrictions on the form of admitted automorphisms can be derived. While it might be a complicated problem to obtain a complete list of megaideals of $\mathfrak{g}^{\sim}$, for the practical application of the megaideal-based method the more tractable problem of finding a set of megaideals that gives maximal restrictions on point transformations should be tackled. Thus, it is for example not essential to

[6] Although the stronger condition $G^{\sim}_{*}\mathfrak{i} = \mathfrak{i}$ is always satisfied, the inverse inclusion $G^{\sim}_{*}\mathfrak{i} \supseteq \mathfrak{i}$ is trivial in view of the presence of the identical automorphism in $G^{\sim}_{*}$, gives no constraints on elements of $G^{\sim}$ and, therefore, can be neglected.

[7] The invariance condition $\mathfrak{T}^{-1}\mathfrak{i} \subseteq \mathfrak{i}$ of $\mathfrak{i}$ with respect to $\mathfrak{T}^{-1}$ implies $\mathfrak{i} \subseteq \mathfrak{T}\mathfrak{i}$, which gives, together with the condition $\mathfrak{T}\mathfrak{i} \subseteq \mathfrak{i}$, the equality $\mathfrak{T}\mathfrak{i} = \mathfrak{i}$.

consider megaideals that are sums of other megaideals, since they give weaker constraints than their individual summands jointly.

In [9, 14, 15, 18] it was shown how new megaideals can be computed from known ones. For the sake of completeness of the present paper, we collect the statements on megaideals from these papers. The central observation for practical computations is that many megaideals of a Lie algebra $\mathfrak{g}$ can be constructed without prior knowledge of the automorphism group $\mathrm{Aut}(\mathfrak{g})$ using the following obvious assertions:

Proposition 1. *Let $\mathfrak{i}_1$ and $\mathfrak{i}_2$ be megaideals of a Lie algebra $\mathfrak{g}$. Then we have:*

(i) The improper subalgebras of $\mathfrak{g}$ (i.e., the zero subspace and $\mathfrak{g}$ itself) are megaideals of $\mathfrak{g}$.

(ii) The sum $\mathfrak{i}_1+\mathfrak{i}_2$, the intersection $\mathfrak{i}_1\cap\mathfrak{i}_2$ and the Lie product $[\mathfrak{i}_1, \mathfrak{i}_2]$ of megaideals are megaideals.

(iii) If $\mathfrak{i}_2$ is a megaideal of $\mathfrak{i}_1$ and $\mathfrak{i}_1$ is a megaideal of $\mathfrak{g}$ then $\mathfrak{i}_2$ is a megaideal of $\mathfrak{g}$. Thus, megaideals of megaideals are again megaideals.

(iv) The centralizer (resp. the normalizer) of a megaideal is a megaideal.

(v) All elements of the derived, upper and lower central series of a Lie algebra are its megaideals. It thus follows that the center and the derivative of $\mathfrak{g}$ are megaideals.

(vi) The radical $\mathfrak{r}$ and nil-radical $\mathfrak{n}$ (i.e., the maximal solvable and nilpotent ideals, respectively) of $\mathfrak{g}$ as well as different Lie products, sums and intersections involving $\mathfrak{g}$, $\mathfrak{r}$ and $\mathfrak{n}$ ($[\mathfrak{g}, \mathfrak{r}]$, $[\mathfrak{r}, \mathfrak{r}]$, $[\mathfrak{g}, \mathfrak{n}]$, $[\mathfrak{r}, \mathfrak{n}]$, $[\mathfrak{n}, \mathfrak{n}]$, etc.) are megaideals of $\mathfrak{g}$.

In order to find more megaideals without computing automorphisms, we also apply an assertion, which in general has no clear interpretation in terms of distinguished object related to the structure of the corresponding Lie algebra. It was first proved in [14], but for the sake of reference we repeat the proof here.

Proposition 2. *If $\mathfrak{i}_0$, $\mathfrak{i}_1$ and $\mathfrak{i}_2$ are megaideals of $\mathfrak{g}$ then the set $\mathfrak{s}$ of elements from $\mathfrak{i}_0$ whose commutators with arbitrary elements from $\mathfrak{i}_1$ belong to $\mathfrak{i}_2$ is also a megaideal of $\mathfrak{g}$.*

Proof. It is clear that the set $\mathfrak{s}$ is a linear subspace of $\mathfrak{g}$. Consider an arbitrary element $z_0 \in \mathfrak{s}$. Hence $z_0 \in \mathfrak{i}_0$ and $[z_0, z_1] \in \mathfrak{i}_2$ for any $z_1 \in \mathfrak{i}_1$. Then for any $\mathfrak{T} \in \mathrm{Aut}(\mathfrak{g})$ and any $z_1 \in \mathfrak{i}_1$ we have $\mathfrak{T}z_0 \in \mathfrak{i}_0$ and $[\mathfrak{T}z_0, z_1] = [\mathfrak{T}z_0, \mathfrak{T}\mathfrak{T}^{-1}z_1] = \mathfrak{T}[z_0, \mathfrak{T}^{-1}z_1] \in \mathfrak{i}_2$ as $\mathfrak{T}^{-1}z_1 \in \mathfrak{i}_1$ and thus $[z_0, \mathfrak{T}^{-1}z_1] \in \mathfrak{i}_2$. This means that $\mathfrak{T}z_0 \in \mathfrak{s}$. Therefore, $\mathfrak{s}$ is a megaideal of $\mathfrak{g}$. □

Assertion (iv) of Proposition 1 is a particular case of Proposition 2, where $\mathfrak{i}_0 = \mathfrak{g}$ and $\mathfrak{i}_2 = \{0\}$ for the centralizer of $\mathfrak{i}_1$, and $\mathfrak{i}_0 = \mathfrak{g}$ and $\mathfrak{i}_2 = \mathfrak{i}_1$ for the normalizer of $\mathfrak{i}_1$.

Remark 3. If $\mathfrak{m}$ is a finite-dimensional megaideal and its automorphism group $\mathrm{Aut}(\mathfrak{m})$ is already known, all megaideals of $\mathfrak{m}$ can be found by direct computation according to the definition of megaideals as subspaces of $\mathfrak{m}$ invariant with respect to $\mathrm{Aut}(\mathfrak{m})$. In the course of calculating the automorphisms we can use knowledge about structural megaideals of $\mathfrak{m}$ such as the center, the radical, the nilradical, elements of the derived, lower central and upper central series of $\mathfrak{m}$ and different megaideals related to structural megaideals via certain operations.

Theorem 1 implies that any transformation $\mathcal{T}$ from $G^{\sim}$ satisfies the condition $\mathcal{T}_*\mathfrak{i} = \mathfrak{i}$ for each megaideal $\mathfrak{i}$ of $\mathfrak{g}^{\sim}$. In the course of the derivation of constraints for $\mathcal{T}$, this condition is interpreted in the following way: Let the megaideal $\mathfrak{i}$ consist of the vector fields Q_γ, where γ runs through a parameter set $\Gamma_\mathfrak{i}$, and be spanned by the vector fields $Q_{\gamma'}$ parameterized by γ' running through a subset $\Gamma'_\mathfrak{i}$ of $\Gamma_\mathfrak{i}$. The condition $\mathcal{T}_*\mathfrak{i} = \mathfrak{i}$ implies that for any $\gamma' \in \Gamma'_\mathfrak{i}$, there exists a $\gamma \in \Gamma_\mathfrak{i}$ such that $\mathcal{T}_*Q_{\gamma'} = Q_\gamma$.

Similarly to the automorphism-based method, the megaideal-based counterpart can also be split in a few algorithmic steps. Some steps are the same or almost the same as for the automorphism-based method. Even so, for the convenience of further application we describe each step in its entirety. We also re-interpret the condition for megaideal invariance in a manner more convenient for practical use.

(i) Given a class of (systems of) differential equations $\mathcal{L}|_{\mathcal{S}}$, find the equivalence algebra $\mathfrak{g}^\sim$ of this class by the infinitesimal invariance criterion. Let $\mathfrak{g}^\sim = \{Q_\gamma \mid \gamma \in \Gamma\}$ for some parameter set Γ.

(ii) Fix a set $\{Q_{\gamma'} \mid \gamma' \in \Gamma'\}$ of vector fields spanning $\mathfrak{g}^\sim$, $\mathfrak{g}^\sim = \langle Q_{\gamma'} \mid \gamma' \in \Gamma'\rangle$ with $\Gamma' \subset \Gamma$, and compute commutation relations for all pairs of spanning vector fields.

(iii) Using Propositions 1 and 2, Remark 3 and other tools, construct as wide a list of megaideals of the equivalence algebra $\mathfrak{g}^\sim$ as possible. Optimize the list by the exclusion of inessential megaideals, which give no new constraints for automorphisms of $\mathfrak{g}^\sim$ as compared with other megaideals. In particular, megaideals being sums of other megaideals are not essential.

(iv) Take the general form of an appropriate point transformation $\mathcal{T}$ in the space of x, $u_{(p)}$ and θ and push-forward the vector fields $Q_{\gamma'}$, $\gamma' \in \Gamma'$. For each γ' we choose, from the above list, the minimal megaideal $\mathfrak{i}_{\gamma'} = \{Q_\gamma \mid \gamma \in \Gamma_{\mathfrak{i}_{\gamma'}} \subset \Gamma\}$ containing $Q_{\gamma'}$ and set $\mathcal{T}_*Q_{\gamma'} = Q_{\gamma_{\gamma'}}$, where the parameter $\gamma_{\gamma'}$ satisfies the constraints singling out the set $\Gamma_{\mathfrak{i}_{\gamma'}}$ from the entire set Γ. Consequently equating the corresponding vector-field components of right- and left-hand sides in the equation $\mathcal{T}_*Q_{\gamma'} = Q_{\gamma_{\gamma'}}$ for each γ' leads to a system of differential equations for components of the transformation $\mathcal{T}$.[8] The solution of this system results in an intermediate form for $\mathcal{T}$.

(v) Use the derived form of $\mathcal{T}$ within the framework of the direct method in order to complete the system of constraints for $\mathcal{T}$ and to produce the final form for $\mathcal{T}$ by solving the completed system.

4. Complete equivalence group of the class of nonlinear wave equations

In order to clarify all the algorithmic steps, in this section we demonstrate in detail the computation of the usual equivalence group $G^\sim$ of the class of nonlinear wave equations (1) using the megaideal-based method. The preliminary presentation of this result is contained in the arXiv preprint of [4].

4.1. Equivalence algebra

The equivalence algebra of class (1) including both linear and nonlinear equations was first computed in [22]. It was then shown in [4], that the class consisting of only nonlinear equations admits the same equivalence algebra. Here we use both the representation of the equivalence algebra and the notation introduced in [4]. The equivalence algebra $\mathfrak{g}^\sim$ of the class (1) is spanned by the vector fields

$$\begin{aligned}
&\mathcal{D}^u = u\partial_u + u_x\partial_{u_x} + g\partial_g, \quad \mathcal{D}^t = t\partial_t - 2f\partial_f - 2g\partial_g, \quad \mathcal{P}^t = \partial_t,\\
&\mathcal{D}(\varphi) = \varphi\partial_x - \varphi_x u_x\partial_{u_x} + 2\varphi_x f\partial_f + \varphi_{xx}u_x f\partial_g,\\
&\mathcal{G}(\psi) = \psi\partial_u + \psi_x\partial_{u_x} - \psi_{xx}f\partial_g, \quad \mathcal{F}^1 = t\partial_u, \quad \mathcal{F}^2 = t^2\partial_u + 2\partial_g,
\end{aligned} \tag{2}$$

where $\varphi = \varphi(x)$ and $\psi = \psi(x)$ run through the set of smooth functions of x. It has been pointed out in [4] that while the elements of the equivalence algebra of class (1) are vector

[8] Similarly to the automorphism-based method, the system obtained involves the parameters $\gamma_{\gamma'}$, $\gamma' \in \Gamma'$. Therefore, integrating the system requires solving a compatibility problem in order to find values of the parameters for which the system is consistent with respect to components of $\mathcal{T}$. In addition to that the group $G^\sim_*$ may be a proper subgroup of Aut($\mathfrak{g}^\sim$), for the megaideal-based version there may be one more reason for the system to be inconsistent for some parameter values: In general, even the complete megaideal hierarchy cannot properly capture delicate constraints for elements of Aut($\mathfrak{g}^\sim$).

fields in the space of $(t, x, u_{(2)}, f, g)$, for practical purposes it suffices to present only their projection to the space of (t, x, u, u_x, f, g). The reason for this is that the vector field components corresponding to u_t, u_x, u_{tt}, u_{tx} and u_{xx}, can be obtained via prolongation from the components corresponding to t, x and u. At the same time, we explicitly include the components associated with the derivative u_x in the representation of vector fields (2) spanning $\mathfrak{g}^\sim$ since the arbitrary elements f and g depend on u_x. Therefore, for the proper computation of commutation relations, the components with ∂_{u_x} are crucially needed.

For the construction of megaideals, it is convenient to also recall the nonvanishing commutation relations between the vector fields of $\mathfrak{g}^\sim$. They read

$$[\mathcal{G}(\psi), \mathcal{D}^u] = \mathcal{G}(\psi), \quad [\mathcal{F}^1, \mathcal{D}^u] = \mathcal{F}^1, \quad [\mathcal{F}^2, \mathcal{D}^u] = \mathcal{F}^2,$$
$$[\mathcal{D}^t, \mathcal{F}^1] = \mathcal{F}^1, \quad [\mathcal{D}^t, \mathcal{F}^2] = 2\mathcal{F}^2,$$
$$[\mathcal{P}^t, \mathcal{D}^t] = \mathcal{P}^t, \quad [\mathcal{P}^t, \mathcal{F}^1] = \mathcal{G}(1), \quad [\mathcal{P}^t, \mathcal{F}^2] = 2\mathcal{F}^1,$$
$$[\mathcal{D}(\varphi^1), \mathcal{D}(\varphi^2)] = \mathcal{D}(\varphi^1\varphi^2_x - \varphi^1_x\varphi^2), \quad [\mathcal{D}(\varphi), \mathcal{G}(\psi)] = \mathcal{G}(\varphi\psi_x).$$

Therefore, steps 1 and 2 of the megaideal-based method were in fact realized in [4, 22].

4.2. Megaideals of equivalence algebra

In order to compute the complete equivalence group of the class (1) by the megaideal-based method (cf. Section 4.3), we need to know a set of megaideals of the equivalence algebra $\mathfrak{g}^\sim$ of this class.

Let $\mathfrak{g} = \mathfrak{g}^\sim$ for the sake of notational simplicity. Using Proposition 1, it is easy to compute the following megaideals of $\mathfrak{g}^\sim$:

$$\mathfrak{g}' = \langle \mathcal{P}^t, \mathcal{D}(\varphi), \mathcal{G}(\psi), \mathcal{F}^1, \mathcal{F}^2 \rangle, \quad \mathfrak{g}'' = \langle \mathcal{D}(\varphi), \mathcal{G}(\psi), \mathcal{F}^1 \rangle, \quad \mathfrak{g}''' = \langle \mathcal{D}(\varphi), \mathcal{G}(\psi) \rangle,$$
$$\mathrm{C}_{\mathfrak{g}}(\mathfrak{g}''') = \langle \mathcal{D}^t, \mathcal{P}^t, \mathcal{G}(1), \mathcal{F}^1, \mathcal{F}^2 \rangle, \quad \mathrm{C}_{\mathfrak{g}'}(\mathfrak{g}''') = \langle \mathcal{P}^t, \mathcal{G}(1), \mathcal{F}^1, \mathcal{F}^2 \rangle,$$
$$\mathrm{C}_{\mathfrak{g}'}(\mathfrak{g}'') = \langle \mathcal{G}(1), \mathcal{F}^1, \mathcal{F}^2 \rangle, \quad \mathrm{Z}_{\mathfrak{g}''} = \langle \mathcal{G}(1), \mathcal{F}^1 \rangle, \quad \mathrm{Z}_{\mathfrak{g}'} = \langle \mathcal{G}(1) \rangle,$$
$$\mathrm{R}_{\mathfrak{g}} = \langle \mathcal{D}^u, \mathcal{D}^t, \mathcal{P}^t, \mathcal{G}(\psi), \mathcal{F}^1, \mathcal{F}^2 \rangle, \quad \mathrm{R}_{\mathfrak{g}'''} = \langle \mathcal{G}(\psi) \rangle,$$

where $\mathfrak{a}'$, $\mathrm{R}_\mathfrak{a}$, $\mathrm{Z}_\mathfrak{a}$ and $\mathrm{C}_\mathfrak{a}(\mathfrak{b})$ denote the derivative, the radical and the center of a Lie algebra $\mathfrak{a}$ and the centralizer of a subalgebra $\mathfrak{b}$ in $\mathfrak{a}$, respectively. We present proofs only for the last two equalities.

The linear span $\mathfrak{s}_1 = \langle \mathcal{D}^u, \mathcal{D}^t, \mathcal{P}^t, \mathcal{G}(\psi), \mathcal{F}^1, \mathcal{F}^2 \rangle$ obviously is a solvable ideal of $\mathfrak{g}$. Moreover, it is the maximal solvable ideal of $\mathfrak{g}$. Indeed, suppose that $\mathfrak{s}_1 \subsetneq \mathfrak{i}$ and $\mathfrak{i}$ is an ideal of $\mathfrak{g}$. Then there exists a smooth function ζ of x which does not identically vanish such that the vector field $\mathcal{D}(\zeta)$ belongs to $\mathfrak{i}$. As $\mathfrak{i}$ is an ideal of $\mathfrak{g}$, for an arbitrary smooth function φ of x the commutator $[\mathcal{D}(\zeta), \mathcal{D}(\varphi)] = \mathcal{D}(\zeta\varphi_x - \zeta_x\varphi)$ belongs to $\mathfrak{i}$. If ζ is not a constant function, we define the following series of operators:

$$R^{0k} = k^{-1}[\mathcal{D}(\zeta), \mathcal{D}(\zeta^{k+1})], \quad R^{jk} = k^{-1}[R^{j-1,1}, R^{j-1,k+1}], \quad j, k = 1, 2, \dots .$$

It is possible to prove by induction that $R^{j-1,k} = \mathcal{D}(\zeta^k(\zeta\zeta_x)^{2^j-1}) \neq 0$, $j, k = 1, 2, \dots$. Moreover, as $R^{0k} \in \mathfrak{i}$, we have $R^{jk} \in \mathfrak{i}^{(j)}$, $j, k = 1, 2, \dots$, i.e., $\mathfrak{i}^{(j)} \neq \{0\}$ for any nonnegative integer j. This means that the ideal $\mathfrak{i}$ is not solvable. If ζ is a constant function, we can set $\zeta \equiv 1$. We choose any smooth function φ of x with $\varphi_{xx} \not\equiv 0$ and denote φ_x by $\tilde\zeta$. As the commutator $[\mathcal{D}(1), \mathcal{D}(\varphi)] = \mathcal{D}(\tilde\zeta)$ belongs to $\mathfrak{i}$, the consideration for the previous case again implies that the ideal $\mathfrak{i}$ is not solvable. Therefore, $\mathfrak{s}_1$ is really the maximal solvable ideal of $\mathfrak{g}$, i.e., $\mathrm{R}_\mathfrak{g} = \mathfrak{s}_1$.

The linear span $\mathfrak{s}_2 = \langle \mathcal{G}(\psi) \rangle$ is an Abelian and, therefore, solvable ideal of $\mathfrak{g}'''$. The maximality of this solvable ideal is proved in the same way as for $\mathfrak{s}_1$. Hence $\mathrm{R}_{\mathfrak{g}'''} = \mathfrak{s}_2$.

Journal of Physics: Conference Series **621** (2015) 012001 doi:10.1088/1742-6596/621/1/012001

The same megaideals can be obtained in different ways. For example, $\langle \mathcal{G}(1)\rangle = \mathrm{Z}_{\mathfrak{g}'} = \mathrm{Z}_{\mathfrak{g}'''}$.

To find one more megaideal which will be used in the course of the computation of the complete equivalence group of the class (1) by the algebraic method in Section 4.3, we should apply a more sophisticated technique than before. There are two ways to do this.

The way directly based on the definition of megaideals is to calculate the automorphism group $\mathrm{Aut}(\mathfrak{m})$ of the finite-dimensional megaideal $\mathfrak{m} = \mathrm{C}_{\mathfrak{g}}(\mathfrak{g}''') = \langle \mathcal{D}^t, \mathcal{P}^t, \mathcal{G}(1), \mathcal{F}^1, \mathcal{F}^2\rangle$ and then determine megaideals of $\mathfrak{m}$ as subspaces of $\mathfrak{m}$ which are invariant with respect to $\mathrm{Aut}(\mathfrak{m})$. In the course of calculating the automorphisms we use the knowledge about simple megaideals of $\mathfrak{m}$,

$$\mathfrak{m}' = \langle \mathcal{P}^t, \mathcal{G}(1), \mathcal{F}^1, \mathcal{F}^2\rangle, \quad \mathfrak{m}'' = \langle \mathcal{G}(1), \mathcal{F}^1\rangle, \quad \mathrm{Z}_{\mathfrak{m}} = \langle \mathcal{G}(1)\rangle, \quad \mathrm{C}_{\mathfrak{m}}(\mathfrak{m}'') = \langle \mathcal{G}(1), \mathcal{F}^1, \mathcal{F}^2\rangle.$$

The presence of the above set of nested megaideals is equivalent to that for any automorphism A of $\mathfrak{m}$, its matrix $(a_{ij})_{i,j=1}^5$ in the basis $\{\mathcal{G}(1), \mathcal{F}^1, \mathcal{F}^2, \mathcal{P}^t, \mathcal{D}^t\}$ is upper triangular with nonzero diagonal elements. In particular,

$$A\mathcal{P}^t = a_{14}\mathcal{G}(1) + a_{24}\mathcal{F}^1 + a_{34}\mathcal{F}^2 + a_{44}\mathcal{P}^t,$$
$$A\mathcal{D}^t = a_{15}\mathcal{G}(1) + a_{25}\mathcal{F}^1 + a_{35}\mathcal{F}^2 + a_{45}\mathcal{P}^t + a_{55}\mathcal{D}^t,$$

where $a_{44}a_{55} \neq 0$. As $[\mathcal{P}^t, \mathcal{D}^t] = \mathcal{P}^t$ and $A \in \mathrm{Aut}(\mathfrak{m})$, we should have $[A\mathcal{P}^t, A\mathcal{D}^t] = A\mathcal{P}^t$. Collecting coefficients of basis elements in the last equality, we derive a system of equations with respect to a's which implies, in view of the condition $a_{44} \neq 0$, that $a_{55} = 1$, $a_{34} = 0$, $a_{24} = a_{44}a_{35}$ and $a_{14} = a_{44}a_{25} - a_{45}a_{24}$. As $a_{34} = 0$, we get that the span $\langle \mathcal{P}^t, \mathcal{G}(1), \mathcal{F}^1\rangle$ is a megaideal of $\mathfrak{m}$ and, therefore, of $\mathfrak{g}^\sim$.

The other way for finding the megaideal $\langle \mathcal{P}^t, \mathcal{G}(1), \mathcal{F}^1\rangle$, which is based on Proposition 2, allows us to avoid the above computation of automorphisms of the megaideal $\mathfrak{m}$ and, thus, is much simpler: Choosing $\mathfrak{i}_0 = \mathfrak{i}_1 = \mathrm{C}_{\mathfrak{g}'}(\mathfrak{g}''') = \langle \mathcal{P}^t, \mathcal{G}(1), \mathcal{F}^1, \mathcal{F}^2\rangle$ and $\mathfrak{i}_2 = \mathrm{Z}_{\mathfrak{g}'} = \langle \mathcal{G}(1)\rangle$ in Proposition 2, we obtain the megaideal $\mathfrak{s} = \langle \mathcal{P}^t, \mathcal{G}(1), \mathcal{F}^1\rangle$.

4.3. Calculation of equivalence group

The careful study of megaideals of the equivalence algebra $\mathfrak{g}^\sim$ in Section 4.2 according to the third step of the megaideal-based procedure supplies us with a sufficient store of megaideals in order to commence the computations directly concerned with equivalence transformations.

Theorem 2. *The equivalence group $G^\sim$ of the class* (1) *consists of the transformations*

$$\tilde t = c_1 t + c_0, \quad \tilde x = \varphi(x), \quad \tilde u = c_2 u + c_4 t^2 + c_3 t + \psi(x), \quad \tilde u_{\tilde x} = \frac{c_2 u_x + \psi_x}{\varphi_x},$$
$$\tilde f = \frac{\varphi_x^2}{c_1^2} f, \quad \tilde g = \frac{1}{c_1^2}\left(c_2 g + \frac{c_2 u_x + \psi_x}{\varphi_x}\varphi_{xx} f - \psi_{xx} f + 2c_4\right), \tag{3}$$

where $c_0, \ldots, c_4$ are arbitrary constants satisfying the condition $c_1c_2 \neq 0$ and φ and ψ run through the set of smooth functions of x, $\varphi_x \neq 0$.

Proof. The group $G^\sim$ consists of nondegenerate point transformations in the joint space of variables t, x and u, the first derivatives u_t and u_x and the arbitrary elements f and g, which are projectable to the variable space and whose components for first derivatives are defined via the first prolongation of their projections to the variable space. Thus, the general form of a transformation $\mathcal{T}$ from $G^\sim$ is

$$\tilde t = T(t, x, u), \quad \tilde x = X(t, x, u), \quad \tilde u = U(t, x, u),$$
$$\tilde u_{\tilde t} = U^t(t, x, u, u_t, u_x), \quad \tilde u_{\tilde x} = U^x(t, x, u, u_t, u_x),$$
$$\tilde f = F(t, x, u, u_t, u_x, f, g), \quad \tilde g = G(t, x, u, u_t, u_x, f, g),$$

where U^t and U^x are determined via T, X and U and the nondegeneracy condition should be additionally satisfied. To obtain the constrained form of $\mathcal{T}$, we will act by the push-forward $\mathcal{T}_*$ induced by $\mathcal{T}$ on the vector fields (2) additionally including the terms with ∂_{u_t} and use megaideals of the equivalence algebra $\mathfrak{g}^\sim$ of the class (1) and restrictions on automorphisms of $\mathfrak{g}^\sim$ found in Section 4.2. The principal part of the consideration is based on objects and properties of $\mathfrak{g}^\sim$ that are related to the finite-dimensional megaideal $\mathfrak{m} = \mathrm{C}_{\mathfrak{g}}(\mathfrak{g}''') = \langle \mathcal{D}^t, \mathcal{P}^t, \mathcal{G}(1), \mathcal{F}^1, \mathcal{F}^2\rangle$. Recall that megaideals being sums of other megaideals are not essential for the computation since they give weaker constraints than their summands. For example, the megaideal $\mathfrak{g}''$ is the sum of $\mathfrak{g}'''$ and $\mathrm{Z}_{\mathfrak{g}'}$ and hence we do not use it in the further consideration. A list of essential megaideals is in fact exhausted by the spans

$$\langle\mathcal{G}(1)\rangle,\quad \langle\mathcal{G}(1),\mathcal{F}^1\rangle,\quad \langle\mathcal{G}(1),\mathcal{F}^1,\mathcal{F}^2\rangle,\quad \langle\mathcal{P}^t,\mathcal{G}(1),\mathcal{F}^1\rangle,\quad \langle\mathcal{D}^t,\mathcal{P}^t,\mathcal{G}(1),\mathcal{F}^1,\mathcal{F}^2\rangle,$$
$$\langle\mathcal{G}(\psi)\rangle,\quad \langle\mathcal{D}(\varphi),\mathcal{G}(\psi)\rangle.$$

We apply the condition of preserving megaideals following their ordering by inclusion. In other words, for each vector field Q from (2) we take the megaideal that contains Q and is minimal among the above listed megaideals. As a result, we derive an optimized set of constraints for $\mathcal{T}_*$ as automorphism of $\mathfrak{g}$,

$$\mathcal{T}_*\mathcal{G}(1) = a_{11}\tilde{\mathcal{G}}(1), \tag{4a}$$
$$\mathcal{T}_*\mathcal{F}^1 = a_{12}\tilde{\mathcal{G}}(1) + a_{22}\tilde{\mathcal{F}}^1, \tag{4b}$$
$$\mathcal{T}_*\mathcal{F}^2 = a_{13}\tilde{\mathcal{G}}(1) + a_{23}\tilde{\mathcal{F}}^1 + a_{33}\tilde{\mathcal{F}}^2, \tag{4c}$$
$$\mathcal{T}_*\mathcal{P}^t = a_{14}\tilde{\mathcal{G}}(1) + a_{24}\tilde{\mathcal{F}}^1 + a_{44}\tilde{\mathcal{P}}^t, \tag{4d}$$
$$\mathcal{T}_*\mathcal{D}^t = a_{15}\tilde{\mathcal{G}}(1) + a_{25}\tilde{\mathcal{F}}^1 + a_{35}\tilde{\mathcal{F}}^2 + a_{45}\tilde{\mathcal{P}}^t + \tilde{\mathcal{D}}^t, \tag{4e}$$
$$\mathcal{T}_*\mathcal{G}(\hat{\psi}) = \tilde{\mathcal{G}}(\tilde{\psi}^{\hat{\psi}}), \tag{4f}$$
$$\mathcal{T}_*\mathcal{D}(\hat{\varphi}) = \tilde{\mathcal{G}}(\tilde{\psi}^{\hat{\varphi}}) + \tilde{\mathcal{D}}(\tilde{\varphi}^{\hat{\varphi}}), \tag{4g}$$

where a's are constants, $a_{11}a_{22}a_{33}a_{44} \neq 0$ and $\hat{\psi}$ and $\hat{\varphi}$ are arbitrary smooth functions of x. The constants a's completed with $a_{55} = 1$ and $a_{ij} = 0$, $1 \leqslant i < j \leqslant 5$, form a matrix of an automorphism of the megaideal $\mathfrak{m}$. Tildes over vector fields on the right hand sides of the above equations mean that these vector fields are written in terms of the transformed variables. The parameter-functions $\tilde{\psi}^{\hat{\psi}}$, $\tilde{\psi}^{\hat{\varphi}}$ and $\tilde{\varphi}^{\hat{\varphi}}$ are smooth functions of $\tilde{x}$ associated with the parameter-functions $\hat{\psi}$ or $\hat{\varphi}$, which is indicated by the corresponding superscripts. We will derive constraints for $\mathcal{T}$, consequently equating the corresponding vector-field components of right- and left-hand sides in each of the conditions (4) and taking into account constraints obtained in previous steps. As the components of vector fields and of the transformation $\mathcal{T}$ associated with the derivatives u_t and u_x are defined via first-order prolongation involving the similar values related to the variables t, x and u, the coefficients of ∂_{u_t} and ∂_{u_x} give no essentially new equations in comparison with the coefficients of ∂_t, ∂_x and ∂_u. This is why we will not equate the coefficients of ∂_{u_t} and ∂_{u_x}. To have a proper representation of the final result, we will re-denote certain values in an appropriate way.

Thus, the equation (4a) implies that $T_u = X_u = 0$, $U_u = c_2$ and $F_u = G_u = 0$, where the nonvanishing constant a_{11} is re-denoted by c_2. Then we derive from Eq. (4b) that $tU_u = a_{22}T + a_{12}$, i.e., $T = c_1t + c_0$ where $c_1 = c_2/a_{22} \neq 0$ and $c_0 = -a_{12}/a_{22}$, and $F_{u_t} = G_{u_t} = 0$. The consequence $t^2U_u = a_{33}T^2 + a_{23}T + a_{13}$ of Eq. (4c) gives only relations between a's. In particular, $a_{33} = c_2/c_1^2$. Then the other consequences of Eq. (4c) are $F_g = 0$ and $G_g = c_2/c_1^2$. The essential consequences of Eq. (4d) are exhausted by $X_t = 0$, $U_t = a_{24}T + a_{14}$ and $F_t = G_t = 0$. Therefore, $X = \varphi(x)$ and $U = c_2 + c_4t^2 + c_3t + \psi(x)$, where $\varphi_x \neq 0$, $c_4 = a_{24}c_1/2$ and $c_3 = a_{14} + a_{24}c_0$.

As we have already derived the precise expressions for the components of $\mathcal{T}$ corresponding to the variables (cf. Eq. (3)), at this point we could terminate the computation of equivalence transformations by the algebraic method and calculate the expressions for F and G by the direct method. At the same time, all the determining equations for transformations from the equivalence group $G^\sim$ of the class (1) follow from restrictions for automorphisms of the equivalence algebra $\mathfrak{g}^\sim$. This is not a common situation when the algebraic method is applied. Usually it gives only a part of the determining equations simplifying the subsequent application of the direct method. See, e.g., the computations of the complete point symmetry groups of the barotropic vorticity equation and quasi-geostrophic two-layer model in [9, Section 3] and [17, Section 4], respectively. This is why we complete the consideration of the equivalence group $G^\sim$ within the framework of the algebraic method.

From Eq. (4e) we obtain in particular that $tU_t = a_{35}T^2 + a_{25}T + a_{15}$, $fF_f = F$ and $fG_f + gG_g = G - a_{35}$, where $a_{35} = 2c_4/c_1^2$ in view of the first of these consequences.

The equation (4f) implies the equations

$$\hat\psi U_u = \tilde\psi^{\hat\psi}, \quad \hat\psi_x F_{u_x} = 0, \quad \hat\psi_x G_{u_x} - \hat\psi_{xx} f G_g = \tilde\psi^{\hat\psi}_{\tilde x\tilde x} F. \tag{5}$$

The first and second equations of (5) are equivalent to $\tilde\psi^{\hat\psi} = c_2\hat\psi$ and $F_{u_x} = 0$. Then we can express the derivative $\tilde\psi^{\hat\psi}_{\tilde x\tilde x}$ via derivatives of $\hat\psi$, $\tilde\psi^{\hat\psi}_{\tilde x\tilde x} = c_2\varphi_x^{-3}(\varphi_x\hat\psi_{xx} - \varphi_{xx}\hat\psi_x)$, substitute the expression into the third equation of (5) and split with respect to the derivatives $\hat\psi_x$ and $\hat\psi_{xx}$, as the function $\hat\psi$ is arbitrary. As a result, we obtain $F = c_1^{-2}{\varphi_x}^2 f$ and $G_{u_x} = c_2\varphi_x^{-3}\varphi_{xx}F$, i.e., $G_{u_x} = c_2c_1^{-2}\varphi_x^{-1}\varphi_{xx}f$. The expression for F coincides with the transformation component for f presented in the theorem.

The last essential equation for G is given by Eq. (4g). Collecting coefficients of ∂_x, ∂_u and ∂_g in Eq. (4g), we have that $\tilde\varphi^{\hat\varphi}(\tilde x) = \varphi_x\hat\varphi$, $\tilde\psi^{\hat\varphi}(\tilde x) = \psi_x\hat\varphi$ and

$$\hat\varphi G_x - \hat\varphi_x u_x G_{u_x} + 2\hat\varphi_x f G_f + \hat\varphi_{xx} u_x f G_g = \tilde\varphi^{\hat\varphi}_{\tilde x\tilde x}\tilde u_{\tilde x} F - \tilde\psi^{\hat\varphi}_{\tilde x\tilde x} F, \tag{6}$$

respectively. We proceed in a way analogous to the previous step. Namely, we express the derivatives $\tilde\varphi^{\hat\varphi}_{\tilde x\tilde x}$ and $\tilde\psi^{\hat\varphi}_{\tilde x\tilde x}$ via derivatives of $\hat\varphi$, substitute the expressions into Eq. (6) and split with respect to derivatives of $\hat\varphi$ because the function $\hat\psi$ is arbitrary. Equating the coefficients of $\hat\varphi_x$ leads to the equation

$$fG_f = c_1^{-2}\varphi_x^{-1}(c_2u_x + \psi_x\varphi_{xx} - \psi_{xx}\varphi_x).$$

The simultaneous integration of all the equations obtained for G precisely results in the transformation component for g from the theorem, which completes step 4 of the megaideal-based method.

In order to complete the proof, we should realize the last procedure step, which is in fact trivial for the class (1). We should just check by the direct computation of expressions for transformed derivatives that any transformation of the form (3) maps any equation from the class (1) to an equation from the same class. □

After comparing the equivalence algebra $\mathfrak{g}^\sim$ and the equivalence group $G^\sim$, the following corollary is evident:

Corollary 1. *A complete list of discrete equivalence transformations of the class* (1) *that are independent up to combining with each other and with continuous equivalence transformations of this class is exhausted by the transformations*

$$(t, x, u, u_x, f, g) \mapsto (-t, x, u, u_x, f, g),$$
$$(t, x, u, u_x, f, g) \mapsto (t, -x, u, -u_x, f, g),$$
$$(t, x, u, u_x, f, g) \mapsto (t, x, -u, -u_x, f, -g).$$

Theorem 2 implies that any transformation $\mathcal{T}$ from $G^{\sim}$ of the class (1) can be represented as the composition

$$\mathcal{T}=\mathcal{D}^t(c_1)\mathcal{P}^t(c_0)\mathcal{D}(\varphi)\mathcal{D}^u(c_2)\mathcal{F}^1(c_4)\mathcal{F}^2(c_3)\mathcal{G}(\psi),$$

where the above parameterized equivalence transformations are

$$\begin{array}{lllllll}
\mathcal{P}^t(c_0)\colon & \tilde t=t+c_0, & \tilde x=x, & \tilde u=u, & \tilde u_{\tilde x}=u_x, & \tilde f=f, & \tilde g=g,\\
\mathcal{D}^t(c_1)\colon & \tilde t=c_1t, & \tilde x=x, & \tilde u=u, & \tilde u_{\tilde x}=u_x, & \tilde f=c_1^{-2}f, & \tilde g=c_1^{-2}g,\\
\mathcal{D}(\varphi)\colon & \tilde t=t, & \tilde x=\varphi, & \tilde u=u, & \tilde u_{\tilde x}=u_x/\varphi_x, & \tilde f=\varphi_x^2f, & \tilde g=g+\varphi_{xx}u_xf/\varphi_x,\\
\mathcal{D}^u(c_2)\colon & \tilde t=t, & \tilde x=x, & \tilde u=c_2u, & \tilde u_{\tilde x}=c_2u_x, & \tilde f=f, & \tilde g=c_2g,\\
\mathcal{F}^1(c_3)\colon & \tilde t=t, & \tilde x=x, & \tilde u=u+c_3t, & \tilde u_{\tilde x}=u_x, & \tilde f=f, & \tilde g=g,\\
\mathcal{F}^2(c_4)\colon & \tilde t=t, & \tilde x=x, & \tilde u=u+c_4t^2, & \tilde u_{\tilde x}=u_x, & \tilde f=f, & \tilde g=g+2c_4,\\
\mathcal{G}(\psi)\colon & \tilde t=t, & \tilde x=x, & \tilde u=u+\psi, & \tilde u_{\tilde x}=u_x+\psi_x, & \tilde f=f, & \tilde g=g-\psi_{xx}f,
\end{array}$$

and the nondegeneracy requires that $c_1c_2\varphi_x\neq0$. These transformations are shifts and scalings in t, arbitrary transformations in x, scalings of u, gauging transformations of u with square polynomials in t and arbitrary functions of x.

5. Conclusion

In this paper we have extended the algebraic method for the computation of complete point symmetry groups of single systems of differential equations to the computation of equivalence groups of classes of such systems. In general, the equivalence group may include both discrete and continuous equivalence transformations. Unlike the infinitesimal method, which aims at finding the corresponding equivalence algebras from which the continuous parts of equivalence groups are computed, the algebraic method allows also constructing discrete equivalence transformations. We have developed two versions of this method, automorphism-based and megaideal-based, which may be combined depending on the problem under consideration [14]. The effectiveness of both versions rests on the fact that each equivalence transformation induces, via push-forwarding the vector fields that constitute the corresponding equivalence algebra $\mathfrak{g}^{\sim}$, an automorphism of the algebra $\mathfrak{g}^{\sim}$. This imposes substantial restrictions on the functional form of equivalence transformations and strongly simplifies, or even makes trivial, the further application of the direct method.

Each of the versions has own advantages and disadvantages. We briefly recall some of them.

The automorphism-based version ensures the maximal use of properties of equivalence algebras within the framework of the algebraic method. One more benefit of this version is that most continuous equivalence transformations, which are in fact known as easily obtainable from elements of the corresponding equivalence algebra $\mathfrak{g}^{\sim}$, can be factored out in the course of the computation. At the same time, the main ingredient of the automorphism-based version is finding the entire automorphism group of the algebra $\mathfrak{g}^{\sim}$, which places a strong limitation on the dimension and/or structure of $\mathfrak{g}^{\sim}$. This is why usually the automorphism-based version is applied to classes of differential equations whose equivalence algebras are low-dimensional.

The megaideal-based version rests on the invariance of megaideals of equivalence algebras under the associated automorphisms and hence it does not require the explicit computation of the automorphism groups of equivalence algebras, which makes it suitable for finding the equivalence groups of classes with infinite-dimensional equivalence algebras. A disadvantage of the megaideal-based version is that the equivalence algebra $\mathfrak{g}^{\sim}$ of a class of differential equations $\mathcal{L}|_{\mathcal{S}}$ may admit properties that induce essential constraints for equivalence transformations of $\mathcal{L}|_{\mathcal{S}}$ but cannot be interpreted in terms of megaideals of $\mathfrak{g}^{\sim}$.

Sometimes disadvantages of either of the versions of the algebraic method can be reduced by combining these versions.

We have restricted ourselves to the computation of the equivalence group of a single physically relevant example, a class of nonlinear wave equations arising in nonlinear elasticity. The equivalence algebra of this class is infinite dimensional, which thus gives a prototypical example for the effectiveness of the megaideal-based method. We plan to write a more extensive paper with further examples on the use of both versions of the algebraic method in the future.

Important further theoretical developments of the present method may include an extension to the computation of extended equivalence groups, which also play a role in the theory of group classification. The study of equivalence groupoids of classes of differential equations using the algebraic method will be another central milestone.

Acknowledgements

AB is a recipient of an APART Fellowship of the Austrian Academy of Sciences. This research was supported by the Austrian Science Fund (FWF), project P25064 (EDSCB and ROP). The authors are grateful to Vyacheslav Boyko for productive discussions and useful suggestions. ROP is also grateful for the hospitality and financial support provided by the University of Cyprus.

References

[1] Ovsiannikov L V 1982 *Group Analysis of Differential Equations* (New York: Acad. Press)
[2] Popovych R O 2006 Classification of admissible transformations of differential equations *Collection of Works of Institute of Mathematics* vol 3 (Kyiv: Institute of Mathematics) p 239
[3] Popovych R O, Kunzinger M and Eshraghi H 2010 Admissible transformations and normalized classes of nonlinear Schrödinger equations *Acta Appl. Math.* **109** 315 (*Preprint* arXiv:math-ph/0611061)
[4] Bihlo A, Dos Santos Cardoso-Bihlo E M and Popovych R O 2012 Complete group classification of a class of nonlinear wave equations *J. Math. Phys.* **53** 123515 (*Preprint* arXiv:1106.4801)
[5] Akhatov I S, Gazizov R K and Ibragimov N K 1991 Nonlocal symmetries. Heuristic approach *J. Math. Sci.* **55** 1401
[6] Kingston J G and Sophocleous C 1991 On point transformations of a generalised Burgers equation *Phys. Lett.* A **155** 15
[7] Kingston J G 1991 On point transformations of evolution equations *J. Phys.* A **24** L769
[8] Kingston J G and Sophocleous C 1998 On form-preserving point transformations of partial differential equations *J. Phys.* A: *Math. Gen.* **31** 1597
[9] Bihlo A and Popovych R O 2011 Point symmetry group of the barotropic vorticity equation *Proc. 5th Int. Workshop on Group Analysis of Differential Equations and Integrable Systems (Protaras, 2010)* (Nicosia: University of Cyprus) p 15 (*Preprint* arXiv:1009.1523)
[10] Hydon P E 1998 Discrete point symmetries of ordinary differential equations *Proc. R. Soc. Lond. Ser.* A *Math. Phys. Eng. Sci.* **454** 1961
[11] Hydon P E 2000 How to construct the discrete symmetries of partial differential equations *Eur. J. Appl. Math.* **11** 515
[12] Hydon P E 2000 *Symmetry Methods for Differential Equations* (Cambridge: Cambridge University Press)
[13] Gray R J 2013 Ph.D. Thesis *Automorphisms of Lie Algebras* (Surrey: University of Surrey)
[14] Dos Santos Cardoso-Bihlo E M and Popovych R O 2013 Complete point symmetry group of the barotropic vorticity equation on a rotating sphere *J. Engrg. Math.* **82** 31 (*Preprint* arXiv:1206.6919)
[15] Popovych R O, Boyko V M, Nesterenko M O and Lutfullin M W 2003 Realizations of real low-dimensional Lie algebras *J. Phys.* A **36** 7337; see arXiv:math-ph/0301029v7 for an extended and revised version
[16] Hilgert J and Neeb K H 2011 *Structure and Geometry of Lie Groups* (Springer Science & Business Media)
[17] Bihlo A and Popovych R O 2011 Lie symmetry analysis and exact solutions of the quasi-geostrophic two-layer problem *J. Math. Phys.* **52** 033103 (*Preprint* arXiv:1010.1542)
[18] Dos Santos Cardoso-Bihlo E M and Popovych R O On the ineffectiveness of constant rotation in the primitive equations (*Preprint* arXiv:1503.04168)
[19] Jacobson N 1979 *Lie Algebras* (New York: Dover)
[20] Fisher D J, Gray R J and Hydon P E 2013 Automorphisms of real Lie algebras of dimension five or less *J. Phys.* A **46** 225204
[21] Christodoulakis T, Papadopoulos G O and Dimakis A 2003 Automorphisms of real four-dimensional Lie algebras and the invariant characterization of homogeneous 4-spaces *J. Phys.* A **36** 427; corrigendum: *J. Phys.* A 2003 **36** 2379

[22] Ibragimov N H, Torrisi M and Valenti A 1991 Preliminary group classification of equations $v_{tt} = f(x, v_x)v_{xx} + g(x, v_x)$ *J. Math. Phys.* **32** 2988
[23] Harin A O 1993 On a countable-dimensional subalgebra of the equivalence algebra for equations $v_{tt} = f(x, v_x)v_{xx} + g(x, v_x)$ *J. Math. Phys.* **34** 3676
[24] Ibragimov N H and Khabirov S V 2000 Contact transformation group classification of nonlinear wave equations *Nonlin. Dyn.* **22** 61
[25] Khabirov S V 2009 A property of the determining equations for an algebra in the group classification problem for wave equations *Sibirsk. Mat. Zh.* **50** 647–668 [in Russian]; English transl. in *Sib. Math. J.* **50** 515
[26] Ibragimov N H, Torrisi M and Valenti A 2004 Differential invariants of nonlinear equations $v_{tt} = f(x, v_x)v_{xx} + g(x, v_x)$ *Commun. Nonlinear Sci. Numer. Simul.* **9** 69
[27] Dos Santos Cardoso-Bihlo E M, Bihlo A and Popovych R O 2011 Enhanced preliminary group classification of a class of generalized diffusion equations *Commun. Nonlinear Sci. Numer. Simulat.* **16** 3622 (*Preprint* arXiv:1012.0297)
[28] Popovych R O and Bihlo A 2012 Symmetry preserving parameterization schemes *J. Math. Phys.* **53** 073102 (*Preprint* arXiv:1010.3010)
[29] Meleshko S V 1994 Group classification of equations of two-dimensional gas motions *Prikl. Mat. Mekh.* **58** 56 [in Russian]; English. transl. in *J. Appl. Math. Mech.* **58** 629
[30] Olver P J 2000 *Application of Lie Groups to Differential Equations* (New York: Springer)

Equivalence groupoids of classes of linear ordinary differential equations and their group classification

Vyacheslav M Boyko[1], **Roman O Popovych**[1,2] **and Nataliya M Shapoval**[3]

[1] Institute of Mathematics of NAS of Ukraine, 3 Tereshchenkivs'ka Str., 01601 Kyiv-4, Ukraine
[2] Wolfgang Pauli Institut, Universität Wien, Oskar-Morgenstern-Platz 1, A-1090 Wien, Austria
[3] Faculty of Mechanics and Mathematics, Taras Shevchenko National University of Kyiv, 4e Academician Glushkov Ave., 03127 Kyiv, Ukraine

E-mail: boyko@imath.kiev.ua, rop@imath.kiev.ua, natalya.shapoval@gmail.com

Abstract. Admissible point transformations of classes of rth order linear ordinary differential equations (in particular, the whole class of such equations and its subclasses of equations in the rational form, the Laguerre–Forsyth form, the first and second Arnold forms) are exhaustively described. Using these results, the group classification of such equations is carried out within the algebraic approach in three different ways.

1. Introduction

The study of Lie symmetries of ordinary differential equations (ODEs) has a long history, and "Lie theory" was just started as a systematical and elegant approach to integration of various classes of ODEs. The first results on possible dimensions of the maximal Lie invariance algebras of ODEs of any fixed order were obtained by Sophus Lie, see, e.g., [1, pp 294–301] and a modern treatment in [2, Section 2]. Namely, Sophus Lie proved that the dimension of the maximal Lie invariance algebra of an rth order ODE is infinite for $r = 1$, not greater than 8 for $r = 2$ and not greater than $r + 4$ for $r \geqslant 3$. He also showed that each ODE of order $r = 1$ is similar with respect to a point transformation to the elementary equation $x' = 0$ and that, for equations of order $r \geqslant 2$, the maximal dimension of invariance algebras is reached for equations that are reduced by point transformations to the elementary equation $x^{(r)} = 0$; cf. [3, Theorem 14] and [4, Theorems 6.39 and 6.43].

In spite of the fact that transformational and, in particular, symmetry properties of linear ODEs were intensively investigated (see, e.g., detailed reviews [5–7] and textbooks [4, 8–10]), in the present paper we consider them from another side, describing the whole set of admissible transformations between such equations. This creates a basis for the group classification of linear ODEs within the framework of the algebraic approach, which is quite effective for solving group classification problems for both ordinary and partial differential equations; see, e.g., [11–15] and references therein. Previously, in [16, 17], the group classification of linear ODEs was carried out within the framework of the classical infinitesimal approach, which led to cumbersome calculations. Although the infinitesimal approach is the most commonly used in group analysis of differential equations, it is efficient only for classes of simple structure. In [4, pp 217–218] the solution of the group classification problem of linear ODEs was related to Wilczynski's result [18] on relative invariants of the Laguerre–Forsyth form of these equations. The similar problem on the classification of linear ODEs up to contact transformations as well as the associated equivalence problem were considered in detail in [19–21].

The main purpose of the present paper is to carry out the complete group classification of the class $\mathcal{L}$ of rth order ($r \geqslant 3$) linear ODEs in more elegant algebraic ways, using subalgebra analysis of the equivalence algebra associated with $\mathcal{L}$. This properly works since the class $\mathcal{L}$ is (pointwise) normalized (in the usual sense), i.e., transformations from its (usual) point equivalence (pseudo)group[1] $G^{\sim}$ generate all admissible point transformations[2] between equations from $\mathcal{L}$. The group classification of the class of second-order linear ODEs is trivial since its equivalence group acts transitively. Note that the equivalence group $G^{\sim}$ of the class $\mathcal{L}$ with $r \geqslant 2$ was first found by Stäckel [22].[3] The set of admissible transformations of any class of differential equations possesses the groupoid structure and is called the *equivalence groupoid* of this class [14, 15]. See, e.g., [13–15, 26] for the definition of normalized classes and other related notions. So, we can say that the equivalence groupoid $\mathcal{G}^{\sim}$ of the class $\mathcal{L}$ with $r \geqslant 3$ is generated by its equivalence group $G^{\sim}$.

In Section 2 we begin the study of the class $\mathcal{L}$ of rth order ($r \geqslant 2$) linear ODEs with the description of its equivalence groupoid in terms of its equivalence group and normalization. As a rule, we present only the components of equivalence transformations that correspond to the dependent and independent variables. One can gauge arbitrary elements of the class $\mathcal{L}$ by parameterized families of transformations from $G^{\sim}$, which induces mappings of the class $\mathcal{L}$ onto its subclasses. Two such gauges for arbitrary elements related to the subleading-order derivatives are well known. They result in the rational form with the first subleading coefficient being equal to zero (the subclass $\mathcal{L}_1$) and Laguerre–Forsyth form with the first two subleading coefficients being equal to zero (the subclass $\mathcal{L}_2$). It appears that for $r \geqslant 3$ both the subclasses $\mathcal{L}_1$ and $\mathcal{L}_2$ are also normalized with respect to their equivalence groups. Then we study two gauges for arbitrary elements related to the lowest-order derivatives, which give the first and second Arnold forms. The corresponding subclasses are even not semi-normalized and hence these gauges are not convenient for symmetry analysis. Having the chain of nested normalized classes $\mathcal{L} \supset \mathcal{L}_1 \supset \mathcal{L}_2$ for $r \geqslant 3$ and the associated chain of classes of homogeneous equations, which are peculiarly semi-normalized, we can classify Lie symmetries of rth order linear ODEs within the algebraic approach in three different ways, which is done in Section 3. In the final section we summarize results of the paper and discuss their connection with possible approaches to solving group classification problems for classes of systems of linear differential equations.

2. Equivalence groupoids of classes of linear ODEs

Consider the class $\mathcal{L}$ of rth order linear ODEs, which have the form

$$x^{(r)} + a_{r-1}(t)x^{(r-1)} + \cdots + a_1(t)x^{(1)} + a_0(t)x = b(t), \tag{1}$$

where a_{r-1}, ..., a_1, a_0 and b are arbitrary smooth functions of t, $x = x(t)$ is the unknown function, $x^{(k)} = d^k x/dt^k$, $k = 1, \ldots, r$, $r \geqslant 2$. Below we also use the notation $x' = dx/dt$

[1] There exist other names for this notion, e.g., "structure invariance group" [10]. The attribute "usual" and the prefix "pseudo-" are usually omitted for usual equivalence pseudogroups. We will also say "normalization" without attributes in the case of pointwise normalization in the usual sense.

[2] An *admissible (point) transformation* of a class of differential equations is a triple of the form $(\mathcal{E}, \tilde{\mathcal{E}}, \mathcal{T})$. Here $\mathcal{E}$ and $\tilde{\mathcal{E}}$ are equations from the class or, equivalently, the corresponding values of the arbitrary elements parameterizing the class. The transformational part $\mathcal{T}$ of the admissible transformation is a point transformation mapping the equation $\mathcal{E}$ to the equation $\tilde{\mathcal{E}}$.

[3] Recall also the contribution by Halphen [23], Laguerre [24], Forsyth [25] and Wilczynski [18] in the study of point transformations between linear ODEs.

and $x'' = d^2x/dt^2$ for the first and second derivatives, respectively. The subscripts t and x denote differentiation with respect to the corresponding variables. We assume that all variables, functions and other values are either real or complex, i.e., the underlying field $\mathbb{F}$ is either $\mathbb{F} = \mathbb{R}$ or $\mathbb{F} = \mathbb{C}$, respectively. We work within the local approach.

2.1. General class

There are two different cases for the structure of the equivalence groupoid of the entire class $\mathcal{L}$ depending on the value r, namely $r = 2$ and $r \geqslant 3$. We begin with the case $r = 2$.

Proposition 1. *The equivalence group $G^{\sim}$ of the class $\mathcal{L}$ with $r = 2$ consists of the transformations whose projections to the variable space*[4] *have the form*

$$\tilde{t} = T(t), \quad \tilde{x} = X_1(t)x + X_0(t), \tag{2}$$

where T, X_1 and X_0 are arbitrary smooth functions of t with $T_tX_1 \neq 0$.

Proof. Suppose that a point transformation $\mathcal{T}$ of the general form

$$\tilde{t} = T(t,x), \quad \tilde{x} = X(t,x), \tag{3}$$

where the Jacobian $J = |\partial(T,X)/\partial(t,x)|$ does not vanish, $J \neq 0$, connects two fixed second-order linear ODEs $\mathcal{E}$ and $\tilde{\mathcal{E}}$. We substitute the expressions for the new variables (which are with tildes) and the corresponding derivatives in terms of the old variables (which are without tildes) into $\tilde{\mathcal{E}}$. The equality obtained should be identically satisfied on solutions of the equation $\mathcal{E}$. Therefore, additionally substituting the expression for x'' implied by $\mathcal{E}$, we can split the equality with respect to the derivative x'. Collecting the coefficients of $(x')^3$, we obtain the equation

$$X_{xx}T_x - X_xT_{xx} + \tilde{a}_1X_xT_x^2 + (\tilde{a}_0X - \tilde{b})T_x^3 = 0.$$

As $\tilde{a}_1$, $\tilde{a}_0$ and $\tilde{b}$ are the only arbitrary elements involved in this equation and we study the equivalence group, we can vary the arbitrary elements and hence split with respect to them. Hence $T_x = 0$, i.e., $T = T(t)$. Then the terms with $(x')^2$ give the equation $T_tX_{xx} = 0$. As the condition $J \neq 0$ reduces to the inequality $T_tX_x \neq 0$, we have $X = X_1(t)x + X_0(t)$ with $X_1 \neq 0$. The other determining equations, which are derived by the additional splitting with respect to x' and x, define the transformation components for the arbitrary element as functions of the variables and the arbitrary elements. □

Proposition 2. *The equivalence groupoid of the class $\mathcal{L}$ of second-order linear ODEs is generated by compositions of transformations from the equivalence group $G^{\sim}$ of this class with transformations from the point symmetry group of the equation $x'' = 0$. Therefore, the class $\mathcal{L}$ is semi-normalized but not normalized.*

Proof. The free particle equation $x'' = 0$ admits point symmetry transformations that are truly fractional linear with respect to x or whose components for t depend on x. Each of these properties is not consistent with the transformation form (2). This is why there exists admissible transformations in the class $\mathcal{L}$ that are not generated by its equivalence transformations, i.e., this class is not normalized.

[4] Recall that for all equivalence transformations we present only the components corresponding to the variables t and x since each pair of these components completely determine the corresponding equivalence transformation. As the form (2) is the most general for relevant equivalence transformations, the transformation components for arbitrary elements can be derived using Faà di Bruno's formula and the general Leibniz rule, and thus they are quite cumbersome.

It is commonly known that any second-order linear ODE $\mathcal{E}$ is locally reduced to the equation $x''=0$ by an equivalence transformation, so-called Arnold transformation,

$$\tilde t=\frac{\varphi_2(t)}{\varphi_1(t)},\quad \tilde x=\frac{x-\varphi_0(t)}{\varphi_1(t)},\tag{4}$$

where φ_0 is a particular solution of the equation $\mathcal{E}$, φ_1 and φ_2 are linearly independent solutions of the corresponding homogeneous equation, see, e.g., [27, p 43] or [28]. In other words, the class $\mathcal{L}$ is a single orbit of its equivalence group $G^\sim$. Any class with this property is semi-normalized. We show this in detail. Consider two fixed equations $\mathcal{E}_1$ and $\mathcal{E}_2$ from the class $\mathcal{L}$ with $r=2$ and a point transformation $\mathcal{T}$ linking these equations. Let $\mathcal{T}_1$ and $\mathcal{T}_2$ be the projections of elements of $G^\sim$ that map the equations $\mathcal{E}_1$ and $\mathcal{E}_2$, respectively, to $x''=0$. Then the transformation $\mathcal{T}_0:=\mathcal{T}_2\mathcal{T}\mathcal{T}_1^{-1}$ belongs to the point symmetry group of the equation $x''=0$. This implies the representation $\mathcal{T}=\mathcal{T}_2^{-1}\mathcal{T}_0\mathcal{T}_1$. Roughly speaking, $\mathcal{T}$ is the composition of the equivalence transformations $\mathcal{T}_1$ and $\mathcal{T}_2^{-1}$ and the symmetry transformation $\mathcal{T}_0$ of the equation $x''=0$. It is obvious that any transformation possessing such a representation maps the equation $\mathcal{E}_1$ to the equation $\mathcal{E}_2$. The above representation can be rewritten as $\mathcal{T}=\check{\mathcal{T}}\hat{\mathcal{T}}$, where $\check{\mathcal{T}}=\mathcal{T}_2^{-1}\mathcal{T}_1$ is an equivalence transformation of the class $\mathcal{L}$ and $\hat{\mathcal{T}}=\mathcal{T}_1^{-1}\mathcal{T}_0\mathcal{T}_1$ is a symmetry transformation of the equation $\mathcal{E}_1$, which means that the class $\mathcal{L}$ is semi-normalized. □

Remark 3. The equivalence group $G^\sim$ induces all fiber-preserving admissible transformations of the class $\mathcal{L}$ with $r=2$. This directly follows from the fact that imposing the constraint $T_x=0$ for admissible transformations also implies the condition $X_{xx}=0$.

In what follows we consider the class $\mathcal{L}$ with $r\geqslant 3$. Although the projections of transformations from the equivalence group to the variable space in the case $r\geqslant 3$ coincide with that in the case $r=2$, the corresponding equivalence groupoids have different structures.

Proposition 4. *The equivalence group $G^\sim$ of the class $\mathcal{L}$, where $r\geqslant 3$, consists of the transformations whose projections to the variable space have the form* (2). *This group generates the entire equivalence groupoid of the class $\mathcal{L}$, i.e., the class $\mathcal{L}$ is normalized.*

Proof. In order to study admissible transformations in the class $\mathcal{L}$, we consider a pair of equations from this class, namely an equation $\mathcal{E}$ of the form (1) and an equation $\tilde{\mathcal{E}}$ of the same form, where all variables, derivatives and arbitrary elements are with tildes, and assume that these equations are connected by a point transformation $\mathcal{T}$ of the general form (3). At first we express derivatives with tildes in terms of the variables without tildes,

$$\tilde x^{(k)}=\left(\frac{1}{DT}D\right)^k X,$$

where $D=\partial_t+x'\partial_x+x''\partial_{x'}+\cdots$ is the total derivative operator with respect to the variable t. After substituting the expressions for the variables and derivatives with tildes into $\tilde{\mathcal{E}}$, we derive an equation in the variables without tildes. It should be an identity on the manifold determined by $\mathcal{E}$ in the rth order jet space with the independent variable t and the dependent variable x. The coefficient of $x''x^{(r-1)}$ in this equation equals

$$-\frac{J}{(DT)^{r+2}}T_x\left(3+\frac{(r-2)(r+3)}{2}\right)=0,$$

and hence $T_x=0$, i.e., the function T does not depend on the variable x, $T=T(t)$. The nondegeneracy condition $J\neq 0$ is simplified to $T_tX_x\neq 0$. Taking into account the condition $T_x=0$, we collect coefficients of $x'x^{(r-1)}$, which gives $rT_t^{-r}X_{xx}=0$. This equation implies that

$X_{xx}=0$, i.e., X is a linear function of x, $X=X_1(t)x+X_0(t)$. Therefore, the transformation $\mathcal{T}$ has the form (2). The other determining equations, which are obtained by splitting with respect to derivatives of x after substituting for $x^{(r)}$ in view of $\mathcal{E}$, establish the relation between arbitrary elements of the initial and the transformed equations.

The transformation $\mathcal{T}$ maps any equation from the class $\mathcal{L}$ to another equation from the same class, and its prolongation to the arbitrary elements a_{r-1}, …, a_0 and b, which is given by the above relation, is a point transformation in the joint space of the variables and the arbitrary elements. Hence such prolongations of the transformations of the form (2) constitute the (usual) equivalence group $G^{\sim}$. □

Consider the corresponding subclass $\widehat{\mathcal{L}}$ of rth order ($r\geqslant 2$) linear homogeneous ODEs, which is singled out from the class $\mathcal{L}$ by the constraint $b=0$. The arbitrary element b can be gauge to zero by equivalence transformations. Namely, the class $\mathcal{L}$ is mapped to its subclass $\widehat{\mathcal{L}}$ by a family of point transformations with $T=t$, $X^1=1$ and X^0 being a particular solution of the initial equation, and thus these transformations are parameterized by b.

Corollary 5. *The equivalence group $\widehat{G}^{\sim}$ of the subclass $\widehat{\mathcal{L}}$ is obtained from the equivalence group $G^{\sim}$ of the class $\mathcal{L}$ by setting $X_0=0$ and neglecting the transformation component for b.*

Corollary 6. *The subclass $\widehat{\mathcal{L}}$ with $r=2$ is a single orbit of the elementary equation $x''=0$ under the action of the equivalence group $\widehat{G}^{\sim}$. Hence the subclass $\widehat{\mathcal{L}}$ is semi-normalized but not normalized.*

Corollary 7. *Given any equation $\mathcal{E}$ from the subclass $\widehat{\mathcal{L}}$ with $r\geqslant 3$, a point transformation maps $\mathcal{E}$ to another equation from the same subclass if and only this transformation has the form* (2), *where the ratio X_0/X_1 is a solution of $\mathcal{E}$.*

In other words, the transformational part of any admissible transformation within the class $\widehat{\mathcal{L}}$ with $r\geqslant 3$ can be represented as the composition of a linear superposition symmetry transformation of the initial equation with the projection of an element of the equivalence group $\widehat{G}^{\sim}$ to the variable space. For all equations from $\widehat{\mathcal{L}}$, the associated groups of linear superposition symmetry transformations are of the same structure. In particular, they are commutative and r-dimensional. Therefore, although the class of $\widehat{\mathcal{L}}$ is not normalized, it is semi-normalized in a quite specific way, which is a particular case of so-called *uniform semi-normalization.*[5] For short, in similar situations we will say that a class is uniformly semi-normalized with respect to linear superposition of solutions.

2.2. Rational form

Using parameterized families of transformations from the equivalence group $G^{\sim}$, we can gauge arbitrary elements of the class $\mathcal{L}$. For example, we can set $a_{r-1}=0$. This gauge can be realized by the parameterized family of projections of equivalence transformations to the (t,x)-space

$$\tilde t=t,\quad \tilde x=\exp\left(\frac{1}{r}\int a_{r-1}(t)dt\right)x, \tag{5}$$

which maps the class $\mathcal{L}$ onto the subclass $\mathcal{L}_1$ of equations in the *rational form*

$$x^{(r)}+a_{r-2}(t)x^{(r-2)}+\dots+a_1(t)x'+a_0(t)x=b(t), \tag{6}$$

where we omitted tildes over the variables and the arbitrary elements. This form was used in [16, 17] for the group classification of linear ODEs within the framework of the infinitesimal approach.

[5] Similar properties are known for classes of homogeneous linear PDEs whose corresponding classes of (in general, inhomogeneous) linear PDEs are normalized [11].

Proposition 8. *The equivalence group $G_1^{\sim}$ of the subclass $\mathcal{L}_1$ consists of the transformations whose projections to the variable space have the form*

$$\tilde{t} = T(t), \quad \tilde{x} = C(T_t(t))^{\frac{r-1}{2}}x + X_0(t), \tag{7}$$

where T and X_0 are arbitrary smooth functions of t with $T_t \neq 0$,[6] *and C is an arbitrary nonzero constant. The subclass $\mathcal{L}_1$ with $r = 2$ is semi-normalized. If $r \geqslant 3$, then the group $G_1^{\sim}$ generates the equivalence groupoid of this subclass, i.e., it is normalized.*

Proof. In the case $r = 2$ we follow the proof of Proposition 1 and derive the form (2) for equivalence transformations of the subclass $\mathcal{L}_1$. Then further collecting coefficients of x' and x gives the equation

$$\frac{X_1}{T_t^2}\left(\frac{X_{1,t}}{X_1} - \frac{1}{2}\frac{T_{tt}}{T_t}\right) = 0,$$

which is integrated to the relation $X_1 = CT_t^{\frac{1}{2}}$ with an arbitrary nonzero constant C, as well as the equivalence transformation components for the arbitrary elements a_0 and b of the subclass $\mathcal{L}_1$ with $r = 2$. The semi-normalization of this subclass is proved in the same way as Proposition 2.

In the case $r \geqslant 3$ we describe the entire equivalence groupoid. Suppose that an equation $\mathcal{E}$ of the form (6) and an equation $\tilde{\mathcal{E}}$ of the same form, where all variables, derivatives and arbitrary elements are with tildes, are connected by a point transformation $\mathcal{T}$. In view of Proposition 4 this transformation has the form (2). We express the variables with tildes and the corresponding derivatives in terms of the variables and derivatives without tildes, substitute the expressions obtained into $\tilde{\mathcal{E}}$ and collect terms containing the derivative $x^{(r-1)}$, which gives

$$r\frac{X_1}{T_t^r}\left(\frac{X_{1,t}}{X_1} - \frac{r-1}{2}\frac{T_{tt}}{T_t}\right)x^{(r-1)}.$$

The coefficient of $x^{(r-1)}$ vanishes only if $X_1 = CT_t^{\frac{r-1}{2}}$, where C is an arbitrary nonzero constant. Therefore, the point transformation $\mathcal{T}$ has the form (7). Analogously to the end of the proof of Proposition 4, the transformations of this form when prolonged to the arbitrary elements $a_{r-2}, \ldots, a_0$ and b constitute the equivalence group $G_1^{\sim}$ of the subclass $\mathcal{L}_1$. □

The status and properties of the corresponding subclass $\widehat{\mathcal{L}}_1$ of homogeneous equations within the class $\mathcal{L}_1$ are the same as those of the subclass $\widehat{\mathcal{L}}$ within the class $\mathcal{L}$.

Corollary 9. *The equivalence group $\widehat{G}_1^{\sim}$ of the subclass $\widehat{\mathcal{L}}_1$ is derived from the equivalence group $G_1^{\sim}$ of the class $\mathcal{L}_1$ by setting $X_0 = 0$ and neglecting the transformation component for b.*

The equivalence groupoid of the subclass $\widehat{\mathcal{L}}_1$ is exhaustively described by the following two assertions depending on the value of r.

Corollary 10. *If $r = 2$, then the subclass $\widehat{\mathcal{L}}_1$ is semi-normalized since it is a single orbit of the elementary equation $x'' = 0$ under the action of the group $\widehat{G}_1^{\sim}$.*

Corollary 11. *For each equation $\mathcal{E}$ from the subclass $\widehat{\mathcal{L}}_1$ with $r \geqslant 3$, a point transformation is the transformational part of an admissible transformation in $\widehat{\mathcal{L}}_1$ with $\mathcal{E}$ as source if and only if it is of the form (7), where the product $T_t^{-\frac{r-1}{2}}X_0$ is a solution of $\mathcal{E}$. This means that this subclass is uniformly semi-normalized with respect to linear superposition of solutions.*

[6] For even r, the power of T_t is half-integer and hence the absolute value of T_t should be substituted instead of T_t in the real case or a branch of square root should be fixed in the complex case.

2.3. Laguerre–Forsyth form

Transformations from $G_1^{\sim}$ are parameterized by an arbitrary function $T = T(t)$ with $T_t \neq 0$. Hence we can set $a_{r-2} = 0$ in the equation (6) by a transformation from the group $G_1^{\sim}$, where the parameter-function T is a solution of the equation

$$T_{ttt}T_t - \frac{3}{2}T_{tt}^{\ 2} + \frac{12}{r(r^2-1)}a_{r-2}T_t^{\ 4} = 0.$$

Thus, a family of such equivalence transformations parameterized by the arbitrary element a_{r-2} maps the subclass $\mathcal{L}_1$ onto the narrower subclass $\mathcal{L}_2$ of equations in the *Laguerre–Forsyth form*

$$x^{(r)} + a_{r-3}(t)x^{(r-3)} + \cdots + a_1(t)x' + a_0(t)x = b(t). \tag{8}$$

Note that, in contrast to the transformation (5), the above map does not preserve the corresponding subclass of linear ODEs with constant coefficients.

Proposition 12. *The equivalence group $G_2^{\sim}$ of the subclass $\mathcal{L}_2$ with $r \geqslant 2$ consists of the transformations whose projections to the variable space have the form*

$$\tilde t = \frac{\alpha t + \beta}{\gamma t + \delta}, \quad \tilde x = \frac{C}{(\gamma t + \delta)^{r-1}}x + X_0(t), \tag{9}$$

where α, β, γ, δ and C are arbitrary constants with $\alpha\delta - \beta\gamma \neq 0$ and $C \neq 0$ that are defined up to obvious rescaling (so, only four constants among them are essential), and X_0 is an arbitrary smooth function of t. The subclass $\mathcal{L}_2$ is semi-normalized in the case $r = 2$. If $r \geqslant 3$, then the group $G_2^{\sim}$ generates the equivalence groupoid of this subclass, i.e., it is normalized.

Proof. We should again consider the cases $r = 2$ and $r \geqslant 3$ separately.

For $r = 2$ we repeat the proof of Proposition 1 and the corresponding part of the proof of Proposition 8 and derive the form (2) for equivalence transformations of the subclass $\mathcal{L}_2$. Then further collecting coefficients of x gives the equation

$$\frac{T_{ttt}}{T_t} - \frac{3}{2}\left(\frac{T_{tt}}{T_t}\right)^2 = 0. \tag{10}$$

In other words, the Schwarzian derivative of the function T vanishes, i.e., the function T is fractional linear,

$$T(t) = \frac{\alpha t + \beta}{\gamma t + \delta}, \tag{11}$$

where α, β, γ, δ are arbitrary constants with $\alpha\delta - \beta\gamma \neq 0$ that are defined up to nonvanishing constant multiplier. The rest of terms in the determining equation results in the equivalence transformation component for the arbitrary element b of the subclass $\mathcal{L}_2$ with $r = 2$. The semi-normalization of this subclass is proved in the same way as Proposition 2.

In order to describe the entire equivalence groupoid the subclass $\mathcal{L}_2$ in the case $r \geqslant 3$, we suppose that a point transformation $\mathcal{T}$ links equations $\mathcal{E}$ and $\tilde{\mathcal{E}}$ from the class $\mathcal{L}_2$. (We assume that all the values in the equation $\tilde{\mathcal{E}}$ are with tildes.) Thus, $\mathcal{T}$ has the form (7). We express the derivatives of $\tilde x$ with respect to $\tilde t$ in terms of (t, x), substitute these expressions into the equation $\tilde{\mathcal{E}}$ and then substitute the expression for $x^{(r)}$ implied by $\mathcal{E}$. Collecting the coefficients of $x^{(r-2)}$ gives the equation (10), i.e., the function T is of the form (11). We substitute the expression for T into (7) and obtain transformations that map any equation from the subclass $\mathcal{L}_2$ to another equation from the same subclass, and the new arbitrary elements functionally depend on the old variables and the old arbitrary elements. Therefore, prolongations of these transformations to the arbitrary elements $a_{r-3}, \ldots, a_0$ and b constitute the equivalence group $G_2^{\sim}$ of the subclass $\mathcal{L}_2$. □

Denote by $\widehat{\mathcal{L}}_2$ the subclass of homogeneous equations within the class $\mathcal{L}_2$. The case $r = 2$ is singular for transformational properties of $\widehat{\mathcal{L}}_2$ since then the only element of this subclass is the free particle equation $x'' = 0$. This implies that the subclass $\widehat{\mathcal{L}}_2$ is normalized, and its equivalence group coincides with the point symmetry group of the equation $x'' = 0$, which consists of fractional linear transformations in the space of (t, x). The case $r \geqslant 3$ for $\widehat{\mathcal{L}}_2$ is similar to ones for $\widehat{\mathcal{L}}$ and $\widehat{\mathcal{L}}_1$.

Corollary 13. *If $r \geqslant 3$, the equivalence group $\widehat{G}_2^\sim$ of the subclass $\widehat{\mathcal{L}}_2$ is derived from the equivalence group $G_2^\sim$ of the class $\mathcal{L}_2$ by setting $X_0 = 0$ and neglecting the transformation component for b. The subclass $\widehat{\mathcal{L}}_2$ is uniformly semi-normalized with respect to linear superposition of solutions. More precisely, the equivalence groupoid of this subclass is constituted by triples of the form $(\mathcal{E}, \tilde{\mathcal{E}}, \mathcal{T})$, where the equation-source $\mathcal{E}$ runs through the entire subclass $\widehat{\mathcal{L}}_2$, the transformational part $\mathcal{T}$ is of the form (9) with the product $(\gamma t+\delta)^{r-1}X_0$ being an arbitrary solution of $\mathcal{E}$, and the equation-target is defined by $\tilde{\mathcal{E}} = \mathcal{T}(\mathcal{E})$.*

Remark 14. In fact, the equivalence and admissible transformations for the classes $\widehat{\mathcal{L}}$, $\widehat{\mathcal{L}}_1$ and $\widehat{\mathcal{L}}_2$ were known in the literature for a long time due to Stäckel, Laguerre, Forsyth et al.; see, e.g., [10, Section 4.1] and [18, Chapter I and § II.4]. At the same time, we rigorously formulate these results and explicitly describe the associated equivalence groupoids by proving that these classes are normalized if $r \geqslant 3$ and semi-normalized if $r = 2$.

2.4. First Arnold form

The above gauges of arbitrary elements of the class $\mathcal{L}$, which are commonly used, concern the subleading coefficients a_{r-1} and a_{r-2}, but this is not a unique possibility. Instead of a_{r-1} and a_{r-2} one can gauge the lowest coefficients a_0 and a_1. We can set $a_0 = 0$ in any equation from the class $\mathcal{L}$ by an Arnold transformation

$$\tilde{t} = t, \quad \tilde{x} = \frac{x}{\varphi_1(t)},$$

where φ_1 is a nonzero solution of the corresponding homogeneous equation. As a result, we obtain the subclass $\mathcal{A}_1$ of the class $\mathcal{L}$ that is constituted by the equations of the form

$$x^{(r)} + a_{r-1}(t)x^{(r-1)} + \cdots + a_1(t)x^{(1)} = b(t). \tag{12}$$

Following the (eponymous) Arnold Principle,[7] we call this form the first Arnold form.

Proposition 15. *The equivalence groupoid of the class $\mathcal{A}_1$, where $r \geqslant 3$, is constituted by the admissible transformations whose equations-sources exhaust the whole class $\mathcal{A}_1$ and whose transformational parts are of the form*

$$\tilde{t} = T(t), \quad \tilde{x} = \frac{x}{\psi_1(t)} + X_0(t), \tag{13}$$

where T and X_0 are arbitrary smooth functions of t with $T_t \neq 0$, and $\psi_1 = \psi_1(t)$ is a nonzero solution of the homogeneous equation associated with the corresponding equation-source.

Proof. Let $\mathcal{T}$ be a point transformation between equations $\mathcal{E}$ and $\tilde{\mathcal{E}}$ of the form (12). Then the transformation $\mathcal{T}$ is of the general form (2). As the equation $\tilde{\mathcal{E}}$ belongs to the class $\mathcal{A}_1$, the function $\tilde{\psi}_1 \equiv 1$ is a solution of the corresponding homogeneous equation. Hence $\psi_1 = 1/X_1$ is a solution of the homogeneous equation associated with $\mathcal{E}$. Therefore, the transformation $\mathcal{T}$ is of the form (13). □

[7] The Arnold Principle states that if a notion bears a personal name, then this name is not the name of the discoverer. The Berry Principle extends the Arnold Principle by stating the following: the Arnold Principle is applicable to itself.

Remark 16. In the case $r = 2$, the same assertion is true for the fiber-preserving part of the equivalence groupoid of the class $\mathcal{A}_1$; cf. Remark 3.

Proposition 17. *The equivalence group $G^{\sim}_{\mathcal{A}_1}$ of the class $\mathcal{A}_1$, where $r \geqslant 2$, consists of the transformations whose projections to the variable space have the form*

$$\tilde{t} = T(t), \quad \tilde{x} = Cx + X_0(t), \tag{14}$$

where T and X_0 are arbitrary smooth functions of t with $T_t \neq 0$ and C is an arbitrary nonzero constant.

Proof. The projection of any transformation from the group $G^{\sim}_{\mathcal{A}_1}$ to the variable space is of the form (13). For $r \geqslant 3$, this is an obvious consequence of Proposition 15. In the case $r = 2$, similarly to the proof of Proposition 1, we can first show that projections of equivalence transformations to the variable space are fiber-preserving and then take into account Remark 16.

Note that constant functions are solutions of any homogeneous equation from the class $\mathcal{A}_1$. Therefore, the group $G^{\sim}_{\mathcal{A}_1}$ contains the transformations whose restrictions on the space of (t, x) have the form (14). Moreover, only these transformations are in the group $G^{\sim}_{\mathcal{A}_1}$. Indeed, consider a transformation $\mathcal{T}$ of the form (13) with $\psi_1 \neq \text{const}$. Then there exists a homogeneous equation from $\mathcal{A}_1$ that is not satisfied by the function ψ_1. Hence the coefficient $\tilde{a}_0$ of the corresponding transformed equation is nonzero. This means that the transformation $\mathcal{T}$ is not a projection of an element of the group $G^{\sim}_{\mathcal{A}_1}$. □

By $\widehat{\mathcal{A}}_1$ we denote the subclass consisting of homogeneous equations of the form (12), i.e., singled out from the class $\mathcal{A}_1$ by the constraint $b = 0$.

Corollary 18. *The equivalence group $\widehat{G}^{\sim}_{\mathcal{A}_1}$ of the subclass $\widehat{\mathcal{A}}_1$ with $r \geqslant 2$ is obtained from the group $G^{\sim}_{\mathcal{A}_1}$ by setting $X_0 = 0$ and neglecting the transformation component for the arbitrary element b. A point transformation $\mathcal{T}$ relates two equations from this subclass if and only if it has the form (13), where the product $\psi_1 X_0$ is a solution of the corresponding initial equation.*

Corollary 19. *The classes $\mathcal{A}_1$ and $\widehat{\mathcal{A}}_1$ with $r \geqslant 3$ are not semi-normalized.*

Proof. There exists an equation $\mathcal{E}$ in $\widehat{\mathcal{A}}_1 \subset \mathcal{A}_1$ whose point symmetry group consists only of transformations related to the linearity and the homogeneity of $\mathcal{E}$.[8] These transformations are of the form $\tilde{t} = t$, $\tilde{x} = \widehat{C}x + \widehat{X}^0(t)$, where $\widehat{C}$ is an arbitrary nonzero constant and $\widehat{X}_0$ is an arbitrary solution of $\mathcal{E}$. The compositions of symmetry transformations of $\mathcal{E}$ and projections of transformations from $G^{\sim}_{\mathcal{A}_1}$ (resp. $\widehat{G}^{\sim}_{\mathcal{A}_1}$) to the variable space are at most of the form (14). At the same time, the equation $\mathcal{E}$ has nonconstant solution. This implies that some admissible transformations of $\mathcal{E}$, which are of the form (13), are not generated by the above compositions. □

Remark 20. Analogously to Proposition 2, the classes $\mathcal{A}_1$ and $\widehat{\mathcal{A}}_1$ with $r = 2$ are semi-normalized since are orbits of the elementary equation $x'' = 0$ with respect to the equivalence groups $G^{\sim}_{\mathcal{A}_1}$ and $\widehat{G}^{\sim}_{\mathcal{A}_1}$, respectively.

[8] A common fact is that similar equations have similar point symmetry groups. Moreover, any admissible transformation in the class $\bar{\mathcal{L}}$ maps point symmetries associated to the linearity and homogeneity of the corresponding initial equation to symmetries of the same kind that are admitted by the target equation. Hence it suffices to find an equation $\bar{\mathcal{E}}$ with trivial point symmetries in the class $\widehat{\mathcal{L}}_2$. In view of Corollary 13, possible point transformations of $\bar{\mathcal{E}}$ within the class $\widehat{\mathcal{L}}_2$ are exhausted, up to point symmetries associated to the linearity and the homogeneity of $\bar{\mathcal{E}}$, by the transformations of the form (9), where $C = 1$ and $X^0 = 0$. Equations that are not invariant with respect to any of such transformations exist even among equations from $\widehat{\mathcal{L}}_2$ with polynomial coefficients.

2.5. Second Arnold form

Using a transformation of the form (4) with $\varphi_0 = 0$, we can set additionally $a_1 = 0$ in any equation from the class $\mathcal{A}_1$. In this way the class $\mathcal{A}_1$ (as well as the entire class $\mathcal{L}$) is mapped onto its subclass $\mathcal{A}_2$ that consists of the equations of the form

$$x^{(r)} + a_{r-1}(t)x^{(r-1)} + \cdots + a_2(t)x^{(2)} = b(t), \tag{15}$$

called the second Arnold form.

Proposition 21. *The equivalence groupoid of the subclass $\mathcal{A}_2$, where $r \geqslant 3$, is constituted by the admissible transformations whose equations-sources exhaust the whole class $\mathcal{A}_2$ and whose transformational parts are of the form*

$$\tilde{t} = \frac{\psi_2(t)}{\psi_1(t)}, \quad \tilde{x} = \frac{x}{\psi_1(t)} + X_0(t), \tag{16}$$

where $\psi_1 = \psi_1(t)$ and $\psi_2 = \psi_2(t)$ are arbitrary linearly independent solutions of the homogeneous equation associated with the corresponding equation-source, and X_0 is an arbitrary smooth function of t.

Proof. Suppose that a point transformation $\mathcal{T}$ connects two equations $\mathcal{E}$ and $\tilde{\mathcal{E}}$ from the class $\mathcal{A}_2$. Then the transformation $\mathcal{T}$ has the form (13). Note that the function $\tilde{\psi}_2 = \tilde{t}$ is a solution of the homogeneous equation associated with $\tilde{\mathcal{E}}$. Hence the function $\psi_2 = \psi_1 T$ is a solution of the homogeneous equation corresponding to $\mathcal{E}$, i.e., $T = \psi_2/\psi_1$. As a result, the transformation $\mathcal{T}$ is of the form (16). □

Proposition 22. *The equivalence group $G^{\sim}_{\mathcal{A}_2}$ of the subclass $\mathcal{A}_2$, where $r \geqslant 3$, consists of the transformations whose projections to the variable space have the form*

$$\tilde{t} = \frac{\alpha t + \beta}{\gamma t + \delta}, \quad \tilde{x} = \frac{x}{\gamma t + \delta} + X_0(t), \tag{17}$$

where α, β, γ and δ are arbitrary constants with $\alpha\delta - \beta\gamma \neq 0$.

Proof. Any transformation from the group $G^{\sim}_{\mathcal{A}_2}$ generates a family of elements from the equivalence groupoid of the subclass $\mathcal{A}_2$, and hence its projection to the variable space has the form (16). Since all linear functions of t are solutions of any homogeneous equation from the subclass $\mathcal{A}_2$, then the group $G^{\sim}_{\mathcal{A}_2}$ contains the transformations whose restrictions on the variable space are of the form (17). We prove that there are no other transformations in the group $G^{\sim}_{\mathcal{A}_2}$, i.e., a transformation $\mathcal{T}$ of the form (16) does not induce an equivalence transformation of the class $\mathcal{A}_2$ if at least one of the parameter-functions ψ_1 or ψ_2 is a nonlinear function of t. First consider the case where ψ_1 is nonlinear. We take a homogeneous equation $\mathcal{E}$ from the subclass $\mathcal{A}_2$ that is not satisfied by ψ_1. Then the corresponding transformed equation $\tilde{\mathcal{E}}$ possesses no constant solutions and hence its coefficient $\tilde{a}_0$ is nonzero. This means that the equation $\tilde{\mathcal{E}}$ does not belong to the subclass $\mathcal{A}_2$, which gives the necessary statement. Now consider the complementary case, namely, where the function ψ_1 is linear and the function ψ_2 is nonlinear. Then there is a homogeneous equation $\mathcal{E}$ of the form (15) that is not satisfied by ψ_2. As ψ_1 is a solution of the equation $\mathcal{E}$, the coefficient $\tilde{a}_0$ is equal to zero in the corresponding transformed equation $\tilde{\mathcal{E}}$. Moreover, the function $\tilde{\psi}_2 \equiv \tilde{t}$ is not a solution of the equation $\tilde{\mathcal{E}}$, so its coefficient $\tilde{a}_1$ is nonzero. Therefore, the equation $\tilde{\mathcal{E}}$ is not contained in the subclass $\mathcal{A}_2$, which completes the proof. □

Transformational properties of the subclass $\widehat{\mathcal{A}}_2$ of homogeneous equations from $\mathcal{A}_2$ with $r \geqslant 3$ are similar to ones of the class $\mathcal{A}_1$.

Corollary 23. *The equivalence group $\hat{G}^{\sim}_{\mathcal{A}_2}$ of the subclass $\hat{\mathcal{A}}_2$ with $r \geqslant 3$ is obtained from the group $G^{\sim}_{\mathcal{A}_2}$ by setting $X_0 = 0$ and neglecting the transformation component for the arbitrary element b. A point transformation relates two equations from this subclass if and only if it has the form* (16), *where the product $\psi_1 X_0$ is a solution of the corresponding initial equation.*

Corollary 24. *The classes $\mathcal{A}_2$ and $\hat{\mathcal{A}}_2$ with $r \geqslant 3$ are not semi-normalized.*

Proof. Following the proof of Corollary 19, consider an admissible transformation $(\mathcal{E}_1, \mathcal{E}_2, \mathcal{T})$ in the class $\hat{\mathcal{A}}_2 \subset \mathcal{A}_2$, where the point symmetry group of $\mathcal{E}_1$ consists only of transformations related to the linearity and the homogeneity of $\mathcal{E}_1$, in the representation (16) for $\mathcal{T}$ the solution ψ_1 of $\mathcal{E}_1$ is not affine in t, $\psi_1'' \neq 0$, and $\mathcal{E}_2 = \mathcal{T}(\mathcal{E}_1)$. At the same time, the compositions of symmetry transformations of $\mathcal{E}_1$ with projections of transformations from $G^{\sim}_{\mathcal{A}_2}$ (resp. $\hat{G}^{\sim}_{\mathcal{A}_2}$) to the variable space are at most of the form (17). Therefore, the admissible transformation $(\mathcal{E}_1, \mathcal{E}_2, \mathcal{T})$ is not generated by one of the above compositions. □

Remark 25. For $r = 2$, the classes $\mathcal{A}_2$ and $\hat{\mathcal{A}}_2$ coincides with the classes $\mathcal{L}_2$ and $\hat{\mathcal{L}}_2$, respectively.

3. Group classification

Any inhomogeneous linear ODE can be reduced to the corresponding homogeneous equation by a transformation from the equivalence group $G^{\sim}$. The class $\mathcal{L}$ is normalized and its subclass $\hat{\mathcal{L}}$ of homogeneous equations is semi-normalized. Therefore, for solving the group classification problem for the class $\mathcal{L}$ it suffices to solve the similar problem for the subclass $\hat{\mathcal{L}}$.

As remarked earlier, any second-order linear ODE can be reduced to the elementary equation $x'' = 0$ (see, e.g., [29]), which admits the eight-dimensional Lie invariance algebra

$$\langle \partial_t,\ \partial_x,\ t\partial_t,\ x\partial_t,\ t\partial_x,\ x\partial_x,\ tx\partial_t + x^2\partial_x,\ t^2\partial_t + tx\partial_x \rangle.$$

This gives the exhaustive group classification of second-order linear ODEs.

Let $r \geqslant 3$. It is also a classical result by Sophus Lie [1, pp 296–298] that the dimension of Lie invariance algebras for rth order ODEs with $r \geqslant 3$ is not greater than $r + 4$. Much later this result was partially reproved in [16, 17] only for linear ODEs.

Consider an rth order linear homogeneous ODE $\mathcal{E}$. Due to linearity leading to the linear superposition principle, this equation admits the r-dimensional abelian Lie invariance algebra $\mathfrak{g}^{\mathcal{E}}_{\rm a}$ spanned by the vector fields

$$\varphi_1(t)\partial_x,\ \varphi_2(t)\partial_x,\ \dots,\ \varphi_r(t)\partial_x, \tag{18}$$

where the functions $\varphi_i = \varphi_i(t)$, $i = 1, \dots, r$, form a fundamental set of solutions of the equation $\mathcal{E}$. By virtue of homogeneity, the equation $\mathcal{E}$ also admits the one-parameter symmetry group of scale transformations generated by the vector field $x\partial_x$. Therefore, each equation $\mathcal{E}$ from the subclass $\hat{\mathcal{L}}$ admits the $(r+1)$-dimensional Lie invariance algebra

$$\mathfrak{g}^{\mathcal{E}}_0 = \langle x\partial_x,\ \varphi_1(t)\partial_x,\ \varphi_2(t)\partial_x,\ \dots,\ \varphi_r(t)\partial_x \rangle. \tag{19}$$

Corollary 7 implies that any Lie symmetry operator Q of the equation $\mathcal{E}$ is of the general form $Q = \tau(t)\partial_t + (\xi_1(t)x + \xi_0(t))\partial_x$, where τ, ξ_1 and ξ_0 are smooth functions of t, and ξ_0 is additionally a solution of $\mathcal{E}$. Hence the maximal Lie invariance algebra $\mathfrak{g}^{\mathcal{E}}$ of $\mathcal{E}$ contains the subalgebra $\mathfrak{g}^{\mathcal{E}}_0$ as an ideal.

Now we carry out the group classification of linear ODEs in three different ways, which respectively involve the Laguerre–Forsyth form (8), the rational form (6) and Lie's classification of realizations of finite-dimensional Lie algebras in the space of two variables.

3.1. The first way: Laguerre–Forsyth form

The group classification of the class $\widehat{\mathcal{L}}$ can be further reduced to the group classification of its subclass $\widehat{\mathcal{L}}_2$ of rth order homogeneous linear ODEs in the Laguerre–Forsyth form, which is singled out from $\widehat{\mathcal{L}}$ by the constraint $a_{r-1} = a_{r-2} = 0$. Indeed, both the arbitrary elements a_{r-1} and a_{r-2} can be gauged to zero by a family of point transformations, which are parameterized by these arbitrary elements and are associated with equivalence transformations. Moreover, equations from the class $\widehat{\mathcal{L}}$ are $\widehat{G}^{\sim}$-equivalent if and only if their images in the class $\widehat{\mathcal{L}}_2$ are $\widehat{G}_2^{\sim}$-equivalent.

In view of Corollary 13, the class $\widehat{\mathcal{L}}_2$ is not normalized. At the same time, it is uniformly semi-normalized with respect to linear superposition of solutions (i.e., with respect to the symmetry groups of its equations related to linear superposition of their solutions). This suffices for using an advanced version of the algebraic method to group classification of the class $\widehat{\mathcal{L}}_2$. The equivalence group $\widehat{G}_2^{\sim}$ of $\widehat{\mathcal{L}}_2$ consists of the transformations whose projections to the variable space are of the form (9) with $X_0 = 0$. The scalings of x constitute the kernel group $\widehat{G}_2^{\cap}$ of $\widehat{\mathcal{L}}_2$ since they are only common point symmetry transformations for all equations from the class $\widehat{\mathcal{L}}_2$. By trivial prolongation to the arbitrary elements, the group $\widehat{G}_2^{\cap}$ is embedded into $\widehat{G}_2^{\sim}$ as a normal subgroup. The factor group $\widehat{G}_2^{\sim}/\widehat{G}_2^{\cap}$ can by identified with the subgroup H of $\widehat{G}_2^{\sim}$ singled out by the constraint $C = 1$, where additionally the expression $\alpha\delta - \beta\gamma$ is equal to 1 or ± 1 in the complex or real case, respectively. Up to point symmetry transformations associated with the linear superposition principle and homogeneity, i.e., related to the Lie algebra $\mathfrak{g}_0^{\mathcal{E}}$, all admissible transformations of the equation $\mathcal{E}$ within $\widehat{\mathcal{L}}_2$ are exhausted by the projections of elements from H to the variable space. The subgroup H is isomorphic to the projective general linear group PGL$(2,\mathbb{F})$ of fractional linear transformations of t.

Therefore, all possible Lie symmetry extensions within the class $\widehat{\mathcal{L}}_2$ are necessarily associated with subgroups of PGL$(2,\mathbb{F})$. In infinitesimal terms, the maximal Lie invariance algebra of the equation $\mathcal{E}$ is a semidirect sum of the algebra $\mathfrak{g}_0^{\mathcal{E}}$ and a subalgebra of the realization of $\mathfrak{sl}(2,\mathbb{F})$ spanned by the vector fields

$$\mathcal{P} = \partial_t, \quad \mathcal{D} = t\partial_t + \tfrac{1}{2}(r-1)x\partial_x, \quad \mathcal{K} = t^2\partial_t + (r-1)tx\partial_x.$$

Subalgebras of $\mathfrak{sl}(2,\mathbb{F})$ are well known (see, e.g., [30] or the appendix in the arXiv version of [31]). Up to internal automorphisms of $\mathfrak{sl}(2,\mathbb{F})$, a complete list of subalgebras of $\mathfrak{sl}(2,\mathbb{F})$ is exhausted by the zero subalgebra $\{0\}$, the one-dimensional subalgebras $\langle\mathcal{P}\rangle$, $\langle\mathcal{D}\rangle$ and, only for $\mathbb{F} = \mathbb{R}$, $\langle\mathcal{P}+\mathcal{K}\rangle$,[9] the two-dimensional subalgebra $\langle\mathcal{P}, \mathcal{D}\rangle$ and the entire realization $\langle\mathcal{P}, \mathcal{D}, \mathcal{K}\rangle$. The equivalence of subalgebras of $\mathfrak{sl}(2,\mathbb{F})$ well agrees with the similarity of equations within the class $\widehat{\mathcal{L}}_2$.

The zero subalgebra $\{0\}$ corresponds to the general case with no extension.

For one-dimensional extensions of algebras of the form (19) by the subalgebras $\langle\mathcal{P}\rangle$, $\langle\mathcal{D}\rangle$ and $\langle\mathcal{P}+\mathcal{K}\rangle$, the corresponding equations from the class $\widehat{\mathcal{L}}_2$ respectively take the forms

$$x^{(r)} + c_{r-3}x^{(r-3)} + \cdots + c_1x' + c_0x = 0, \tag{20}$$

$$x^{(r)} + c_{r-3}t^{-3}x^{(r-3)} + \cdots + c_1t^{-r+1}x' + c_0t^{-r}x = 0, \tag{21}$$

$$x^{(r)} + q_{r-3}(t)x^{(r-3)} + \cdots + q_1(t)x' + q_0(t)x = 0, \tag{22}$$

[9] The subalgebra $\langle\mathcal{P}+\mathcal{K}\rangle$ is equivalent to $\langle\mathcal{P}\rangle$ if $\mathbb{F} = \mathbb{C}$. This is why the part of the consideration corresponding to the subalgebra $\langle\mathcal{P}+\mathcal{K}\rangle$ can be then just neglected in the complex case.

where $c_0, \dots, c_{r-3}$ are arbitrary constants,

$$q_{r-3}(t) = \frac{c_{r-3}}{(1+t^2)^3},$$

$$q_m(t) = \frac{c_m}{(1+t^2)^{r-m}} - \frac{(m+1)(r-m-1)}{(1+t^2)^{r-m}} \int (1+t^2)^{r-m-1} q_{m+1}(t)dt, \quad m = r-4, \dots, 0,$$

and the integral denotes a fixed antiderivative. However, by the point transformations[10]

$$\tilde t = \ln t,\ \tilde x = xt^{-(r-1)/2} \quad \text{and} \quad \tilde t = \arctan t,\ \tilde x = x\big(1+t^2\big)^{-(r-1)/2}$$

the maximal Lie invariance algebras of equations of the form (21) (known as the Euler–Cauchy equation, or just Euler's equation) and the form (22) are reduced to the algebras looking as

$$\langle \partial_{\tilde t},\ \tilde x\partial_{\tilde x},\ \tilde\varphi_1(\tilde t)\partial_{\tilde x},\ \tilde\varphi_2(\tilde t)\partial_{\tilde x},\ \dots,\ \tilde\varphi_r(\tilde t)\partial_{\tilde x}\rangle. \tag{23}$$

Moreover, the above transformations map these equations to constant-coefficient equations from the class (6), where $a_{r-2} = -\frac{1}{24}r(r^2-1)$ for (21) and $a_{r-2} = \frac{1}{6}r(r^2-1)$ for (22). Additionally scaling t, we can set $a_{r-2} = -1$ and $a_{r-2} = 1$, respectively. Thus, any equation from $\widehat{\mathcal L}_2$ that admits an $(r{+}2)$-dimensional Lie invariance algebra is equivalent to a homogeneous equation with constant coefficients from the class (6), in which $a_{r-2} = 0$, $a_{r-2} = -1$ and $a_{r-2} = 1$ for (20), (21) and (22), respectively.

If an equation from $\widehat{\mathcal L}_2$ possesses the $(r{+}3)$-dimensional Lie invariance algebra

$$\langle \partial_t,\ t\partial_t + \tfrac12(r-1)x\partial_x,\ x\partial_x,\ \varphi_1(t)\partial_x,\ \varphi_2(t)\partial_x,\ \dots,\ \varphi_r(t)\partial_x\rangle,$$

then it has the form $x^{(r)} = 0$ and hence its maximal Lie invariance algebra is $(r{+}4)$-dimensional,

$$\langle \partial_t,\ t\partial_t + \tfrac12(r-1)x\partial_x,\ t^2\partial_t + (r-1)tx\partial_x,\ x\partial_x,\ \varphi_1(t)\partial_x,\ \varphi_2(t)\partial_x,\ \dots,\ \varphi_r(t)\partial_x\rangle. \tag{24}$$

As the functions $\varphi_1, \dots, \varphi_r$ form a fundamental set of solutions of the elementary equation, we can choose $\varphi_i = t^{i-1}$, $i = 1, \dots, r$. Thus, there is no rth order linear ODE whose maximal Lie invariance algebra is $(r{+}3)$-dimensional. Moreover, if such an equation admits an $(r{+}4)$-dimensional Lie invariance algebra, then it is similar to the elementary equation $x^{(r)} = 0$ with respect to a point transformation of the form (2).

3.2. The second way: rational form

Consider now the subclass $\widehat{\mathcal L}_1$ of homogeneous ODEs of the rational form (6). Again, the group classification of the class $\widehat{\mathcal L}$ reduces to that of the subclass $\widehat{\mathcal L}_1$ since the gauge $a_{r-1} = 0$ singling out the subclass $\widehat{\mathcal L}_1$ from $\widehat{\mathcal L}$ is realized by a family of equivalence transformations. Moreover, the subclass $\widehat{\mathcal L}_1$ is uniformly semi-normalized with respect to linear superposition of solutions.

Each equation $\mathcal E$ from $\widehat{\mathcal L}_1$ possesses the $(r{+}1)$-dimensional Lie invariance algebra $\mathfrak g_0^{\mathcal E}$, which is an ideal of the maximal Lie invariance algebra $\mathfrak g^{\mathcal E}$ of $\mathcal E$. At the same time, in contrast to the Laguerre–Forsyth form, the equivalence group $G_1^\sim$ of $\widehat{\mathcal L}_1$ is parameterized by an arbitrary function $T = T(t)$. Corollary 11 implies that the algebra $\mathfrak g^{\mathcal E}$ is contained in the algebra $\langle \mathcal R(\tau)\rangle + \mathfrak g_0^{\mathcal E}$, where plus denotes the sum of vector spaces,

$$\mathcal R(\tau) = \tau(t)\partial_t + \tfrac12(r-1)\tau_t(t)x\partial_x,$$

[10] For the first transformation, the absolute value of t should be substituted instead of t in the real case or branches of the ln and power functions should be fixed in the complex case.

and the parameter τ runs through the set of smooth functions of t. It is easy to see that $[\mathcal{R}(\tau^1), \mathcal{R}(\tau^2)] = \mathcal{R}(\tau^1\tau^2_t - \tau^2\tau^1_t)$. Hence $\mathfrak{g}^{\mathcal{E}}_1 := \langle\mathcal{R}(\tau)\rangle \cap \mathfrak{g}^{\mathcal{E}}$ is a (finite-dimensional) subalgebra of $\mathfrak{g}^{\mathcal{E}}$. Each vector field $\mathcal{R}(\tau)$ is completely defined by its projection $\mathrm{pr}_t\mathcal{R}(\tau)$ to the space of the variable t, $\mathrm{pr}_t\mathcal{R}(\tau) = \tau(t)\partial_t$. In other words, the algebras $\mathfrak{g}^{\mathcal{E}}_1$ and $\mathrm{pr}_t\mathfrak{g}^{\mathcal{E}}_1$ are isomorphic. Moreover, the corresponding projection of the equivalence group of the subclass $\widehat{\mathcal{L}}_1$ coincides with the group of all local diffeomorphisms in the space of the variable t.

As a result, the group classification of the class $\widehat{\mathcal{L}}_1$ reduces to the classification of realizations of finite-dimensional Lie algebras by vector fields in the space of the single variable t. The latter classification is well known and was done by Sophus Lie himself. A complete list of inequivalent realizations on the line is exhausted by the algebras $\{0\}$, $\langle\partial_t\rangle$, $\langle\partial_t, t\partial_t\rangle$ and $\langle\partial_t, t\partial_t, t^2\partial_t\rangle$, which gives $\{0\}$, $\langle\mathcal{P}\rangle$, $\langle\mathcal{P}, \mathcal{D}\rangle$ and $\langle\mathcal{P}, \mathcal{D}, \mathcal{K}\rangle$ as possible inequivalent Lie symmetry extensions within the class $\mathcal{L}_1$. Here $\mathcal{P} = \mathcal{R}(1)$, $\mathcal{D} = \mathcal{R}(t)$ and $\mathcal{K} = \mathcal{R}(t^2)$ are the same operators as those in the first way. If the equation $\mathcal{E}$ admits the two-dimensional extension $\langle\mathcal{P}, \mathcal{D}\rangle$, then it coincides with the elementary equation $x^{(r)} = 0$, which possesses the three-dimensional extension $\langle\mathcal{P}, \mathcal{D}, \mathcal{K}\rangle$. This is why the two-dimensional extension is improper. Finally, we have three inequivalent cases of Lie symmetry extensions in the class $\mathcal{L}$, which are

- the general case with no extension,
- general constant-coefficient equations admitting the one-dimensional extension $\langle\mathcal{P}\rangle$, and
- the elementary equation $x^{(r)} = 0$ possessing the three-dimensional extension $\langle\mathcal{P}, \mathcal{D}, \mathcal{K}\rangle$.

3.3. The third way: general form

The group classification of the entire class $\widehat{\mathcal{L}}$ can also be obtained directly from Lie's classification of realizations of finite-dimensional Lie algebras in the spaces of two real or complex variables [29, 32]. A modern treatment of these results on realizations was presented, e.g., in [3, 4, 33, 34]. In order to solve the group classification problem for the class $\widehat{\mathcal{L}}$, from Lie's list of realizations we select candidates for the maximal invariance algebras of equations from $\widehat{\mathcal{L}}$. All candidates should satisfy the following obvious properties, which are preserved by point transformations:

- The maximal Lie invariance algebra $\mathfrak{g}^{\mathcal{E}}$ of each rth order linear ODE $\mathcal{E}$ ($r \geqslant 3$) contains the (r+1)-dimensional almost abelian ideal $\mathfrak{g}^{\mathcal{E}}_0$. More precisely, the ideal $\mathfrak{g}^{\mathcal{E}}_0$ is the semidirect sum of an r-dimensional abelian ideal $\mathfrak{g}^{\mathcal{E}}_{\mathrm{a}}$ of the whole algebra $\mathfrak{g}^{\mathcal{E}}$ and the linear span of one more vector field whose adjoint action on the ideal $\mathfrak{g}^{\mathcal{E}}_{\mathrm{a}}$ is the identity operator.
- Moreover, the ideal $\mathfrak{g}^{\mathcal{E}}_0$ is an intransitive Lie algebra of vector fields, $\mathrm{rank}\,\mathfrak{g}^{\mathcal{E}}_0 = 1$.

The above properties are satisfied by realization families 21, 23, 26 and 28 from [33, Table 1] (or realization families 3.2, 1.6, 1.9 and 1.11 from [4, pp 472–473], or realization families 49, 51, 54 and 56 from [34, Table 1.1], respectively). As in the previous two ways, among rth order linear ODEs, only the elementary equation $x^{(r)} = 0$ admits the third realization. At the same time, this equation also possesses the fourth realization, which is of greater dimension than the third one. This is why the third realization should be neglected.

The equivalence within the chosen families of realizations well conforms with the point equivalence of linear ODEs. Indeed, given two such realizations that are equivalent with respect to a point transformation $\mathcal{T}$ and whose (r+1)-dimensional almost abelian ideals are of the form (19), the transformation $\mathcal{T}$ should have the form (2), where X_0/X_1 is a linear combinations of the parameters-functions $\varphi_1, \ldots, \varphi_r$ of the initial realization. The proof of this claim is based on two facts. The first fact is trivial: The mapping generated by a point transformation between realizations of a Lie algebra by vector fields is a Lie algebra isomorphism and, in particular, establishes a bijection between the corresponding nilradicals. The second fact is that nilradicals of appropriate realizations are spanned by vector fields $\varphi_1(t)\partial_x, \ldots, \varphi_r(t)\partial_x$, where $\varphi_i = \varphi_i(t)$,

$i = 1, \dots, r$, are linearly independent functions. This fact is obvious for realizations reducing to the forms (19) and (24). Suppose that it is not the case for a realization reducing to the form (23), where tildes over all values are omitted. Then for some constant ν the corresponding nilradical includes the vector field $\partial_t + \nu x \partial_x$, and commutation relations within the nilradical imply the existence of a constant nilpotent matrix (μ_{ij}) such that $\varphi_i' - \nu\varphi_i = \mu_{i1}\varphi_1 + \dots + \mu_{ir}\varphi_r$. The constant ν can be set to zero by a point transformation $\bar t = t$, $\bar x = e^{\nu t}x$. As the functions φ_i are linearly independent, up to their linear combining the matrix (μ_{ij}) can be assumed to coincide with the $r \times r$ nilpotent Jordan block, and hence we can set $\varphi_i = t^{i-1}$. At the same time, the only equation that belongs to the class $\widehat{\mathcal{L}}$ and is invariant with respect to the algebra $\langle \partial_t, t\partial_t, \dots, t^{r-1}\partial_t \rangle$ is the elementary equation $x'' = 0$, but the maximal Lie invariance algebra of this equation is of higher dimension.

As a result, using the algebraic method of group classification we have reproved the following assertion in three different ways (see, e.g., [4, 10, 16, 17, 21]):

Proposition 26. *The dimension of the maximal Lie invariance algebra $\mathfrak{g}^{\mathcal{E}}$ of an rth order ($r \geqslant 3$) linear ODE $\mathcal{E}$ takes a value from $\{r+1, r+2, r+4\}$. If $\dim \mathfrak{g}^{\mathcal{E}} \geqslant r+2$, then the equation $\mathcal{E}$ is similar to a linear ODE with constant coefficients. In the case $\dim \mathfrak{g}^{\mathcal{E}} = r+4$ the equation $\mathcal{E}$ is reduced by a point transformation of the form (2) to the elementary equation $x^{(r)} = 0$.*

4. Conclusion

In this paper we exhaustively describe the equivalence groupoid of the class $\mathcal{L}$ of rth order linear ODEs as well as equivalence groupoids of its subclasses $\mathcal{L}_1$, $\mathcal{L}_2$, $\mathcal{A}_1$ and $\mathcal{A}_2$ associated with the rational, the Laguerre–Forsyth, the first and second Arnold forms, i.e., the classes of equations of the form (1), (6), (8), (12) and (15), respectively. The corresponding classes $\widehat{\mathcal{L}}$, $\widehat{\mathcal{L}}_1$, $\widehat{\mathcal{L}}_2$, $\widehat{\mathcal{A}}_1$ and $\widehat{\mathcal{A}}_2$ of homogeneous equations are also studied from the point of view of admissible transformations.

The case $r = 2$ is singular for all the above classes. Each of the classes $\mathcal{L}$, $\widehat{\mathcal{L}}$, $\mathcal{L}_1$, $\widehat{\mathcal{L}}_1$, $\mathcal{L}_2 = \mathcal{A}_2$, $\mathcal{A}_1$ and $\widehat{\mathcal{A}}_1$ with $r = 2$ is an orbit of the elementary equation $x'' = 0$ with respect to the equivalence group of this class. This is why its equivalence groupoid is generated by the compositions of transformations from its equivalence group with transformations from the point symmetry group of the equation $x'' = 0$. Hence the class is semi-normalized, cf. the proof of Proposition 2. The class $\widehat{\mathcal{L}}_2 = \widehat{\mathcal{A}}_2$ is constituted by the single equation $x'' = 0$ and, thus, is normalized.

For $r \geqslant 3$, the equivalence groupoids of the classes $\mathcal{L}$, $\mathcal{L}_1$ and $\mathcal{L}_2$ are generated by the corresponding (usual) equivalence groups. In other words, each of these classes is normalized, see Propositions 4, 8 and 12. The associated subclasses of homogeneous equations are uniformly semi-normalized with respect to the linear superposition symmetry groups of their equations. This allows us to classify Lie symmetries of linear ODEs using the algebraic tools in three different ways. The purpose of the presentation of various ways for carrying out the known classification is to demonstrate advantages and disadvantages of each of them, which is important, e.g., to effectively apply the algebraic approach to systems of linear ODEs. Thus, the classification based on the Laguerre–Forsyth form, which is associated with a maximal gauge of arbitrary elements, is just reduced to the classification of subalgebras of the algebra $\mathfrak{sl}(2,\mathbb{F})$, which is finite-dimensional (more precisely, three-dimensional). The use of the rational form leads to involving the classification of all possible realizations of finite-dimensional Lie algebras on the line. At the same time, the single classification case of constant-coefficient equations in the rational form is split in the Laguerre–Forsyth form into three (resp. two) cases over the real (resp. complex) field, and two (resp. one) of them are related to variable-coefficient equations.

If we neglect the possibility of gauging arbitrary elements and consider general linear ODEs, we need to classify specific realizations of specific Lie algebras in the space of two variables.

The structure of the equivalence groupoids of the classes $\mathcal{A}_1$, $\widehat{\mathcal{A}}_1$, $\mathcal{A}_2$ and $\widehat{\mathcal{A}}_2$ associated with the Arnold forms, where $r \geqslant 3$, is more complicated since these classes are even not semi-normalized. This is why they are not usable for the group classification of the class $\mathcal{L}$, although these are the forms that are involved in reduction of order of linear ODEs.

In contrast to single linear ODEs, results concerning group properties of normal systems of second-order linear ODEs are very far from to be completed, not to mention general systems of linear ODEs; see a more detailed discussion in [35]. Only recently the group classification of systems of second-order linear ODEs with commuting constant-coefficient matrices was considered for various particular cases of the number of equations (two, three or four) and of the structure of the coefficient matrices in a series of papers [36–39] and was then exhaustively solved in [35]. In spite of a number of publications on the subject, the group classification of systems of linear second-order ODEs with noncommuting constant-coefficient matrices or with general nonconstant coefficients was carried out only for the cases of two and three equations [40–42]. The consideration of a greater number of equations or equations of higher and different orders within the framework of the classical infinitesimal approach requires cumbersome computations. Hence there is a demand for the development of new more powerful algebraic and geometric tools, which, for instance, involve a deep investigation of associated equivalence groupoids and other related algebraic structures.

Acknowledgments
The research of ROP was supported by the Austrian Science Fund (FWF), project P25064. VMB and ROP thank the University of Cyprus for the hospitality and financial support.

References
[1] Lie S 1893 *Vorlesungen über continuierliche Gruppen mit geometrischen und anderen Adwendungen* (B.G. Teubner, Leipzig)
[2] González-Gascón F and González-López A 1983 Symmetries of differential equations. IV *J. Math. Phys.* **24** 2006
[3] Olver P J 1994 Differential invariants and invariant differential equations *Lie Groups Appl.* **1** 177
[4] Olver P J 1995 *Equivalence, Invariants, and Symmetry* (Cambridge University Press, Cambridge)
[5] Ibragimov N H 1992 Group analysis of ordinary differential equations and the invariance principle in mathematical physics (on the occasion of the 150th anniversary of the birth of Sophus Lie) *Russ. Math. Surv.* **47** no 4, 89
[6] Mahomed F M 2007 Symmetry group classification of ordinary differential equations: survey of some results *Math. Methods Appl. Sci.* **30** 1995
[7] Schwarz F 2000 Solving second-order differential equations with Lie symmetries *Acta Appl. Math.* **60** 39
[8] Berkovich L M 2002 *Factorization and Transformations of the Differential Equations. Methods and Applications* (RCD, Moscow) (in Russian)
[9] Ibragimov N H 1999 *Elementary Lie Group Analysis and Ordinary Differential Equations* (*Wiley Series in Mathematical Methods in Practice* vol 4) (John Wiley & Sons, Ltd., Chichester)
[10] Schwarz F 2008 *Algorithmic Lie Theory for Solving Ordinary Differential Equations* (*Pure and Applied Mathematics* vol 291) (Chapman & Hall/CRC, Boca Raton, FL)
[11] Popovych R O, Kunzinger M and Ivanova N M 2008 Conservation laws and potential symmetries of linear parabolic equations *Acta Appl. Math.* **100** 113 (*Preprint* arXiv:0706.0443)
[12] Vaneeva O O, Popovych R O and Sophocleous C 2009 Enhanced group analysis and exact solutions of variable coefficient semilinear diffusion equations with a power source *Acta Appl. Math.* **106** 1 (*Preprint* arXiv:0708.3457)
[13] Popovych R O, Kunzinger M and Eshraghi H 2010 Admissible transformations and normalized classes of nonlinear Schrödinger equations *Acta Appl. Math.* **109** 315 (*Preprint* arXiv:math-ph/0611061)
[14] Popovych R O and Bihlo A 2012 Symmetry preserving parameterization schemes *J. Math. Phys.* **53** 073102 (*Preprint* arXiv:1010.3010)

[15] Bihlo A, Dos Santos Cardoso-Bihlo E and Popovych R O 2012 Complete group classification of a class of nonlinear wave equations *J. Math. Phys.* **53** 123515 (*Preprint* arXiv:1106.4801)
[16] Krause J and Michel L 1991 Classification of the symmetries of ordinary differential equations *Group Theoretical Methods in Physics (Moscow, 1990)* (*Lecture Notes in Phys.* vol 382) (Springer, Berlin) p 251
[17] Mahomed F M and Leach P G L 1990 Symmetry Lie algebras of nth order ordinary differential equations *J. Math. Anal. Appl.* **151** 80
[18] Wilczynski E J 1906 *Projective Differential Geometry of Curves and ruled Surfaces* (B.G. Teubner, Leipzig)
[19] Yumaguzhin V A 2002 Classification of linear ordinary differential equations. I *Differ. Equ.* **38** 1132
[20] Yumaguzhin V A 2002 Classification of linear ordinary differential equations. II *Differ. Equ.* **38** 1724
[21] Yumaguzhin V A 2002 Contact classification of linear ordinary differential equations. I *Acta Appl. Math.* **72** 155
[22] Stäckel P 1893 Über Transformationen von Differentialgleichungen *J. Reine Angew. Math.* **111** 290
[23] Halphen G H Sur les invariants différentielles Thése présentée à la Faculté de Sciences de Paris, 1878
[24] Laguerre E 1879 Sur les équations différentielles linéaires du troisième ordre *C. R. Acad. Sci. Paris* **88** 116
[25] Forsyth A R 1888 Invariants, covariants and quotient-derivatives associated with linear differential equations *Philos. Trans. Roy. Soc. London, Ser.* A **179** 377
[26] Vaneeva O O, Popovych R O and Sophocleous C 2014 Equivalence transformations in the study of integrability *Phys. Scr.* **89** 038003 (*Preprint* arXiv:1308.5126)
[27] Arnold V I 1988 *Geometrical Methods in the Theory of Ordinary Differential Equations* 2nd ed (*Grundlehren der Mathematischen Wissenschaften* vol 250) (Springer-Verlag, New York)
[28] González-López A 1988 Symmetries of linear systems of second-order ordinary differential equations *J. Math. Phys.* **29** 1097
[29] Lie S 1888 Classification und Integration von gewöhnlichen Differentialgleichungen zwischen xy, die eine Gruppe von Transformationen gestatten *Math. Ann.* **32** 213
[30] Patera J and Winternitz P 1977 Subalgebras of real three- and four-dimensional Lie algebras *J. Math. Phys.* **18** 1449
[31] Popovych R O, Boyko V M, Nesterenko M O and Lutfullin M W 2003 Realizations of real low-dimensional Lie algebras *J. Phys.* A **36** 7337 (*Preprint* arXiv:math-ph/0301029, extended and revised version)
[32] Lie S 1888, 1890, 1893 *Theorie der Transformationsgruppen, Bds. 1–3* (B.G. Teubner, Leipzig)
[33] González-López A, Kamran N and Olver P J 1992 Lie algebras of vector fields in the real plane *Proc. London Math. Soc. (3)* **64** 339
[34] Nesterenko M O 2006 Transformation groups on real plane and their differential invariants *Int. J. Math. Math. Sci.* Art. ID 17410 (*Preprint* arXiv:math-ph/0512038)
[35] Boyko V M, Popovych R O and Shapoval N M 2013 Lie symmetries of systems of second-order linear ordinary differential equations with constant coefficients *J. Math. Anal. Appl.* **397** 434 (*Preprint* arXiv:1203.0387)
[36] Wafo Soh C 2010 Symmetry breaking of systems of linear second-order ordinary differential equations with constant coefficients *Commun. Nonlinear Sci. Numer. Simul.* **15** 139
[37] Meleshko S V 2011 Comment on "Symmetry breaking of systems of linear second-order ordinary differential equations with constant coefficients" *Commun. Nonlinear Sci. Numer. Simul.* **16** 3447
[38] Campoamor-Stursberg R 2011 Systems of second-order linear ODE's with constant coefficients and their symmetries *Commun. Nonlinear Sci. Numer. Simul.* **16** 3015
[39] Campoamor-Stursberg R 2012 Systems of second-order linear ODE's with constant coefficients and their symmetries II. The case of non-diagonal coefficient matrices *Commun. Nonlinear Sci. Numer. Simul.* **17** 1178
[40] Wafo Soh C and Mahomed F M 2000 Symmetry breaking for a system of two linear second-order ordinary differential equations *Nonlinear Dynam.* **22** 121
[41] Moyo S, Meleshko S V and Oguis G F 2013 Complete group classification of systems of two linear second-order ordinary differential equations *Commun. Nonlinear Sci. Numer. Simul.* **18** 2972
[42] Suksern S, Moyo S and Meleshko S V Application of group analysis to classification of systems of three second-order ordinary differential equations *Preprint* arXiv:1310.5203

Properties of the sums of resonances found in the singularity analysis of certain classes of ordinary differential equations

Kyriakos Charalambous[1] **and Peter Leach**[2]

[1]Department of Mathematics and Statistics, University of Cyprus, P.O. Box 20537, 1678 Nicosia, Cyprus

[2]Department of Mathematics and Institute of Systems Science, Durban University of Technology, P.O. Box 1334, Durban 4000, Republic of South Africa

E-mail: kyriakosnj20@gmail.com, leach@ucy.ac.cy

Abstract. In the singularity analysis of ordinary differential equations it often happens that there are various possible leading-order behaviours. It is observed that the sums of the resonances in certain instances are the same. We illustrate this with certain examples from differential sequences and higher-order equations of Bureau symbol $P2$. For the Riccati Differential Sequence we prove two generic properties dealing with the occurrecne of the coefficients of the leading-order term and the variation of the sum of the resonances. An explanation of observed discrepancies in the cases of fourth- and fifth-order equations of Bureau symbol $P2$ is presented.

1. Introduction

There are three elements to the singularity analysis of ordinary differential equations. They are the determination of the leading-order behaviour, the determination of the resonances and the establishment of consistency at successive resonances and with the terms in the equation which are not included in the leading-order elements. See for example the standard references of Tabor [1] and Ramani *et al* [2]. By way of example consider the equation $y'' = y^2+1$. We substitute $y = \alpha(x-x_0)^p$ to determine the leading-order behaviour and obtain $\alpha p(p-1)(x-x_0)^{p-2} = \alpha^2(x-x_0)^{2p}+1$ from which it is evident that balance is obtained when $p = -2$ and $\alpha = 6$ with the first and the second terms. To determine the resonances, i.e. the values of the exponents at which the second arbitrary constant of integration appears, we substitute $y = 6(x-x_0)^{-2} + \mu(x-x_0)^{-2+s}$ into the dominant terms and find that μ is arbitrary for $s = -1, 6$. The substitution of the series up to the resonance at 6 into the full equation shows that there is consistency. The resonance, $s = -1$, is generic and is associated with the arbitrary location of the pole at $x = x_0$.

When one makes a study of a number of equations with various possibilities for the coefficient of the leading-order term, there are some occasions for which the sum of the resonances follows some pattern or is even the same. Here we examine several systems of equations for which one or other of these phenomena occurs. One example is found in the Riccati Differential Sequence [3] and another set of examples comprises higher-order equations of Painlevé-type as discussed by Cosgrove [4]. In Section 2 we investigate these properties for members of the Riccati Differential Sequence and in Section 3 we examine the same properties for fourth- and fifth-order equations of Bureau symbol $P2$, i.e. the singularity is a second-order pole. For the former we infer general properties based on the examples of earlier members of the sequence, formulate conjectures based on these inferences and then prove them for general n. In the latter examples we find that there are some inconsistencies and we explain how they arise.

2. The Riccati Differential Sequence

The Riccati Differential Sequence, $R_n = 0$, is defined through the Rodrigues Formula [5]

$$R_n = \exp\left[-\int y(x)\mathrm{d}x\right] D^n \left\{\exp\left[\int y(x)\mathrm{d}x\right]\right\}, \quad n = 0, 1, 2, \ldots . \tag{1}$$

This leads to a recursion relation. We write the Rodrigues Formula, (1), as

$$\begin{aligned} R_{n+1} &= \exp\left[-\int y(x)\mathrm{d}x\right] \left\{D^{n+1}\left(\exp\left[\int y(x)\mathrm{d}x\right]\right)\right\} \\ &= \exp\left[-\int y(x)\mathrm{d}x\right] D\left\{D^{n}\left(\exp\left[\int y(x)\mathrm{d}x\right]\right)\right\} \\ &= \exp\left[-\int y(x)\mathrm{d}x\right] D\left\{R_n\left(\exp\left[\int y(x)\mathrm{d}x\right]\right)\right\} \\ &= \exp\left[-\int y(x)\mathrm{d}x\right] \left\{R_n' + yR_n\right\}\left(\exp\left[\int y(x)\mathrm{d}x\right]\right) \\ &= R_n' + yR_n, \end{aligned}$$

i.e. the recursion relation is

$$R_{n+1} = R_n' + yR_n$$

and the recursion operator is $D + y$.

An alternate definition makes use of the recursion operator, $D+y$, with the same seed function. We list the first few members of the Sequence, omitting the zeroth and first elements as they are rather trivial,

$$\begin{aligned} R_2 : &\quad y' + y^2 = 0, \\ R_3 : &\quad y'' + 3yy' + y^3 = 0, \\ R_4 : &\quad y^{(3)} + 4yy'' + 3y'^2 + 6y^2y' + y^4 = 0, \\ R_5 : &\quad y^{(4)} + 10y'y'' + 10y^2y'' + 5yy'^3 + 15yy'^2 + 10y^3y' + y^5 = 0, \\ R_6 : &\quad y^{(5)} + 6yy^{(4)} + 15y^2y^{(3)} + 10y''^2 + 60yy'y'' + 20y^3y'' + 15y'^3 + 45y^2y'^2 \\ &\quad + 15y^4y' + y^6 = 0, \end{aligned}$$

where in a standard notation $y^{(n)}$ denotes the nth derivative of $y(x)$ with respect to the independent variable, x. We recognise R_2 as the Riccati Equation [6] and R_3 as the Painlevé–Ince Equation [7] which has been the subject of many studies due to its myriad and interesting properties.

In Table 1 we list the properties relevant to the singularity analysis and the topic of this paper.

Evidently one could continue indefinitely, but the pattern of results has become already quite obvious. Firstly we note that the conventional Painlevé Expansion, that of a series ascending from the singularity occurs only for one possibility of the value of the coefficient of the leading-order term. The possibility of negative resonances was discussed by Fordy and Pickering [8] in terms of a perturbative expansion. Feix *et al* [9] recognised the possibility of ascending and descending series and termed the expressions "Left Painlevé Series" and "Right Painlevé Series". Subsequently Andriopoulos *et al* [10] explained all in terms of the three possible Laurent expansions one could have in the complex plane and supported the explanation with concrete examples.

Table 1. Possible leading-order coefficients (with the exception of a zero), their corresponding resonances and the sum of the resonances. In all cases the singularity is given by $p = -1$.

n	α	s	Σ
2	1	-1	-1
3	1	$-1,\ 1$	0
	2	$-1,\ -2$	-3
4	1	$-1,\ 1,\ 2$	2
	2	$-2,\ -1,\ 1$	-2
	3	$-3,\ -2,\ -1$	-6
5	1	$-1,\ 1,\ 2,\ 3$	5
	2	$-2,\ -1,\ 1,\ 2$	0
	3	$-3,\ -2,\ -1,\ 1$	-5
	4	$-4,\ -3,\ -2,\ -1$	-10
6	1	$-1,\ 1,\ 2,\ 3,\ 4$	9
	2	$-2,\ -1,\ 1,\ 2,\ 3$	3
	3	$-3,\ -2,\ -1,\ 1,\ 2$	-3
	4	$-4,\ -3,\ -2,\ -1,\ 1$	-9
	5	$-5,\ -4,\ -3,\ -2,\ -1$	-15
7	1	$-1,\ 1,\ 2,\ 3,\ 4,\ 5$	14
	2	$-2,\ -1,\ 1,\ 2,\ 3,\ 4$	7
	3	$-3,\ -2,\ -1,\ 1,\ 2,\ 3$	0
	4	$-4,\ -3,\ -2,\ -1,\ 1,\ 2$	-7
	5	$-5,\ -4,\ -3,\ -2,\ -1,\ 1$	-14
	6	$-6,\ -5,\ -4,\ -3,\ -2,\ -1$	-21

From the contents of Table 1 it is evident that for higher members of the Sequence the Left and Right Series are simply the extremes of a number of intermediate series which are indicative of a complete Laurent expansion over some annulus centred upon the singularity. For example in the case of R_7 there are four intermediate series and each one has its own annulus.

If one observes the Check Sum for each particular value of n for which this makes a sensible activity, it is obvious that the decrease is by a common amount for each of the possible coefficients for the leading-order term. Furthermore this constant decrease increases by one as the value of n increases. On the other hand the commencing value of Σ, i.e. the value of the Check Sum for the Right Painlevé Series, increases by unity as n increases by unity. Similarly the magnitude of the final value of Σ, which is the result for the Left Painlevé Series with the exception of $n = 2$, increases by 1 for increasing n.

The observations of the properties listed in Table 1 lead to two conjectures.

Conjecture 1: The coefficients of the leading-order term of R_n for $n > 0$ take the nonzero values $\alpha = 1, 2, \ldots, n-1$.

Conjecture 2: The sum of the resonances for each value of the leading-order coefficient, α, is given by $\sum_{i=1}^{i=n} s_i = \frac{1}{2}\left(n(n+1) - \alpha\beta\right)$.

Proof of Conjecture 1. In terms of the degree of the derivative, y', the ultimate and penultimate terms in R_n are y^n and $\beta_n y^{n-2}y'$, where β_n is some numerical coefficient. From the recursion relation the corresponding terms in R_{n+1} are y^{n+1} and $ny^{n-1}y' + \beta_n y^{n-1}y'$. Thus the penultimate term is $(\beta_n + n)y^{n-1}y'$. When the substitution for the leading-order term, $y \longrightarrow \alpha\chi^p$, is made and p is identified as -1, the two terms are $\alpha^n\chi^{-n} - \alpha^{n-1}(\beta_n + n - 1)\chi^{-n}$ so that

$$\alpha_{n+1} = -(\beta_n + n - 1),$$

i.e. the sum of the roots of the polynomial determining the values of α increases by $n-1$.

From Table 1 we see that the increase for the first few members of the Sequence is indeed $n-1$ and that this increase expresses itself as the value of the additional root. Consequently for R_n the sum of the roots is given by

$$\sum_{i=1}^{i=n} \alpha_i = \tfrac{1}{2}(n-2)(n-1).$$ □

Proof of Conjecture 2. We use the highest and second-highest terms with respect to the degree of the derivative. Without knowing the coefficient we write these two terms for R_n as $y^{(n)} + \gamma yy^{(n-1)}$, where γ is a numerical coefficient. We make the substitution

$$y(x) \longrightarrow \alpha\chi^{-1} + m\chi^{s-1},$$

where, as before, α is the value of the leading-order coefficient. After some rearrangement the coefficient of m in the highest and second-highest terms in the order of the derivative is

$$s^n - \left[\tfrac{1}{2}n(n+1) - \alpha\gamma\right] s^{n-1}.$$

Consequently the sum of the resonances is given by

$$\sum_{i=1}^{i=n} s_i = \tfrac{1}{2}\left(n(n+1) - \alpha\gamma\right),$$

which result is, in a sense, independent of the individual values of the resonances. □

3. Fourth- and fifth-order equations of Bureau Symbol $P2$

Cosgrove has made an extensive study of the singularity properties of higher-order equations. In particular he reported on equations with the so-called Painlevé Property in the Polynomial Class 1 with Bureau Symbol $P2$ [4]. We examine some of those equations in the spirit of the analysis above.

3.1. Equations of the fourth order

The fourth-order equations which we consider are

$$\begin{aligned}
\text{T-I}: &\quad y^{(4)} - 12yy'' - 12y'^2 - (\alpha x + \beta)y' - 2\alpha y + \tfrac{1}{6}(\alpha x + \beta)^2 = 0,\\
\text{T-II}: &\quad y^{(4)} - 3yy'' + 4y'^2 = 0,\\
\text{T-III}: &\quad y^{(4)} - 15yy'' - \tfrac{45}{4}y'^2 + 15y^3 - \alpha y - \beta = 0,\\
\text{T-IV}: &\quad y^{(4)} - 30yy'' + 60y^3 - \alpha y - \beta = 0,\\
\text{T-V}: &\quad y^{(4)} - 20yy'' - 10y'^2 + 40y^3 - \alpha y - \kappa x - \beta = 0 \quad \text{and}\\
\text{T-VI}: &\quad y^{(4)} - 18yy'' - 9y'^2 + 24y^3 - \alpha y^2 - \tfrac{1}{9}\alpha^2 y - \kappa x - \beta = 0,
\end{aligned}$$

where the nomenclature is that of Cosgrove.

In all cases the dominant terms in the equation possess the two Lie point symmetries $\Gamma_1 = \partial_x$ and $\Gamma_2 = x\partial_x - 2y\partial_y$ with the algebra A_2. The structure of the self-similar symmetry, Γ_2, indicates that the leading-order behaviour is χ^{-2}. We can immediately proceed to the next step of the

Journal of Physics: Conference Series **621** (2015) 012003 doi:10.1088/1742-6596/621/1/012003

Table 2. All of the six equations have leading-order exponent -2. We list the coefficient(s) of the leading-order terms, the resonances associated with each coefficient and the Check Sum. Note that we use a as the coefficient of the leading-order terms as Cosgrove uses α as a parameter in some of the equations.

Equation	a	s	Σ
T-I	1	1, 4, 5, 6	14
T-II	60	3, −2, −1, 20	14
T-III	1	1, 3, 5, 7	14
	8	7, −1, 10, 12	14
T-IV	1	1, 2, 3, 10	14
	2	2, −1, 5, 12	14
T-V	1	1, 2, 5, 8	14
	3	3, −1, 8, 10	14
T-VI	1	1, 3, 4, 8	14
	5	5, −1, 8, 12	14

singularity analysis by writing $y \longrightarrow a(x-x_0)^{-2}+\mu(x-x_0)^{s-2}$. In Table 2 we provide the possible coefficients of the leading-order term, resonances and Check Sum.

The sum of the resonances is 14 in each case. Cosgrove excludes T-II as a possible candidate for being integrable in the sense of singularity and in the cases of T-III to T-VI allows only the first set of resonances as indicating integrability as he follows the criterion advanced by Fordy *et al.* In the spirit of Feix *et al* and as developed by Andriopoulos *et al* all cases are integrable either as a Right Painlevé Series over a punctured disc centred upon the singularity or a full Laurent Expansion in an annulus centred upon the singularity. In no case is there the possibility of a Left Painlevé Series. In fact Cosgrove describes T-II as being a fourth-order equivalent to the famous Chazy Barrier Equation [11]. In a comprehensive numerical investigation of the equations governing the Mixmaster Universe Model Bountis and Drossos [12] demonstrated the nature of the barrier in that model. To make the numerical integration one needed to fix the location of the singularity as part of the initial conditions. When one integrates outwards from near the singularity, a barrier of complex structure is certainly found. However, in the case of an equation such as T-II one would expect to have a problem with an integration begun near the singularity but not if one began within the annulus. The surmise is that as one approaches either boundary of the annulus that the numerical integration would show a similar complexity. The problem is where to locate the annulus in the complex plane even if the location of the singularity is fixed.

3.2. Equations of the fifth order

The corresponding equations of the fifth order as listed by Cosgrove are

$$\begin{aligned}
\text{P-I}: &\quad y^{(5)} - 15yy''' - \tfrac{75}{2}y'y'' + 45y^2y' - (\lambda x + \alpha)y' - 2\lambda y = 0,\\
\text{P-II}: &\quad y^{(5)} - 30yy''' - 30y'y'' + 180y^2y' - (\lambda x + \alpha)y' - 2\lambda y = 0,\\
\text{P-IIIa}: &\quad y^{(5)} - 20yy''' - 40y'y'' + 120y^2y' - (\lambda x + \alpha)y' - 2\lambda y - \kappa = 0,\\
\text{P-IIIb}: &\quad y^{(5)} - 20yy''' - 40y'y'' + 120y^2y' - \alpha y'\\
&\quad -\left(y^{(4)} - 20yy'' - 10y'^2 + 40y^3 - \alpha y - \beta\right)/x = 0,\\
\text{P-IVa}: &\quad y^{(5)} - 18yy''' - 36y'y'' + 72y^2y' - 3\lambda y''\\
&\quad -\left(y^{(4)} - 18yy' - 9y'^2 + 24y^3 - 3\lambda y' + \kappa\right)/x\\
&\quad -\tfrac{1}{2}\lambda x(5y''' - 36yy') + \tfrac{1}{2}\lambda^2 x(2xy' + y) = 0,\\
\text{P-IVb}: &\quad y^{(5)} - 18yy''' - 36y'y'' + 72y^2y' - 2\alpha yy' - \tfrac{1}{9}\alpha^2y' - \kappa = 0 \quad \text{and}\\
\text{P-IVc}: &\quad y^{(5)} - 18yy''' - 36y'y'' + 72y^2y' - 2\alpha yy' - \tfrac{1}{9}\alpha^2y'\\
&\quad -\left(y^{(4)} - 18yy'' - 9y'^2 + 24y^3 - \alpha y^2 - \tfrac{1}{9}\alpha^2y - \beta\right)/x = 0.
\end{aligned}$$

The similarity symmetry again indicates that the value of p is -2.

Table 3. We list the coefficient(s) of the leading-order terms, the resonances associated with each coefficient and their Check Sum.

Equation	a	s	Σ
P-I	1	1, 3, 5, 6, 7	20
	8	7, −1, 6, 10, 12	20
P-II	1	1, 2, 3, 6, 10	20
	2	2, −1, 5, 6, 12	20
P-IIIa	1	1, 2, 5, 6, 8	20
	3	3, −1, 6, 8, 10	20
P-IIIb	1	1, 2, 5, 7, 8	21
	3	3, −1, 7, 8, 10	21
P-IVa	1	1, 3, 4, 7, 8	21
	5	5, −1, 7, 8, 12	21
P-IVb	1	1, 3, 4, 6, 8	20
	5	5, −1, 6, 8, 12	20
P-IVc	1	1, 3, 4, 7, 8	21
	5	5, −1, 7, 8, 12	21

We note that the unity of results evinced by the fourth-order equations is not matched by this set of fifth-order equations. The Check Sum is either 20 or 21. Each equation has a Right Painlevé Series and a full Laurent Expansion. For all the Check Sum is unaffected by the nature of the series. The explanation of the two possible case is quite simple when they occur. In the equations for which the Check Sum is 21, there is an additional term contributing to the leading-order behaviour. If we take P-IVc as an example, there is the term $-y^{(4)}/x$ and this adds one to the sum of the resonances either as a Right Painlevé Series or as a full Laurent Expansion. Were the coefficient a number other than -1, the number by which the Check Sum would change would be minus that number.

4. Discussion

The Riccati Differential Sequence represents an example of a set of differential equations of ever-increasing order generated by the Rodrigues Formula (1). An alternate mode of generation is to use the recursion operator $R = D + y$ which is the real form of Dirac's creation operator for the solution of the Schrödinger equation for the simple harmonic oscillator. All members of this sequence are integrable in terms of elementary functions [13]. Here we have been concerned with the singularity analysis of members of this sequence. Admittedly we have presented the analysis of only a very finite selection of equations taken from the lower members of the sequence. Nevertheless the properties are so regular that one may infer the properties of elements of the sequence in general. These properties have been summarised in Table 1 and the chief of these properties were proven in the proofs of Conjectures 1 and 2.

The second class of equations examined is taken from those equations of the fourth- and fifth-order which possess the Painlevé Property with the particular property that the Bureau Symbol is $P2$, i.e. the singularity identified from the leading-order terms is a second-order pole. Here we saw behaviour differing from that observed for the elements of the Riccati Differential Sequence in that across the equations of the same order belonging to a given class the Check Sum was the same no matter the mix of resonances. The exceptions in the class of fifth-order equations were easily explained in terms of subdominant terms being rendered dominant by a multiplier comprising a suitably singular function of x. It is the order of the singularity in this function and not its location which is the critical feature [14].

Finally we emphasise that the singularity analysis does not deal only with a Laurent Expansion commencing at the power of the singularity and thereafter in increasing powers. The nature of the expansion depends upon the region of the complex plane where the expansion is valid. This does open the possibility of numerical experiments to determine "natural barriers" not only from expansion about the singular point but also advancing towards the singularity from afar (Left Painlevé Series) or going both inwards and outwards towards the bounding circles of an annulus (full Laurent Expansion).

Acknowledgments

KC and PGLL thank the University of Cyprus for the kind hospitality afforded them during the seventh GADEIS National Workshop held at Larnaca in June, 2014. PGLL thanks the Durban University of Technology and the National Research Foundation of South Africa for support. The opinions expressed in this paper should not be construed as being those of either institution.

References

[1] Tabor M 1989 *Chaos and Integrability in Nonlinear Dynamics. An Introduction* (New York: John Wiley & Sons)
[2] Ramani A, Grammaticos B and Bountis T 1989 The Painlevé property and singularity analysis of integrable and nonintegrable systems *Phys. Rep.* **180**:3 159
[3] Euler M, Euler N and Leach P G L 2007 The Riccati and Ermakov–Pinney hierarchies *J. Nonlinear Math. Phys.* **14**:2 290
[4] Cosgrove C M 2000 Higher-order Painlevé equations in the polynomial class. I. Bureau Symbol P2 *Stud. Appl. Math.* **104**:1 1
[5] Rodrigues O 1816 De l'attraction des sphéroídes (Correspondence sur l'École Impériale Polytechnique vol 3) (Paris) p 361
[6] Riccati J 1764 *Opere* (Lucca: Jacopo Guisti) p 83
[7] Ince E L 1927 *Ordinary Differential Equations* (London: Longmans, Green and co. ltd)

[8] Fordy A and Pickering A 1991 Analysing negative resonances in the Painlevé test *Phys. Lett.* A **160**:4 347
[9] Feix M R, Géronimi C, Cairó L, Leach P G L, Lemmer R L and Bouquet S É 1997 On the singularity analysis of ordinary differential equations invariant under time translation and rescaling *J. Phys.* A **30**:21 7437
[10] Andriopoulos K and Leach P G L 2006 An interpretation of the presence of both positive and negative nongeneric resonances in the singularity analysis *Phys. Lett.* A **359**:3 199
[11] Chazy J 1911 Sur les équations différentielles du troisième ordre et d'ordre supérieur dont l'intégrale générale a ses points critiques fixes *Acta Math.* **34**:1 317
[12] Bountis T C and Drossos L B 1997 Evidence of a natural boundary and nonintegrability of the mixmaster universe model *Nonlinear Science* **7**:1 45
[13] Andriopoulos K, Leach P G L and Maharaj A 2011 On differential sequences *Appl. Math. Inf. Sci.* **5**:3 525
[14] Andriopoulos K and Leach P G L 2011 Singularity analysis for autonomous and nonautonomous differential equations *Appl. Anal. Discrete Math.* **5**:2 230

Generalized Lotka–Volterra systems connected with simple Lie algebras

Stelios A Charalambides, Pantelis A Damianou and Charalambos A Evripidou

Department of Mathematics and Statistics, University of Cyprus,
P.O. Box 20537, 1678 Nicosia, Cyprus

E-mail: charalambides.stelios@ucy.ac.cy, damianou@ucy.ac.cy, cevrip02@ucy.ac.cy

Abstract. We devise a new method for producing Hamiltonian systems by constructing the corresponding Lax pairs. This is achieved by considering a larger subset of the positive roots than the simple roots of the root system of a simple Lie algebra. We classify all subsets of the positive roots of the root system of type A_n for which the corresponding Hamiltonian systems are transformed, via a simple change of variables, to Lotka–Volterra systems. For some special cases of subsets of the positive roots of the root system of type A_n, we produce new integrable Hamiltonian systems.

1. Introduction

The Volterra model, also known as the KM system is a well-known integrable system defined by

$$\dot{x}_i = x_i(x_{i+1} - x_{i-1}), \quad i = 1, 2, \dots, n, \tag{1}$$

where $x_0 = x_{n+1} = 0$. It was studied by Lotka in [1] to model oscillating chemical reactions and by Volterra in [2] to describe population evolution in a hierarchical system of competing species. It was first solved by Kac and van Moerbeke in [3], using a discrete version of inverse scattering due to Flaschka [4]. In [5] Moser gave a solution of the system using the method of continued fractions and in the process he constructed action-angle coordinates. Equations (1) can be considered as a finite-dimensional approximation of the Korteweg–de Vries (KdV) equation. The Poisson bracket for this system can be thought as a lattice generalization of the Virasoro algebra [6]. The Volterra system is associated with a simple Lie algebra of type A_n. Bogoyavlensky generalized this system for each simple Lie algebra and showed that the corresponding systems are also integrable. See [7, 8] for more details. The generalization in this paper is different from the one of Bogoyavlensky.

The KM-system given by equation (1) is Hamiltonian (see [9, 10]) and can be written in Lax pair form in various ways. We will focus on the symmetric version due to Moser, where the

Journal of Physics: Conference Series **621** (2015) 012004 doi:10.1088/1742-6596/621/1/012004

matrices L and B have the form

$$L = \begin{pmatrix} 0 & a_1 & 0 & \cdots & \cdots & 0 \\ a_1 & 0 & a_2 & \ddots & & \vdots \\ 0 & a_2 & 0 & \ddots & & \vdots \\ \vdots & \ddots & \ddots & \ddots & & 0 \\ \vdots & & & \ddots & \ddots & a_n \\ 0 & \cdots & \cdots & 0 & a_n & 0 \end{pmatrix}, \qquad (2)$$

and

$$B = \begin{pmatrix} 0 & 0 & a_1a_2 & \cdots & \cdots & 0 \\ 0 & 0 & 0 & \ddots & & \vdots \\ -a_1a_2 & 0 & 0 & \ddots & a_2a_3 & \vdots \\ \vdots & -a_2a_3 & \ddots & \ddots & & a_{n-1}a_n \\ \vdots & & & \ddots & \ddots & 0 \\ 0 & \cdots & \cdots & -a_{n-1}a_n & 0 & 0 \end{pmatrix}.$$

In terms of the simple root vectors X_{α_i}, $i = 1, 2, \ldots, n$ the matrices L and B are written as $L = \sum_{i=1}^{n} a_i \left(X_{\alpha_i} + X_{-\alpha_i}\right)$ and $B = \sum_{i=1}^{n-1} a_i a_{i+1} \left(X_{\alpha_i+\alpha_{i+1}} - X_{-\alpha_i-\alpha_{i+1}}\right)$. The matrix equation $\dot{L} = [B, L]$ gives a polynomial (in fact cubic) system of differential equations. The change of variables $x_i = 2a_i^2$ gives equations (1). The purpose of this paper is to show that this Lax pair can be generalized and produce a larger class of Hamiltonian systems which we call generalized Volterra systems.

It is evident from the form of L in the Lax pair, that the position of the variables a_i corresponds to the simple root vectors of a root system of type A_n. On the other hand a non-zero entry of the matrix B occurs at a position corresponding to the sum of two simple roots α_i and α_j. In this paper we generalize the Lax pair of Moser (2) as follows.

Instead of considering the set of simple roots Π, we begin with a subset Φ of the positive roots Δ^+ which contains Π, i.e. $\Pi \subset \Phi \subset \Delta^+$. For each such choice of a set Φ we produce a Lax pair and thus a new Hamiltonian system. We call these cubic systems generalized Volterra systems since in some cases by a simple change of variables we produce Lotka–Volterra systems. In this paper we restrict our attention to some examples in the A_n case. However, this algorithm applies, more generally for each complex simple Lie algebra. In dimension 3 this procedure produces only two systems, the KM system and the periodic KM system. In dimensions 4 and 5 (i.e. the cases of A_3 and A_4) and by allowing the use of complex coefficients this method works in all possible cases and in fact we have verified using Maple that all the resulting systems are Liouville integrable. After the definition of Lotka–Volterra systems in Section 2, we describe our algorithm in Section 3. In Section 4 we give a classification of all cases which give rise to Lotka–Volterra systems via the transformation $a_i \to 2a_i^2$. Finally in Section 5 we present some examples where complex coefficients are used to produce new systems.

2. Lotka–Volterra systems

The KM-system belongs to a large class of the so called Lotka–Volterra systems. The most general form of the Lotka–Volterra equations is

$$\dot{x}_i = \varepsilon_i x_i + \sum_{j=1}^{n} a_{ij} x_i x_j, \quad i = 1, 2, \ldots, n.$$

We may assume that there are no linear terms ($\varepsilon_i = 0$). We also assume that the matrix $A = (a_{ij})$ is skew-symmetric. All these systems can be written in Hamiltonian form using the Hamiltonian function

$$H = x_1 + x_2 + \cdots + x_n.$$

Hamilton's equations take the form $\dot{x}_i = \{x_i, H\} = \sum_{j=1}^{n} \pi_{ij}$ with quadratic functions

$$\pi_{ij} = \{x_i, x_j\} = a_{ij}x_i x_j, \quad i, j = 1, 2, \ldots, n. \tag{3}$$

From the skew symmetry of the matrix $A = (a_{ij})$ it follows that the Jacobi identity is satisfied.

The Poisson tensor (3) is Poisson isomorphic to the constant Poisson structure defined by the constant matrix A, see [11]. If $\mathbf{k} = (k_1, k_2, \ldots, k_n)$ is a vector in the kernel of A then the function

$$f = x_1^{k_1} x_2^{k_2} \cdots x_n^{k_n}$$

is a Casimir. Indeed for an arbitrary function g the Poisson bracket $\{f, g\}$ is

$$\{f, g\} = \sum_{i,j=1}^{n} \{x_i, x_j\} \frac{\partial f}{\partial x_i} \frac{\partial g}{\partial x_j} = \sum_{j=1}^{n} \left(\sum_{i=1}^{n} a_{ij} k_i \right) x_j f \frac{\partial g}{\partial x_j} = 0.$$

If the matrix A has rank r, then there are $n - r$ functionally independent Casimirs. This type of integral can be traced back to Volterra [2]; see also [11–13].

3. The procedure

We recall the following procedure from [14]. Let $\mathfrak{g}$ be any simple Lie algebra equipped with its Killing form $\langle \cdot | \cdot \rangle$. One chooses a Cartan subalgebra $\mathfrak{h}$ of $\mathfrak{g}$, and a basis Π of simple roots for the root system Δ of $\mathfrak{h}$ in $\mathfrak{g}$. The corresponding set of positive roots is denoted by Δ^+. To each positive root α one can associate a triple $(X_\alpha, X_{-\alpha}, H_\alpha)$ of vectors in $\mathfrak{g}$ which generate a Lie subalgebra isomorphic to $\mathfrak{sl}_2(\mathbf{C})$. The set $(X_\alpha, X_{-\alpha})_{\alpha \in \Delta^+} \cup (H_\alpha)_{\alpha \in \Pi}$ is a basis of $\mathfrak{g}$, called a root basis. Let $\Pi = \{\alpha_1, \ldots, \alpha_\ell\}$ and let $X_{\alpha_1}, \ldots, X_{\alpha_\ell}$ be the corresponding root vectors in $\mathfrak{g}$. Define

$$L = \sum_{\alpha_i \in \Pi} a_i (X_{\alpha_i} + X_{-\alpha_i}).$$

To find the matrix B we use the following procedure. For each i, j form the vectors $[X_{\alpha_i}, X_{\alpha_j}]$. If $\alpha_i + \alpha_j$ is a root then include a term of the form $a_i a_j [X_{\alpha_i}, X_{\alpha_j}]$ in B. We make B skew-symmetric by including the corresponding negative root vectors $a_i a_j [X_{-\alpha_i}, X_{-\alpha_j}]$. Finally, we define the system using the Lax pair equation

$$\dot{L} = [B, L].$$

For a root system of type A_n we obtain the KM system.

In this paper we generalize this algorithm as follows. Consider a subset Φ of Δ^+ such that

$$\Pi \subset \Phi \subset \Delta^+.$$

The Lax matrix is easy to construct

$$L = \sum_{\alpha_i \in \Phi} a_i (X_{\alpha_i} + X_{-\alpha_i}).$$

Here we use the following enumeration of Φ which we assume to have m elements. The variables a_j correspond to the simple roots α_j for $j = 1, 2, \ldots, \ell$. We assign the variables a_j for $j = \ell+1, \ell+2, \ldots, m$ to the remaining roots in Φ. To construct the matrix B we use the following algorithm. Consider the set $\Phi \cup \Phi^-$ which consists of all the roots in Φ together with their negatives. Let

$$\Psi = \left\{ \alpha + \beta \,|\, \alpha, \beta \in \Phi \cup \Phi^-, \ \alpha + \beta \in \Delta^+ \right\}.$$

Define

$$B = \sum c_{ij} a_i a_j (X_{\alpha_i + \alpha_j} - X_{-\alpha_i - \alpha_j}), \tag{4}$$

where $c_{ij} = \pm 1$ if $\alpha_i + \alpha_j \in \Psi$ with $\alpha_i, \alpha_j \in \Phi \cup \Phi^-$ and 0 otherwise. In all eight in A_3 cases we are able to make the proper choices of the sign of the c_{ij} so that we can produce a Lax pair. This method produces a Lax pair in all but five out of sixty four cases in A_4. However, when we allow the c_{ij} to take the complex values $\pm i$ we are able to produce a Lax pair in all sixty four cases. By using Maple we were able to check that all these examples in A_3 and A_4 are in fact Liouville integrable. We will not attempt to prove the integrability of these systems in general due to the complexity of their definition.

We have to point out that in [15] there is a similar construction for the case of the Toda lattice where Hamiltonian systems are defined which interpolate between the classical Kostant–Toda lattice and the full Kostant–Toda lattice. In that case there is a simple criterion on the set Φ which ensures the construction of the Lax pair. In our case there is no such simple criterion. In the next proposition we present a sufficient (but not necessary) condition on the subset Φ which gives a consistent Lax pair.

Proposition 1. *Let $\Pi \subset \Phi \subset \Delta^+$ be a subset of the positive roots with the property that whenever $\alpha, \beta, \gamma \in \Phi \cup \Phi^-$ then $\alpha + \beta + \gamma \neq 0$ and if $\alpha + \beta + \gamma \in \Delta^+$ then $\alpha + \beta + \gamma \in \Phi$. Also let B be the matrix constructed using the algorithm described in* (4)*. Then for any choice of the signs $c_{i,j}$ the pair L, B is a Lax pair.*

Proof. Let K be the following subset of the positive roots

$$K = \{\alpha + \beta + \gamma : \alpha, \beta, \gamma \in \Phi \cup \Phi^-, \ \alpha + \beta + \gamma \in \Delta^+\}.$$

It easily follows from the construction of the matrix B that for all possible choices of the signs $c_{i,j}$, the nonzero entries of the bracket $[B, L]$ appear in the positions corresponding to the root vectors X_α, $\alpha \in K$. The condition $\alpha, \beta, \gamma \in \Phi \cup \Phi^- \Rightarrow \alpha + \beta + \gamma \neq 0$ implies that there are no variables in the diagonal of $[B, L]$ while the condition $\alpha, \beta, \gamma \in \Phi \cup \Phi^-$ and $\alpha + \beta + \gamma \in \Delta^+ \Rightarrow \alpha + \beta + \gamma \in \Phi$ implies that $K \subset \Phi$. Since we also have $\Phi \subset K$ we deduce that $\Phi = K$ and therefore the pair L, B is a Lax pair. □

This condition is of course not necessary. For example the KM and the periodic KM systems do not fall in this class. In Theorem 1 we find several other families of subsets Φ which give consistent Lax pairs.

Example 1. Let Φ be the subset of the positive roots of the root system A_n containing all the roots of odd height. We immediately see that if K is the subset of the positive roots described in Proposition 1 then $K = \Phi$ and therefore for all possible choices of the signs $c_{i,j}$ we have a consistent Lax pair. Take for example the case $n = 3$. In that case the subset Φ is given by

$$\Phi = \{\alpha_1, \alpha_2, \alpha_3, \alpha_4 = \alpha_1 + \alpha_2 + \alpha_3\}$$

and this choice gives rise to the Lax matrix

$$L = \sum_{i=1}^{4} a_i \left(X_{\alpha_i} + X_{-\alpha_i} \right).$$

The skew symmetric matrix B constructed using (4) is

$$B = (c_{1,2}a_1a_2 + c_{3,4}a_3a_4)(X_{\alpha_1+\alpha_2} - X_{-\alpha_1-\alpha_2}) + (c_{1,4}a_1a_4 + c_{2,3}a_2a_3)(X_{\alpha_2+\alpha_3} - X_{-\alpha_2-\alpha_3}).$$

Now we easily verify that all 16 possible choices of the signs $c_{i,j}$ give a consistent Lax pair. Of course only half of them give possibly non-isomorphic systems and only one of them gives a Lotka–Volterra system (see Theorem 1), the well known periodic KM system.

Example 2 (A_3 root system)**.** Let E be the hyperplane of $\mathbb{R}^4$ for which the coordinates sum to 0 (i.e. vectors orthogonal to $(1,1,1,1)$). Let Δ be the set of vectors in E of length $\sqrt{2}$ with integer coordinates. There are 12 such vectors in all. We use the standard inner product in $\mathbb{R}^4$ and the standard orthonormal basis $\{\epsilon_1, \epsilon_2, \epsilon_3, \epsilon_4\}$. Then, it is easy to see that $\Delta = \{\epsilon_i - \epsilon_j \mid i \neq j\}$. The vectors

$$\alpha_1 = \epsilon_1 - \epsilon_2, \quad \alpha_2 = \epsilon_2 - \epsilon_3, \quad \alpha_3 = \epsilon_3 - \epsilon_4$$

form a basis of the root system in the sense that each vector in Δ is a linear combination of these three vectors with integer coefficients, either all nonnegative or all nonpositive. For example, $\epsilon_1 - \epsilon_3 = \alpha_1 + \alpha_2$, $\epsilon_2 - \epsilon_4 = \alpha_2 + \alpha_3$ and $\epsilon_1 - \epsilon_4 = \alpha_1 + \alpha_2 + \alpha_3$. Therefore $\Pi = \{\alpha_1, \alpha_2, \alpha_3\}$ and the set of positive roots Δ^+ is given by

$$\Delta^+ = \{\alpha_1, \alpha_2, \alpha_3, \alpha_1 + \alpha_2, \alpha_2 + \alpha_3, \alpha_1 + \alpha_2 + \alpha_3\}.$$

If we take $\Phi = \{\alpha_1, \alpha_2, \alpha_3, \alpha_1 + \alpha_2\}$, then

$$\Phi \cup \Phi^- = \{\alpha_1, \alpha_2, \alpha_3, \alpha_1 + \alpha_2, -\alpha_1, -\alpha_2, -\alpha_3, -\alpha_1 - \alpha_2\}$$

and $\Psi = \{\alpha_1, \alpha_2, \alpha_1 + \alpha_2, \alpha_2 + \alpha_3, \alpha_1 + \alpha_2 + \alpha_3\}$. In this example the variables a_i for $i = 1, 2, 3$ correspond to the three simple roots α_1, α_2, α_3. We associate the variable a_4 to the root $\alpha_4 = \alpha_1 + \alpha_2$. We obtain the following Lax pair

$$L = \sum_{i=1}^{4} a_i (X_{\alpha_i} + X_{-\alpha_i}),$$
$$B = -a_4a_2(X_{\alpha_1} + X_{-\alpha_1}) - a_1a_4(X_{\alpha_2} + X_{-\alpha_2}) + a_1a_2(X_{\alpha_4} + X_{-\alpha_4})$$
$$+ a_2a_3(X_{\alpha_2+\alpha_3} + X_{-\alpha_2-\alpha_3}) - a_3a_4(X_{\alpha_1+\alpha_2+\alpha_3} + X_{\alpha_1-\alpha_2-\alpha_3}).$$

The Lax pair (using the substitution $x_i = a_i^2$ followed by scaling) becomes equivalent to the following equations of motion

$$\begin{aligned}
\dot{x}_1 &= x_1x_2 - x_1x_4, \\
\dot{x}_2 &= -x_1x_2 + x_2x_3 + x_2x_4, \\
\dot{x}_3 &= -x_2x_3 + x_3x_4, \\
\dot{x}_4 &= x_1x_4 - x_2x_4 - x_3x_4.
\end{aligned}$$

The system is integrable. There exist two functionally independent Casimir functions $F_1 = x_1x_3 = \det L$ and $F_2 = x_1x_2x_4$. The additional integral is the Hamiltonian $H = x_1 + x_2 + x_3 + x_4 = \operatorname{tr} L^2$.

The standard quadratic Poisson bracket (3) is given by

$$\pi = \begin{pmatrix} 0 & x_1x_2 & 0 & -x_1x_4 \\ -x_2x_1 & 0 & x_2x_3 & x_2x_4 \\ 0 & -x_3x_2 & 0 & x_3x_4 \\ x_4x_1 & -x_4x_2 & -x_4x_3 & 0 \end{pmatrix}.$$

One can find the Casimirs by computing the kernel of the matrix

$$A = \begin{pmatrix} 0 & 1 & 0 & -1 \\ -1 & 0 & 1 & 1 \\ 0 & -1 & 0 & 1 \\ 1 & -1 & -1 & 0 \end{pmatrix}.$$

The two eigenvectors with eigenvalue 0 are $(1,0,1,0)$ and $(1,1,0,1)$. We obtain the two Casimirs $F_1 = x_1^1 x_2^0 x_3^1 x_4^0 = x_1 x_3$ and $F_2 = x_1^1 x_2^1 x_3^0 x_4^1 = x_1 x_2 x_4$.

Example 3. A Lax pair L, B corresponding to the subset $\Phi = \{\alpha_1, \alpha_2, \alpha_3, \alpha_4 = \alpha_1 + \alpha_2 + \alpha_3\}$ of the root system of type A_3 is

$$L = \sum_{i=1}^{4} a_i \left(X_{\alpha_i} + X_{-\alpha_i}\right),$$
$$B = (a_1 a_2 - a_3 a_4)\left(X_{\alpha_1+\alpha_2} + X_{-\alpha_1-\alpha_2}\right) + (a_2 a_3 - a_1 a_4)\left(X_{\alpha_2+\alpha_3} + X_{-\alpha_2-\alpha_3}\right).$$

Using the substitution $x_i = 2a_i^2$ we obtain the periodic KM-system

$$\begin{aligned} \dot{x_1} &= x_1 x_2 - x_1 x_4, \\ \dot{x_2} &= -x_2 x_1 + x_2 x_3, \\ \dot{x_3} &= x_3 x_4 - x_3 x_2, \\ \dot{x_4} &= x_4 x_1 - x_4 x_3. \end{aligned} \tag{5}$$

The Poisson matrix (which can be read from the right hand side of (5)) is

$$\pi = \begin{pmatrix} 0 & x_1 x_2 & 0 & -x_1 x_4 \\ -x_1 x_2 & 0 & x_2 x_3 & 0 \\ 0 & -x_2 x_3 & 0 & x_3 x_4 \\ x_1 x_4 & 0 & -x_3 x_4 & 0 \end{pmatrix}$$

of rank 2. In addition to the Hamiltonian

$$H = x_1 + x_2 + x_3 + x_4$$

it possesses two Casimirs $C_1 = x_1 x_3$ and $C_2 = x_2 x_4$.

Example 4. Consider the subset

$$\Phi = \{\alpha_1, \alpha_2, \alpha_3, \alpha_4, \alpha_5 = \alpha_2 + \alpha_3\}$$

of the positive roots of the root system of type A_4. The Lax pair corresponding to the subset Φ is given by

$$L = \sum_{i=1}^{5} a_i \left(X_{\alpha_i} + X_{-\alpha_i}\right)$$

and

$$\begin{aligned} B = {} & -a_2 a_5 \left(X_{\alpha_3} - X_{-\alpha_3}\right) - a_3 a_5 \left(X_{\alpha_2} - X_{-\alpha_2}\right) a_1 a_2 \left(X_{\alpha_1+\alpha_2} - X_{-\alpha_1-\alpha_2}\right) \\ & + a_2 a_3 \left(X_{\alpha_5} - X_{-\alpha_5}\right) + a_3 a_4 \left(X_{\alpha_3+\alpha_4} - X_{-\alpha_3-\alpha_4}\right) \\ & - a_1 a_5 \left(X_{\alpha_1+\alpha_2+\alpha_3} - X_{-\alpha_1-\alpha_2-\alpha_3}\right) - a_4 a_5 \left(X_{\alpha_2+\alpha_3+\alpha_4} - X_{-\alpha_2-\alpha_3-\alpha_4}\right). \end{aligned}$$

Using the change of variables $x_i = 2a_i^2$ the corresponding Lotka–Volterra system becomes

$$\begin{aligned}
\dot{x}_1 &= x_1x_2 - x_1x_5,\\
\dot{x}_2 &= -x_2x_5 + x_2x_3 - x_2x_1,\\
\dot{x}_3 &= x_3x_5 + x_3x_4 - x_3x_2,\\
\dot{x}_4 &= x_4x_5 - x_4x_3,\\
\dot{x}_5 &= -x_5x_4 - x_5x_3 + x_5x_1 + x_5x_2\,.
\end{aligned}$$

The associated Poisson matrix is of rank 4.

The constants of motion are

$$\begin{aligned}
H &= x_1 + x_2 + x_3 + x_4 + x_5 \text{ (Hamiltonian)},\\
F &= x_1x_3 + x_1x_4 + x_2x_4,\\
C &= x_2x_3x_5 \text{ (Casimir)}\,.
\end{aligned}$$

4. Classification of Lotka–Volterra systems

In the previous section we presented several examples of cubic systems which (after a simple change of variables) are equivalent to Lotka–Volterra systems. In this section we classify all subsets Φ of the positive roots of A_n which produce, after a suitable change of variables, Lotka–Volterra systems. We also explicitly describe the corresponding Lotka–Volterra systems. We have the following Theorem.

Theorem 1. *Consider the root system of type A_n with simple roots $\Pi = \{\alpha_1, \alpha_2, \ldots, \alpha_n\}$ and positive roots*

$$\Delta^+ = \{\alpha_i + \alpha_{i+1} + \cdots + \alpha_j : 1 \le i \le j \le n\}.$$

The only choices for the subset Φ of Δ^+ so that the corresponding generalized Volterra system transforms into a Lotka–Volterra system, using the substitution $x_i = 2a_i^2$, are the following five:

(1) $\Phi = \Pi$,

(2) $\Phi = \Pi \cup \{\alpha_2 + \alpha_3 + \cdots + \alpha_{n-1}\}$,

(3) $\Phi = \Pi \cup \{\alpha_1 + \alpha_2 + \cdots + \alpha_{n-1}\}$,

(4) $\Phi = \Pi \cup \{\alpha_2 + \alpha_3 + \cdots + \alpha_n\}$,

(5) $\Phi = \Pi \cup \{\alpha_1 + \alpha_2 + \cdots + \alpha_n\}$.

Case (1) gives rise to the KM system while case (5) gives rise to the periodic KM system. Case (2) corresponds to the Lax equation $\dot{L} = [B, L]$ with

$$L = \sum_{i=1}^{n} a_i \left(X_{\alpha_i} + X_{-\alpha_i}\right) + a_{n+1}\left(X_{\alpha_2+\alpha_3+\cdots+\alpha_{n-1}} + X_{-\alpha_2-\alpha_3-\cdots-\alpha_{n-1}}\right).$$

The skew symmetric matrix B is defined using the method described in Section 3 (see also Proposition 2).

After substituting $x_i = 2a_i^2$ for $i = 1, \ldots, n+1$, the Lax pair L, B becomes equivalent to the following equations of motion

$$\begin{aligned}
\dot{x}_1 &= x_1(x_2 - x_{n+1}),\\
\dot{x}_2 &= x_2(x_3 - x_1 - x_{n+1}),\\
\dot{x}_i &= x_i(x_{i+1} - x_{i-1}), \qquad i = 3, 4, \ldots, n-2, n,\\
\dot{x}_{n-1} &= x_{n-1}(x_n - x_{n-2} + x_{n+1}),\\
\dot{x}_{n+1} &= x_{n+1}(x_1 + x_2 - x_{n-1} - x_n).
\end{aligned}$$

It is easily verified that for n even, the rank of the Poisson matrix is n and the function $f = x_2x_3 \cdots x_{n-1}x_{n+1}$ is the Casimir of the system, while for n odd, the rank of the Poisson matrix is $n-1$ and the functions $f_1 = x_1x_3 \cdots x_n = \sqrt{\det L}$ and $f_2 = x_2x_3 \cdots x_{n-1}x_{n+1}$ are the Casimirs.

Case (3) corresponds to the Lax pair whose Lax matrix L is given by

$$L = \sum_{i=1}^{n} a_i \left(X_{\alpha_i} + X_{-\alpha_i}\right) + a_{n+1} \left(X_{\alpha_1+\alpha_2+\cdots+\alpha_{n-1}} + X_{-\alpha_1-\alpha_2-\cdots-\alpha_{n-1}}\right).$$

After substituting $x_i = 2a_i^2$ for $i = 1, \ldots, n+1$, we obtain the following equivalent equations of motion

$$\begin{aligned}
\dot{x}_1 &= x_1(x_2 - x_{n+1}), \\
\dot{x}_i &= x_i(x_{i+1} - x_{i-1}), \qquad i = 2, 3, 4, \ldots, n-2, n, \\
\dot{x}_{n-1} &= x_{n-1}(x_n - x_{n-2} + x_{n+1}), \\
\dot{x}_{n+1} &= x_{n+1}(x_1 - x_n - x_{n-1}).
\end{aligned}$$

For n even, the rank of the Poisson matrix is n and the function $f = x_1x_2 \cdots x_{n-1}x_{n+1}$ is the Casimir, while for n odd, the rank of the Poisson matrix is $n-1$ and the functions $f_1 = x_1x_3x_5 \cdots x_n = \sqrt{\det L}$ and $f_2 = x_1x_2 \cdots x_{n-1}x_{n+1}$ are Casimirs.

The system obtained in case (4) turns out to be isomorphic to the one in case (3). In fact, the change of variables $u_{n+1-i} = -x_i$ for $i = 1, 2, \ldots, n$ and $u_{n+1} = -x_{n+1}$ in case (3) gives the corresponding system of case (4).

Since subsystems of Lotka–Volterra systems are also Lotka–Volterra, in order to prove Theorem 1 it is enough to consider the case where the subset Φ contains the simple roots and only one extra root. The following proposition shows that we have only four possible choices for the extra root in Φ which give rise to a Lotka–Volterra system. Therefore the proof of Theorem 1 is a case by case verification of the 16 possible subsets Φ containing the simple roots and roots given in the following proposition.

Proposition 2. *Let $\Phi = \{\alpha_1, \ldots, \alpha_{n+1}\}$ be the subset of the positive roots of the root system A_n containing the simple roots and the additional extra root α_{n+1}. Suppose that $\alpha_{n+1} = \alpha_k + \alpha_{k+1} + \cdots + \alpha_m$ for some $1 \le k < m \le n$. Then the corresponding generalized Volterra system is transformed, with the substitution $x_i = 2a_i^2$, to a Lotka–Volterra system if and only if the integers k, m satisfy*

$$k \le 2 \quad \text{and} \quad m \ge n-1.$$

We outline the proof of this proposition. We explicitly write down the bracket $[B, L]$ as a linear combination of the root vectors $\{X_\alpha : \alpha \in \Delta^+\}$. Note that there are some signs, $c_{i,j}$, involved in the construction of B which we have to determine so that the corresponding system is transformed to a Lotka–Volterra system. In order to have a Lax pair we set the coefficients of the root vectors $\{X_\alpha : \alpha \notin \Phi\}$ equal to zero and in order to have a Lotka–Volterra system we set the coefficients of the form $c_{i,j}a_ia_ja_k$ for $i \neq j \neq k \neq i$ equal to zero. We end up with a linear system of the coefficients $c_{i,j}$ which in order to have a solution the integers k, m must be of the required form. For a detailed proof see [16].

5. Using complex coefficients

As we noted in the introduction, we may produce more Lotka–Volterra systems by changing the matrix L from symmetric to Hermitian. The aim of this section is to describe this idea of using complex coefficients and give some examples of new Lotka–Volterra systems produced by this new method. Let us begin with an example

Example 5. Consider the case of a root system of type A_4 and the subset of positive roots $\Phi = \{\alpha_1, \alpha_2, \alpha_3, \alpha_4, \alpha_5 = \alpha_1 + \alpha_2\}$. It turns out that if

$$L = \sum_{i=1}^{4} a_i \left(X_{\alpha_i} + X_{-\alpha_i}\right) + a_5 \left(X_{\alpha_5} + X_{-\alpha_5}\right),$$

the corresponding linear system of signs (described in the comments after Proposition 2) does not have a solution while if

$$L = \sum_{i=1}^{4} a_i \left(X_{\alpha_i} + X_{-\alpha_i}\right) + \mathrm{i} a_5 \left(X_{\alpha_5} - X_{-\alpha_5}\right),$$

the corresponding system of signs does have a solution and gives rise to the system

$$\begin{aligned} \dot{a}_1 &= a_1 a_2^2 + a_1 a_5^2, \\ \dot{a}_2 &= a_2 a_3^2 - a_1^2 a_2 - a_2 a_5^2, \\ \dot{a}_3 &= a_3 a_4^2 - a_2^2 a_3 - a_3 a_5^2, \\ \dot{a}_4 &= -a_3^2 a_4, \\ \dot{a}_5 &= a_2^2 a_5 + a_3^2 a_5 - a_1^2 a_5. \end{aligned} \tag{6}$$

This system can be easily transformed to a Lotka–Volterra system which is integrable with one rational Casimir, $\frac{a_1 a_2}{a_5}$, and an extra constant of motion, $\mathrm{tr}(L^4)$.

In general, the idea is to make the matrix L Hermitian by replacing some terms of the form $a_i \left(X_{\alpha_i} + X_{-\alpha_i}\right)$ with $\mathrm{i}a_i \left(X_{\alpha_i} - X_{-\alpha_i}\right)$ where i is the imaginary unit. The construction of the matrix B is the same as in Section 3 with the only difference of B being skew Hermitian. Example 5 suggest that we may replace with $\mathrm{i}a_i \left(X_{\alpha_i} - X_{-\alpha_i}\right)$ all variables in L corresponding to the roots of height 2. But doing so we see that the only possible way to have a consistent Lax pair is to replace with $\mathrm{i}a_i \left(X_{\alpha_i} - X_{-\alpha_i}\right)$ all variables in L corresponding to roots of even height. Therefore we end up with the following alternative method of constructing Lax pairs.

We begin with a subset Φ of the positive roots containing the simple roots. We write $\Phi = \Phi_1 \cup \Phi_2$ where Φ_1 are the roots in Φ of odd height and Φ_2 are the roots in Φ of even height. The Lax matrix is constructed as

$$L = \sum_{\alpha_i \in \Phi_1} a_i (X_{\alpha_i} + X_{-\alpha_i}) + \sum_{\alpha_i \in \Phi_2} \mathrm{i}a_i (X_{\alpha_i} - X_{-\alpha_i}) = \sum_{\alpha_i \in \Phi} b_i (X_{\alpha_i} \pm X_{-\alpha_i}), \tag{7}$$

where the variables b_i are defined as $b_i = a_i$ if $\alpha_i \in \Phi_1$ and $b_i = \mathrm{i}a_i$ if $\alpha_i \in \Phi_2$. Consider the set $\Phi \cup \Phi^-$ which consists of all the roots in Φ together with their negatives. Let

$$\Psi = \left\{\alpha + \beta \,|\, \alpha, \beta \in \Phi \cup \Phi^-,\ \alpha + \beta \in \Delta^+\right\}.$$

We define the upper triangular part of the skew-Hermitian matrix B as

$$\sum c_{ij} b_i b_j X_{\alpha_i + \alpha_j}, \tag{8}$$

where $c_{ij} = \pm 1$ if $\alpha_i + \alpha_j \in \Psi$ with $\alpha_i, \alpha_j \in \Phi \cup \Phi^-$ and 0 otherwise.

An easy consequence of the construction of the matrices L and B is the following lemma.

Lemma 1. *Let Φ be a subset of the positive roots containing the simple roots and L, B the matrices constructed in (7) and (8). Also let K be the subset of the positive roots defined by*

$$K = \{\alpha + \beta + \gamma \colon \alpha, \beta, \gamma \in \Phi \cup \Phi^-,\ \alpha + \beta + \gamma \in \Delta^+\}.$$

Journal of Physics: Conference Series **621** (2015) 012004 doi:10.1088/1742-6596/621/1/012004

Let's write $K = K_1 \cup K_2$ where K_1 are the roots in K of odd height and K_2 are the roots in K of even height. Then the bracket $[B, L]$ is decomposed into $[B, L] = A_1 + \mathrm{i}A_2$ for a symmetric matrix A_1 and a skewsymmetric matrix A_2 where the nonzero entries of A_1 and A_2 appear in the "correct" positions; i.e. those of A_1 in positions corresponding to root vectors X_α, $\alpha \in K_1$ while those of A_2 in positions corresponding to root vectors X_α, $\alpha \in K_2$.

Next we show that this method gives more Lotka–Volterra systems and in general produces more Lax pairs than the one described in Section 3. Since for this method we don't have to worry about the diagonal entries of the bracket $[B, L]$ (see Lemma 1). We end up with the following proposition.

Proposition 3. *Let $\Pi \subset \Phi \subset \Delta^+$ be a subset of the positive roots containing the simple roots with the property that whenever $\alpha, \beta, \gamma \in \Phi \cup \Phi^-$ and $\alpha + \beta + \gamma \in \Delta^+$ then $\alpha + \beta + \gamma \in \Phi$. Also let L, B be the matrices constructed using the algorithm described in (7) and (8). Then for any choice of the signs $c_{i,j}$ the pair L, B is a Lax pair.*

Example 6. Let $k, n \in \mathbb{N}$ with $1 \leq k < n$. If Φ is the subset of the positive roots of the root system A_n containing the simple roots and all the roots of height larger than k then for all possible choices of the signs $c_{i,j}$ we have a consistent Lax pair.

Example 7. Consider the root system of type A_3 with simple roots $\Pi = \{\alpha_1, \alpha_2, \alpha_3\}$ and positive roots

$$\Delta^+ = \Pi \cup \{\alpha_4 = \alpha_1 + \alpha_2,\ \alpha_5 = \alpha_2 + \alpha_3,\ \alpha_6 = \alpha_1 + \alpha_2 + \alpha_3\}.$$

Then all Lax pairs corresponding to $\Phi = \Pi \cup \{\alpha_5, \alpha_6\}$ are given by

$$L = \sum_{i=1}^{3} a_i \left(X_{\alpha_i} + X_{-\alpha_i}\right) + \mathrm{i}a_4 \left(X_{\alpha_5} - X_{-\alpha_5}\right) + a_5 \left(X_{\alpha_6} + X_{-\alpha_6}\right)$$

and

$$\begin{aligned} B = {} & \mathrm{i}c_{4,5}a_4a_5 \left(X_{\alpha_1} + X_{-\alpha_1}\right) + \mathrm{i}c_{3,4}a_3a_4 \left(X_{\alpha_2} + X_{-\alpha_2}\right) + \mathrm{i}c_{4,5}a_2a_4 \left(X_{\alpha_3} + X_{-\alpha_3}\right) \\ & + \left(c_{1,2}a_1a_2 + c_{3,5}a_3a_5\right)\left(X_{\alpha_4} - X_{-\alpha_4}\right) + \left(c_{1,5}a_1a_5 + c_{2,3}a_2a_3\right)\left(X_{\alpha_5} - X_{-\alpha_5}\right) \\ & - \mathrm{i}c_{3,4}a_1a_4 \left(X_{\alpha_6} + X_{-\alpha_6}\right). \end{aligned}$$

We can verify that for all 32 choices of the signs $c_{i,j}$ no one of the corresponding systems is the same as the one produced by the method of Section 3. Therefore this procedure produces systems which in general are different from the ones of Section 3.

Example 8. For the root system of type A_5 with simple roots $\Pi = \{\alpha_i : i = 1, 2, 3, 4, 5\}$ and positive roots

$$\begin{aligned} \Delta^+ = {} & \Pi \cup \{\alpha_{3+i} = \alpha_i + \alpha_{i+1} : i = 1, 2, 3, 4\} \cup \{\alpha_{9+i} = \alpha_i + \alpha_{i+1} + \alpha_{i+2} : i = 1, 2, 3\} \\ & \cup \{\alpha_{12+i} = \alpha_i + \alpha_{i+1} + \alpha_{i+2} + \alpha_{i+3} : i = 1, 2\} \cup \{\alpha_{15} = \alpha_1 + \alpha_2 + \alpha_3 + \alpha_4 + \alpha_5\}, \end{aligned}$$

consider the Lax pair

$$\begin{aligned} L = {} & \sum_{i=1}^{5} a_i \left(X_{\alpha_i} + X_{-\alpha_i}\right) + \mathrm{i}a_6 \left(X_{\alpha_7} - X_{-\alpha_7}\right) + \mathrm{i}a_7 \left(X_{\alpha_9} - X_{-\alpha_9}\right) \\ & + \mathrm{i}a_8 \left(X_{\alpha_{14}} - X_{-\alpha_{14}}\right) + a_9 \left(X_{\alpha_{15}} + X_{-\alpha_{15}}\right) \end{aligned}$$

and

$$\begin{aligned} B = {} & \mathrm{i}a_8a_9\left(X_{\alpha_1}+X_{-\alpha_1}\right)-\mathrm{i}a_3a_6\left(X_{\alpha_2}+X_{-\alpha_2}\right)-\mathrm{i}a_2a_6\left(X_{\alpha_3}+X_{-\alpha_3}\right)\\ & -\mathrm{i}a_5a_7\left(X_{\alpha_4}+X_{-\alpha_4}\right)-\mathrm{i}a_4a_7\left(X_{\alpha_5}+X_{-\alpha_5}\right)-a_1a_2\left(X_{\alpha_6}-X_{-\alpha_6}\right)\\ & +\left(-a_2a_3+a_7a_8\right)\left(X_{\alpha_7}-X_{-\alpha_7}\right)-a_3a_4\left(X_{\alpha_8}-X_{-\alpha_8}\right)+\left(-a_4a_5+a_6a_8\right)\left(X_{\alpha_9}-X_{-\alpha_9}\right)\\ & -\left(\mathrm{i}a_1a_6+\mathrm{i}a_7a_9\right)\left(X_{\alpha_{10}}+X_{-\alpha_{10}}\right)+\left(-\mathrm{i}a_4a_6+\mathrm{i}a_5a_8\right)\left(X_{\alpha_{11}}+X_{-\alpha_{11}}\right)\\ & +\left(\mathrm{i}a_2a_8-\mathrm{i}a_3a_7\right)\left(X_{\alpha_{12}}+X_{-\alpha_{12}}\right)+a_5a_9\left(X_{\alpha_{13}}-X_{-\alpha_{13}}\right)+\left(a_1a_9+a_6a_7\right)\left(X_{\alpha_{14}}-X_{-\alpha_{14}}\right)\\ & +\mathrm{i}a_1a_8\left(X_{\alpha_{15}}+X_{-\alpha_{15}}\right). \end{aligned}$$

This Lax pair corresponds to the subset $\Phi = \Pi \cup \{\alpha_7, \alpha_9, \alpha_{14}, \alpha_{15}\}$ and gives rise to a Lotka–Volterra system. Note that according to Theorem 1 if we construct the matrices L, B using the algorithm of Section 3, the subset Φ does not give a Lotka–Volterra system.

The Poisson matrix for the associated system is given by

$$\pi = \begin{pmatrix} 0 & -a_1a_2 & 0 & 0 & 0 & -a_1a_6 & 0 & a_1a_8 & a_1a_9\\ a_1a_2 & 0 & -a_2a_3 & 0 & 0 & -a_2a_6 & 0 & a_2a_8 & 0\\ 0 & a_2a_3 & 0 & -a_3a_4 & 0 & a_3a_6 & -a_3a_7 & 0 & 0\\ 0 & 0 & a_3a_4 & 0 & -a_4a_5 & a_4a_6 & -a_4a_7 & 0 & 0\\ 0 & 0 & 0 & a_4a_5 & 0 & 0 & a_5a_7 & -a_5a_8 & -a_5a_9\\ a_1a_6 & a_2a_6 & -a_3a_6 & -a_4a_6 & 0 & 0 & -a_6a_7 & a_6a_8 & 0\\ 0 & 0 & a_3a_7 & a_4a_7 & -a_5a_7 & a_6a_7 & 0 & -a_7a_8 & -a_7a_9\\ -a_1a_8 & -a_2a_8 & 0 & 0 & a_5a_8 & -a_6a_8 & a_7a_8 & 0 & a_8a_9\\ -a_1a_9 & 0 & 0 & 0 & a_5a_9 & 0 & a_7a_9 & -a_8a_9 & 0 \end{pmatrix}$$

which has the following Casimirs

$$a_2a_4a_9,\ \frac{a_2a_4a_8}{a_1},\ \frac{a_1a_3a_7}{a_4},\ \frac{a_6}{a_2a_3},\ a_1a_3a_5.$$

The additional integral is given by $H_4 = \operatorname{tr} L^4$.

Acknowledgments

The first author was supported by a University of Cyprus Postdoctoral fellowship. The work of the third author was co-funded by the European Regional Development Fund and the Republic of Cyprus through the Research Promotion Foundation (Project: PENEK/0311/30).

References

[1] Lotka A 1920 Undamped oscillations derived from the law of mass action *J. Amer. Chem. Soc.* **42** 1595
[2] Volterra V 1990 *Leçons sur la théorie mathématique de la lutte pour la vie* [Lessons on the Mathematical Theory of the Struggle for Survival] Reprint of the 1931 original (Les Grands Classiques Gauthier-Villars [Gauthier-Villars Great Classics]) (Sceaux: Éditions Jacques Gabay)
[3] Kac V G 1990 *Infinite-Dimensional Lie Algebras* 3rd ed (Cambridge: Cambridge University Press)
[4] Flaschka H 1974 The Toda lattice. I. Existence of integrals *Phys. Rev.* B **9**:4 1924
[5] Moser J 1975 Three integrable Hamiltonian systems connected with isospectral deformations *Adv. Math.* **16** 197
[6] Faddeev L D and Takhtajan L A 1986 Liouville model on the lattice *Field Theory, Quantum Gravity and Strings (Meudon/Paris, 1984/1985)* (Lecture Notes in Phys. vol 246) (Berlin: Springer) p 166
[7] Bogoyavlensky O I 1988 Five constructions of integrable dynamical systems connected with the Korteweg–de Vries equation *Acta Appl. Math.* **13** 227
[8] Bogoyavlensky O I 1988 Integrable discretizations of the KdV equation *Phys. Lett.* A **134** 34
[9] Damianou P A 1991 The Volterra model and its relation to the Toda lattice *Phys. Lett.* A **155** 126
[10] Faddeev L D and Takhtajan L A 1987 *Hamiltonian Methods in the Theory of Solitons* Springer Series in Soviet Mathematics (Berlin: Springer-Verlag)

[11] Hernández-Bermejo B and Fairén V 2000 Separation of variables in the Jacobi identities *Phys. Lett.* A **271** 258

[12] Plank M 1995 Hamiltonian structures for the n-dimensional Lotka–Volterra equations *J. Math. Phys.* **36** 3520

[13] Bogoyavlenskij O I 2008 Integrable Lotka–Volterra systems *Regul. Chaotic Dyn.* **13** 543

[14] Damianou P A 2013 Lotka–Volterra systems associated with graphs *Proc. 6th Int. Workshop on Group Analysis of Differential Equations and Integrable Systems (Protaras, 2012)* (Nicosia: University of Cyprus) p 30

[15] Damianou P A, Sabourin H and Vanhaecke P 2011 Height-2 Toda systems *Proc. 5th Int. Workshop on Group Analysis of Differential Equations and Integrable Systems (Protaras, 2010)* (Nicosia: University of Cyprus) p 76

[16] Evripidou C A 2014 Ph.D. thesis *Applications of Root Systems in Geometry and Physics* (Nicosia: University of Cyprus)

Integrable discrete nonautonomous quad-equations as Bäcklund auto-transformations for known Volterra and Toda type semidiscrete equations

Rustem N Garifullin and Ravil I Yamilov

Ufa Institute of Mathematics, Russian Academy of Sciences,
112 Chernyshevsky Street, Ufa 450008, Russian Federation

E-mail: `rustem@matem.anrb.ru` and `RvlYamilov@matem.anrb.ru`

Abstract. We construct integrable discrete nonautonomous quad-equations as Bäcklund auto-transformations for known Volterra and Toda type semidiscrete equations, some of which are also nonautonomous. Additional examples of this kind are found by using transformations of discrete equations which are invertible on their solutions. In this way we obtain integrable examples of different types: discrete analogs of the sine-Gordon equation, the Liouville equation and the dressing chain of Shabat. For Liouville type equations we construct general solutions, using a specific linearization. For sine-Gordon type equations we find generalized symmetries, conservation laws and $L-A$ pairs.

1. Introduction

In this paper we consider discrete quad-equations

$$F_{n,m}(u_{n+1,m}, u_{n,m}, u_{n,m+1}, u_{n+1,m+1}) = 0, \quad n, m \in \mathbb{Z}, \tag{1}$$

which may explicitly depend on the discrete variables n, m. These equations are supposed to be polylinear, i.e. the functions $F_{n,m}$ are polynomials of the first order in each of their argument. Most of the known integrable equations of this kind have two generalized symmetries of the form

$$\frac{du_{n,m}}{dt_1} = \phi_{n,m}(u_{n+1,m}, u_{n,m}, u_{n-1,m}), \tag{2}$$

$$\frac{du_{n,m}}{dt_2} = \psi_{n,m}(u_{n,m+1}, u_{n,m}, u_{n,m-1}), \tag{3}$$

see, e.g., [1–4]. Recently a few examples of quad-equations with more complicated generalized symmetries have been found [5–8], but such equations are out of consideration in this paper.

Almost all known equations of the form (1) possessing the symmetries (2), (3) are autonomous. In the essentially more difficult nonautonomous case, we study in this paper, only a few examples are known [4].

The discrete equation (1) can be interpreted as a chain of Bäcklund auto-transformations for the lattice equations (2), (3). Such transformations allow one to construct a new solution $u_{n,m+1}$ in the case of equation (2) and $u_{n+1,m}$ in the case of equation (3), starting from a given solution $u_{n,m}$, see a more detailed comment in [1].

In this paper we start from known integrable equations of the form (2) and look for the discrete equations (1) generating for them chains of the Bäcklund auto-transformations. Such problem has been solved up to now only once by one example and only in the autonomous case [9]. Discrete equations obtained in this way may be not integrable. We select integrable cases by requiring the existence of a second symmetry of the form (3). In such way we can construct integrable discrete equations, using known integrable equations of the Volterra type presented in [10, 11] or using their nonautonomous generalizations given in [12].

More precisely, we are going to use differential-discrete equations of the form:

$$\frac{du_n}{dt} = P_n(u_n)(u_{n+1} - u_{n-1}). \tag{4}$$

It has been shown in [12] that nonlinear integrable equations of this form are described by the following conditions:

$$P_n = \alpha u_n^2 + \beta_n u_n + \gamma_n, \tag{5}$$

where α is an arbitrary constant, and β_n, γ_n are the two-periodic functions:

$$\beta_{n+2} = \beta_n, \quad \gamma_{n+2} = \gamma_n. \tag{6}$$

Up to the transformations

$$\tilde{t} = \eta t, \quad \tilde{u}_n = \mu_n u_n + \nu_n, \quad \tilde{u}_n = u_{n+1},$$

where μ_n and ν_n are the two-periodic functions, we have five cases: the Volterra equation with $P_n = u_n$, its three modifications with $P_n = u_n^2$, $P_n = u_n^2 - 1$, $P_n = u_n^2 - \chi_n$ and an equation with $P_n = \chi_n u_n + \chi_{n+1}$. The last equation is nothing but one of forms of the Toda model, as it has been shown in [12]. Here χ_n is given by

$$\chi_n = \frac{1 + (-1)^n}{2}. \tag{7}$$

We fix the generalized symmetry (2) in the direction n in one of the following five ways:

$$\frac{du_{n,m}}{dt_1} = u_{n,m}(u_{n+1,m} - u_{n-1,m}), \tag{8}$$

$$\frac{du_{n,m}}{dt_1} = u_{n,m}^2(u_{n+1,m} - u_{n-1,m}), \tag{9}$$

$$\frac{du_{n,m}}{dt_1} = (u_{n,m}^2 - 1)(u_{n+1,m} - u_{n-1,m}), \tag{10}$$

$$\frac{du_{n,m}}{dt_1} = (u_{n,m}^2 - \chi_n)(u_{n+1,m} - u_{n-1,m}), \tag{11}$$

$$\frac{du_{n,m}}{dt_1} = (\chi_n u_{n,m} + \chi_{n+1})(u_{n+1,m} - u_{n-1,m}). \tag{12}$$

In all these cases we find all corresponding polylinear discrete equations (1). A more general form of such symmetry is possible:

$$\frac{du_{n,m}}{dt_1} = (\alpha_m u_{n,m}^2 + \beta_{n,m} u_{n,m} + \gamma_{n,m})(u_{n+1,m} - u_{n-1,m}), \tag{13}$$

where $\beta_{n+2,m} = \beta_{n,m}$ and $\gamma_{n+2,m} = \gamma_{n,m}$ for all n, m. In this case we will construct some examples.

As a result we find nonautonomous integrable examples of several different types. According to their symmetry properties, equations (1) are the discrete analogs of the hyperbolic type equations $u_{xy} = f(x, y, u, u_x, u_y)$. Some examples obtained in this paper are of the sine-Gordon type. Such equations have two generalized symmetries (2), (3) and are not Darboux integrable, see definitions in the next section. We also find a few Darboux integrable equations which can be called the discrete analogs of the Liouville equation. For all equations of this type, we construct general solutions. One more new interesting integrable example presented here is a discrete analog of the well-known dressing chain studied in [13–15].

In Section 2 we give some definitions and obtain theoretical results necessary for the paper. In Section 3 we enumerate all the discrete equation (1) corresponding to the differential-discrete equations (8)–(12) and also obtain some examples in the case of equation (13). In Section 4 a discrete analog of the dressing chain is discussed. The problem of construction of the second symmetry (3) for examples obtained in Section 3 is solved in Section 5. In Section 6 some additional examples are found by using special transformations of the discrete equations invertible on their solutions.

2. Theory

In this section we give necessary definitions and derive some conditions for the discrete equations which allow us to make the class (1) of the discrete equations essentially more narrow.

We consider equations of the form (1) which are *polylinear* and *nondegenerate*. It is convenient to formulate definitions in terms of the function

$$F_{n,m}(x_1, x_2, x_3, x_4)$$

which depends on four continuous complex variables x_1, x_2, x_3, x_4 and on two integer discrete variables n, m. An equation of the form (1) is polylinear if

$$\frac{\partial^2 F_{n,m}}{\partial x_i^2} = 0, \quad i = 1, 2, 3, 4,$$

for all $n, m \in \mathbb{Z}$. So, we consider a class of polynomial equations with 16 n, m-dependent coefficients. The nondegeneracy is defined following [2]. If the function $F_{n,m}$ depends on x_4 for all n, m, then we can rewrite the equation $F_{n,m} = 0$ in the form

$$x_4 = f_{n,m}(x_1, x_2, x_3) \tag{14}$$

and we require the function $f_{n,m}$ to depend essentially on all its continuous variables for all n, m. So, we have the following nondegeneracy condition in terms of $F_{n,m}$ and $f_{n,m}$:

$$\frac{\partial F_{n,m}}{\partial x_4}, \frac{\partial f_{n,m}}{\partial x_1}, \frac{\partial f_{n,m}}{\partial x_2}, \frac{\partial f_{n,m}}{\partial x_3} \neq 0 \quad \text{for all} \quad n, m \in \mathbb{Z}. \tag{15}$$

The discrete equation (1) is equivalent to

$$u_{n+1,m+1} = f_{n,m}(u_{n+1,m}, u_{n,m}, u_{n,m+1}). \tag{16}$$

The *compatibility conditions* of equations (16) and (2), (3) have the form:

$$\phi_{n+1,m+1} = \phi_{n+1,m} \frac{\partial f_{n,m}}{\partial u_{n+1,m}} + \phi_{n,m} \frac{\partial f_{n,m}}{\partial u_{n,m}} + \phi_{n,m+1} \frac{\partial f_{n,m}}{\partial u_{n,m+1}}, \tag{17}$$

$$\psi_{n+1,m+1} = \psi_{n+1,m} \frac{\partial f_{n,m}}{\partial u_{n+1,m}} + \psi_{n,m} \frac{\partial f_{n,m}}{\partial u_{n,m}} + \psi_{n,m+1} \frac{\partial f_{n,m}}{\partial u_{n,m+1}}. \tag{18}$$

These relations are obtained by differentiating equation (16) with respect to the times t_1 and t_2 of equations (2), (3). For fixed values of n, m we can express, using equation (16), all functions $u_{n+k,m+l}$ $(kl \neq 0)$ in terms of the functions

$$u_{n+k,m}, u_{n,m+l}, \quad k, l \in \mathbb{Z}, \tag{19}$$

which can be considered as *independent variables*. Equations (17), (18) must be identically satisfied for all values of the independent variables as well as for any $n, m \in \mathbb{Z}$. If equations (16) and (2), (3) are compatible, then (2), (3) are the generalized symmetries of (16) and, on the other hand, equation (16) defines chains of the Bäcklund auto-transformations for equations (2), (3), see [1].

Equation (16) is called *Darboux integrable* if it has two first integrals $W_{n,m}$, $V_{n,m}$ depending of a finite number of the independent variables (19) and satisfying the relations

$$(T_1 - 1)W_{n,m} = 0, \quad (T_2 - 1)V_{n,m} = 0, \quad \text{for all } n, m \in \mathbb{Z}, \tag{20}$$

on the solutions of equation (16). Here T_1, T_2 are the shift operators in the first and second directions, respectively:

$$T_1 h_{n,m} = h_{n+1,m}, \quad T_2 h_{n,m} = h_{n,m+1}.$$

It is easy to show that the first integrals $W_{n,m}$ and $V_{n,m}$ depend only on the independent variables $u_{n,m+i}$ and $u_{n+j,m}$, respectively. Applying the shift operators, we can represent these first integrals as:

$$\begin{aligned} W_{n,m} &= W_{n,m}(u_{n,m}, u_{n,m+1}, \ldots, u_{n,m+k_1}), \\ V_{n,m} &= V_{n,m}(u_{n,m}, u_{n+1,m}, \ldots, u_{n+k_2,m}). \end{aligned} \tag{21}$$

The Darboux integrable equations are linearizable, with linearizing transformations $w_{n,m} = W_{n,m}$, $v_{n,m} = V_{n,m}$, and are analogs of the Liouville equation.

A discrete equation (16) is of the *sine-Gordon type* if it has two generalized symmetries (2), (3) and is not Darboux integrable. Such equations should be integrable by the inverse scattering method. It is difficult to prove that a given equation has no first integrals (21), see [16] for possible difficulties. For examples obtained below, we check that fact for $k_i \leq 4$.

Let us derive two conditions necessary for the compatibility of the discrete equation (16) and an equation of the form

$$\frac{du_{n,m}}{dt_1} = P_{n,m}(u_{n,m})(u_{n+1,m} - u_{n-1,m}). \tag{22}$$

The only restriction here is that $P_{n,m}(x) \neq 0$ for all n, m. The equations (8)–(13) are particular cases of equation (22).

Differentiating the compatibility condition (17) with respect to $u_{n+2,m}$ and applying T_1^{-1}, we obtain the relation

$$T_2 \left(\frac{\partial \phi_{n,m}}{\partial u_{n+1,m}} \right) \frac{\partial f_{n,m}}{\partial u_{n+1,m}} = \frac{\partial \phi_{n,m}}{\partial u_{n+1,m}} T_1^{-1} \left(\frac{\partial f_{n,m}}{\partial u_{n+1,m}} \right). \tag{23}$$

This is nothing but one of so-called integrability conditions obtained in [9]. Here we just present it in the most general non-autonomous case and write it down in a form more convenient for the present paper. In the case of equation (22) it takes the form:

$$P_{n,m+1}(u_{n,m+1}) \frac{\partial f_{n,m}}{\partial u_{n+1,m}} = P_{n,m}(u_{n,m}) T_1^{-1} \left(\frac{\partial f_{n,m}}{\partial u_{n+1,m}} \right). \tag{24}$$

Applying T_1^{-1} to equation (16), we can easily rewrite it in one more form

$$u_{n-1,m+1} = \hat{f}_{n,m}(u_{n-1,m}, u_{n,m}, u_{n,m+1}) \tag{25}$$

equivalent to equation (1). The function $\hat{f}_{n,m}$ essentially depends on all its continuous variables for all n, m. The compatibility condition for (25) and (2) reads:

$$\phi_{n-1,m+1} = \phi_{n-1,m}\frac{\partial \hat{f}_{n,m}}{\partial u_{n-1,m}} + \phi_{n,m}\frac{\partial \hat{f}_{n,m}}{\partial u_{n,m}} + \phi_{n,m+1}\frac{\partial \hat{f}_{n,m}}{\partial u_{n,m+1}}. \tag{26}$$

The following condition analogous to (24) is derived from (26) in a quite similar way:

$$P_{n,m+1}(u_{n,m+1})\frac{\partial \hat{f}_{n,m}}{\partial u_{n-1,m}} = P_{n,m}(u_{n,m})T_1\left(\frac{\partial \hat{f}_{n,m}}{\partial u_{n-1,m}}\right). \tag{27}$$

Theorem 1 *If the discrete equation* (1) *is compatible with* (22)*, then the conditions* (24) *and* (27) *must be satisfied.*

Using the conditions (24) and (27), we can essentially simplify the form of the polylinear discrete equation (1). The resulting form is

$$\begin{aligned}&(\kappa_{1,n,m}u_{n,m} + \kappa_{2,n,m}u_{n,m+1} + \kappa_{3,n,m})u_{n+1,m+1}\\&+(\kappa_{4,n,m}u_{n,m} + \kappa_{5,n,m}u_{n,m+1} + \kappa_{6,n,m})u_{n+1,m}\\&+(\kappa_{7,n,m}u_{n,m} + \kappa_{8,n,m}u_{n,m+1} + \kappa_{9,n,m}) = 0,\end{aligned} \tag{28}$$

where $\kappa_{i,n,m}$ are arbitrary n, m-dependent functions. This form corresponds to the following restrictions

$$\frac{\partial^2 F_{n,m}}{\partial u_{n+1,m+1}\partial u_{n+1,m}} = \frac{\partial^2 F_{n,m}}{\partial u_{n,m+1}\partial u_{n,m}} = 0. \tag{29}$$

Theorem 2 *If a nondegenerate polylinear equation* (1) *is compatible with an equation of the form* (22)*, then it must have the form* (28).

Proof. The relation (24) depends on the following independent variables: $u_{n-1,m}$, $u_{n,m}$, $u_{n,m+1}$, $u_{n+1,m}$, and only the term $\frac{\partial f_{n,m}}{\partial u_{n+1,m}}$ depends on $u_{n+1,m}$. Differentiating (24) with respect to $u_{n+1,m}$ and taking into account the restriction $P_{n,m}(x) \neq 0$, we obtain

$$\frac{\partial^2 f_{n,m}}{\partial u_{n+1,m}^2} = 0. \tag{30}$$

In quite similar way the relation (27) implies

$$\frac{\partial^2 \hat{f}_{n,m}}{\partial u_{n-1,m}^2} = 0. \tag{31}$$

Applying the conditions (30) and (31) to the discrete polylinear equation (1) and using its nondegeneracy, we are led to the form (28). □

3. Bäcklund auto-transformations

Here we describe all polylinear discrete equations (1) compatible with equations (8)–(12). At the end we construct an example corresponding to an equation of the form (13).

We fix one of equations (8)–(12) as the generalized symmetry (2). To find corresponding discrete equation, we use in the first step more simple necessary conditions (24), (27) instead of the compatibility condition (17). The relations (24), (27) are equivalent to a nonlinear algebraic system of equations for eighteen functions $\kappa_{i,n,m}$, $\kappa_{i,n-1,m}$. It is interesting that not only in this system but also in case of the compatibility condition (17) the discrete variable m is not changed. So we can define the dependence of $\kappa_{i,n,m}$ on n only.

Such problem is not difficult in the autonomous case. We just need to solve an algebraic system for nine unknown coefficients k_i, using any computer algebra system like Reduce. In the nonautonomous case, we can only find from that algebraic system a set of solutions which have the form of relations between the functions $k_{i,n,m}$ and $k_{i,n-1,m}$ in a fixed point $n = n_0$. Another difficulty is that there are many divisors of zero $\alpha_n \not\equiv 0$ and $\beta_n \not\equiv 0$, such that $\alpha_n\beta_n \equiv 0$.

Nevertheless, comparing sets of solutions at $n = n_0$ and $n = n_0 + 1$, one can choose consistent pairs of nondegenerate solutions, i.e. such that corresponding discrete equation (28) is nondegenerate, and can determine in this way the dependence of the functions $k_{i,n,m}$ on n. In case of the nonautonomous equations (11) and (12), we need to solve the corresponding algebraic system twice, at $n = 2k_0$ and $n = 2k_0 + 1$, to avoid a dependence on the function χ_n. This the way we find the discrete equations.

In case of the Toda lattice (12), we have checked that there is no polylinear and nondegenerate discrete equation (1) compatible with (12). However, in Section 6 we present an example of the sine-Gordon type, corresponding to (12), which is not polylinear. In case of the Volterra equation (8) we get a positive result.

Theorem 3 *If a polylinear nondegenerate discrete equation* (1) *is compatible with equation* (8), *then, up to the multiplication by a function $\nu_{n,m}$ nonzero for any $n, m \in \mathbb{Z}$, it can be expressed as:*

$$\begin{aligned} &\Omega_{n,m}(u_{n+1,m+1}u_{n,m+1} - u_{n+1,m}u_{n,m}) \\ &\quad + \Omega_{n+1,m}(u_{n+1,m+1} + u_{n,m+1} - u_{n+1,m} - u_{n,m} + k_m) = 0, \\ &\Omega_{n,m} = \frac{1 + \omega_m(-1)^n}{2}, \end{aligned} \tag{32}$$

where $\omega_m^2 = 1$ for all $m \in \mathbb{Z}$ and k_m is an arbitrary m-dependent function.

In the case of equations (9)–(11) the problem of finding discrete equations (1) is easier. We solve an algebraic system, corresponding to equations (24) and (27) in a fixed point n, and see that all possible nondegenerate solutions have the same structure:

$$k_{1,n,m} = k_{5,n,m} = 0, \quad k_{4,n,m} = -k_{2,n,m} \neq 0 \quad \text{for all } n, m \in \mathbb{Z}.$$

Deviding the discrete equation (28) by $k_{2,n,m}$, we obtain a very simple ansatz of the form

$$\begin{aligned} &u_{n+1,m+1}u_{n,m+1} - u_{n+1,m}u_{n,m} + \kappa_{3,n,m}u_{n+1,m+1} + \kappa_{6,n,m}u_{n+1,m} \\ &\quad + \kappa_{7,n,m}u_{n,m} + \kappa_{8,n,m}u_{n,m+1} + \kappa_{9,n,m} = 0. \end{aligned} \tag{33}$$

Now we can interpret algebraic systems of equations for $\kappa_{i,n,m}$, $\kappa_{i,n+1,m}$, $\kappa_{i,n-1,m}$, corresponding to the conditions (24), (27) and (17), as systems of ordinary difference equations for the functions $k_{i,n,m}$ and then we can specify equation (33) with no difficulties.

The resulting discrete equation in the case of (9) is not interesting:

$$u_{n+1,m+1}u_{n,m+1} = u_{n+1,m}u_{n,m}. \tag{34}$$

It can be linearized by the point transformation $v_{n,m} = \log u_{n,m}$.

Theorem 4 *Up to the multiplication by a nonzero function* $\nu_{n,m}$, *a polylinear and nondegenerate discrete equation* (1) *must be of the form:*

$$(u_{n+1,m}+a_{n+1,m})(u_{n,m}-a_{n,m})=(u_{n+1,m+1}+b_{n+1,m+1})(u_{n,m+1}-b_{n,m+1}),\quad a_{n+2,m}=a_{n,m},\quad b_{n+2,m}=b_{n,m},\quad a_{n,m}^2=b_{n,m}^2=1, \tag{35}$$

if it is compatible with equation (10), *and of the form:*

$$(u_{n+1,m}+A_m\chi_{n+1})(u_{n,m}-A_m\chi_n)=(u_{n+1,m+1}+B_{m+1}\chi_{n+1})(u_{n,m+1}-B_{m+1}\chi_n),\quad A_m^2=B_m^2=1, \tag{36}$$

in the case of equation (11).

Let us construct now a generalization of equations (35) and (36), using a pair of Miura type transformations, see [17] and a more close to the discrete quad-equations [18].

Equation (13) with $\alpha_m\neq 0$ for any m can be transformed into

$$\frac{du_{n,m}}{dt_1}=(u_{n,m}^2-a_{n,m}^2)(u_{n+1,m}-u_{n-1,m}),\quad a_{n+2,m}=a_{n,m}\ \text{ for all } n,m. \tag{37}$$

A transformation has the form $\tilde{u}_{n,m}=\mu_m u_{n,m}+\nu_{n,m}$, where $\mu_m\neq 0$, $\nu_{n+2,m}=\nu_{n,m}$ for all n, m. Equation (37) is transformed into the Volterra equation (8) by a Miura type transformation, such that

$$\hat{u}_{n,m}=(u_{n+1,m}+a_{n+1,m})(u_{n,m}-a_{n,m}). \tag{38}$$

Introducing $b_{n,m}$, such that

$$b_{n+2,m}=b_{n,m},\quad b_{n,m}^2=a_{n,m}^2,$$

we have another Miura type transformation of equation (37) into (8):

$$\hat{u}_{n,m-1}=(u_{n+1,m}+b_{n+1,m})(u_{n,m}-b_{n,m}). \tag{39}$$

Excluding $\hat{u}_{n,m}$ we are led to the discrete equation

$$(u_{n+1,m}+a_{n+1,m})(u_{n,m}-a_{n,m})=(u_{n+1,m+1}+b_{n+1,m+1})(u_{n,m+1}-b_{n,m+1}),\quad a_{n+2,m}=a_{n,m},\quad b_{n+2,m}=b_{n,m},\quad a_{n,m}^2=b_{n,m}^2\ \text{ for all } n,m. \tag{40}$$

It can be checked that the quad-equation (40) is compatible with (37).

One can see that equations (35) and (36) correspond to particular cases of the general formula (40). In the second case, we have $a_{n,m}^2=b_{n,m}^2=\chi_n$ and we can represent

$$a_{n,m}=A_m\chi_n,\quad b_{n,m}=B_m\chi_n.$$

4. Discrete analog of the dressing chain

Let us consider a particular case of equation (40), presented in [9], namely:

$$(u_{n+1,m}+\delta_m)(u_{n,m}-\delta_m)=(u_{n+1,m+1}-\delta_{m+1})(u_{n,m+1}+\delta_{m+1}), \tag{41}$$

where δ_m is an arbitrary m-dependent coefficient. It is a complete analogue of the well-known dressing chain, see [13–15]:

$$\frac{d}{dx}(u_{m+1}+u_m)=u_{m+1}^2-u_m^2+\delta_{m+1}-\delta_m. \tag{42}$$

Equation (42) can be constructed by two Miura transformations into the Korteweg–de Vries equation as well as equation (41) is constructed by two discrete Miura transformations into the Volterra equation which also is called the discrete Korteweg–de Vries equation. There is in [9] an $L-A$ pair for equation (41) which is the direct analog of an $L-A$ pair for (42) constructed in [14]. In this section we present a generalization of equation (41) together with its $L-A$ pair.

That generalization is a particular case of equation (40) corresponding to $b_{n,m} = -c_m a_{n,m}$ with $c_m^2 \equiv 1$:

$$\begin{aligned} &(u_{n+1,m+1} - c_{m+1}a_{n+1,m+1})(u_{n,m+1} + c_{m+1}a_{n,m+1}) \\ &= (u_{n+1,m} + a_{n+1,m})(u_{n,m} - a_{n,m}), \quad a_{n+2,m} = a_{n,m}, \quad c_m^2 \equiv 1, \end{aligned} \tag{43}$$

and we lose, e.g., the case $b_{n,m} = (-1)^n a_{n,m}$. In the case when $a_{n,m} \neq 0$ and $c_m = 1$ for all n, m, we can introduce a new function $\hat{u}_{n,m}$: $u_{n,m} = \hat{u}_{n,m} a_{n,m}/\delta_m$, where $\delta_m^2 = a_{n+1,m}a_{n,m}$. The last product does not depend on n, as $a_{n,m}$ is two-periodic with respect to n. This function $\hat{u}_{n,m}$ satisfies an equation of the form (41), i.e. we get nothing new in this case. So, only the case when $c_m = -1$ or $a_{n,m} = 0$ for some n, m is interesting, and we will show examples of this kind in next sections.

An $L-A$ pair for equation (43) reads:

$$\begin{aligned} L_{n,m} &= \begin{pmatrix} \lambda & -(u_{n+1,m} + a_{n+1,m})(u_{n,m} - a_{n,m}) \\ 1 & 0 \end{pmatrix}, \\ A_{n,m} &= \begin{pmatrix} \lambda & a_{n+1,m+1}(u_{n,m+1} + a_{n,m+1})(1 + c_{m+1}) \\ -\frac{a_{n,m+1}(1+c_{m+1})}{u_{n,m+1}-a_{n,m+1}} & \frac{\lambda(u_{n,m+1}+c_{m+1}a_{n,m+1})}{u_{n,m+1}-a_{n,m+1}} \end{pmatrix}, \end{aligned} \tag{44}$$

and these matrices satisfy the standard relation $L_{n,m+1}A_{n,m} = A_{n+1,m}L_{n,m}$. In particular case (41), this $L-A$ pair coincides with one of [9]. A hierarchy of conservation laws for equation (41) has been constructed in [19] by using that $L-A$ pair of [9].

We also can construct conservation laws for (43), using the $L-A$ pair (44) and the same scheme of [19]. Discrete conservation laws are of the form

$$(T_2 - 1)p_{n,m}^{(i)} = (T_1 - 1)q_{n,m}^{(i)} \tag{45}$$

and, for $i = 0, 1, 2$, are given by the following functions $p_{n,m}^{(i)}$ and $q_{n,m}^{(i)}$:

$$\begin{aligned} p_{n,m}^{(0)} &= \log(u_{n+1,m} + a_{n+1,m})(u_{n,m} - a_{n,m}), \\ q_{n,m}^{(0)} &= \log \frac{u_{n,m+1} + c_{m+1}a_{n,m+1}}{u_{n,m+1} - a_{n,m+1}}; \end{aligned} \tag{46}$$

$$\begin{aligned} p_{n,m}^{(1)} &= (u_{n+1,m} + a_{n+1,m})(u_{n,m} - a_{n,m}), \\ q_{n,m}^{(1)} &= -(u_{n,m+1} - a_{n,m+1})(1 + c_{m+1})a_{n+1,m+1}; \end{aligned} \tag{47}$$

$$\begin{aligned} p_{n,m}^{(2)} &= (u_{n+1,m} + a_{n+1,m})(u_{n,m} - a_{n,m})(2u_{n+2,m}u_{n+1,m} + u_{n+1,m}u_{n,m} \\ &\quad - 2u_{n+2,m}a_{n+1,m} + u_{n+1,m}a_{n,m} + a_{n+1,m}u_{n,m} - 3a_{n+1,m}a_{n,m}), \\ q_{n,m}^{(2)} &= -2(1 + c_{m+1})a_{n+1,m+1}(u_{n,m+1} - c_{m+1}a_{n,m+1})(u_{n+1,m}u_{n,m} - u_{n+1,m}a_{n,m} \\ &\quad + a_{n+1,m+1}u_{n,m+1} + a_{n+1,m}u_{n,m} - a_{n+1,m+1}a_{n,m+1} - a_{n+1,m}a_{n,m}). \end{aligned} \tag{48}$$

Equations of the sine-Gordon type possess two hierarchies of generalized symmetries and conservation laws. Equation (43) probably has, in general, only one hierarchy of conservation laws and generalized symmetries, see a comment below. Nevertheless, it has the $L-A$ pair and is integrable for this reason. It deserves further study as a direct discrete analogue of the dressing chain (42).

The discrete equation (41) has the generalized symmetry (37) with $a_{n,m} = \delta_m$ in the direction n. However, in the case when $\delta_m \neq 0$ for all m, we can show that if the second generalized symmetry of the form (3) exists, then the following relation must take place:

$$\delta_m^2 = \delta_0^2 \text{ for all } m.$$

In this case, using rescaling $u_{n,m} = \hat{u}_{n,m}\delta_m$, we can transform this equation into:

$$(u_{n+1,m+1} - 1)(u_{n,m+1} + 1) = (u_{n+1,m} + 1)(u_{n,m} - 1). \tag{49}$$

This is the well-known integrable equation found in [20, 21]. A generalized symmetry in the m-direction can be found in [9], $L-A$ pairs of different forms have been presented, e.g., in [18, 20].

A new sine-Gordon type example of the form (43), which is essentially nonautonomous, will be discussed in Section 6. That equation has two hierarchies of generalized symmetries as well as two hierarchies of conservation laws which can be derived from the $L-A$ pair (44).

5. Second generalized symmetry

Among discrete equations, we consider in this paper, there may be integrable equations which do not have two generalized symmetries of the form (2), (3). Nevertheless, we use here the existence of two generalized symmetries of such form as an integrability criterion. It is constructive and allows us not only to check an equation for integrability, but also to solve some classification problems, as it is demonstrated below, see also [6].

For some of discrete equations found in Section 3, we select in this section cases in which a generalized symmetry of the form (3) exists. The symmetry must be *nondegenerate*, i.e. its right hand side must differ from zero for any n, m.

When searching for generalized symmetries, we use a scheme developed in [2, 9, 22]. To check the Darboux integrability of an equation, we use results of [6, 23]. In both cases some special annihilation operators introduced in [23] play an important role.

5.1. Volterra case

In the case of equation (32) we have two possibilities up to the transformation $\tilde{u}_{n,m} = u_{n+1,m}$. We have $k_m \equiv 0$ in both cases. In the first case $\omega_m = (-1)^m$, therefore $\Omega_{n,m} = \chi_{n+m}$, and equation (32) takes the form:

$$\begin{aligned} &\chi_{n+m}(u_{n+1,m+1}u_{n,m+1} - u_{n+1,m}u_{n,m}) \\ &\quad + \chi_{n+m+1}(u_{n+1,m+1} + u_{n,m+1} - u_{n+1,m} - u_{n,m}) = 0, \end{aligned} \tag{50}$$

where χ_{n+m} is defined by (7). Nonautonomous integrable equations of this kind are known, see [4]. Examples of [4] and equation (50) are essentially different, as corresponding generalized symmetries strongly differ from each other.

The second generalized symmetry (3) in the m-direction of equation (50) reads:

$$\frac{du_{n,m}}{dt_2} = u_{n,m}\left(\frac{C_m}{u_{n,m+1} - u_{n,m}} + \frac{C_{m-1}}{u_{n,m} - u_{n,m-1}}\right), \quad C_m = \alpha m + \beta, \tag{51}$$

where α, β are arbitrary constants. Equation (51) with $\alpha = 0$, $\beta = 1$ is a representative of the well-known complete list of integrable Volterra type equations presented in [10, 11]. Equation (51) with $\alpha = 1$, $\beta = 0$ is its master symmetry found in [24]. It generates generalized symmetries for equation (51) with $\alpha = 0$, $\beta = 1$ and, therefore, for the discrete equation (50), for instance:

$$\begin{aligned}\frac{du_{n,m}}{d\tau_2} = &\frac{u_{n,m+1}u_{n,m}}{(u_{n,m+2} - u_{n,m+1})(u_{n,m+1} - u_{n,m})^2} + \frac{u_{n,m}u_{n,m-1}}{(u_{n,m} - u_{n,m-1})^2(u_{n,m-1} - u_{n,m-2})} \\ &+ \frac{u_{n,m}^2(u_{n,m+1} - u_{n,m-1})}{(u_{n,m+1} - u_{n,m})^2(u_{n,m} - u_{n,m-1})^2}.\end{aligned} \tag{52}$$

It can be proved that equation (50) does not have first integrals (21) with $k_i \leq 4$ and, for this reason, probably is not Darboux integrable.

In the second case, $\omega_m \equiv 1$ in equation (32), hence $\Omega_{n,m} = \chi_n$, and this equation is of the form:

$$\begin{aligned}&\chi_n(u_{n+1,m+1}u_{n,m+1} - u_{n+1,m}u_{n,m}) \\ &\quad + \chi_{n+1}(u_{n+1,m+1} + u_{n,m+1} - u_{n+1,m} - u_{n,m}) = 0.\end{aligned} \tag{53}$$

Its generalized symmetry of the form (3) reads:

$$\frac{du_{n,m}}{dt_2} = c_m \frac{(u_{n,m+1} - u_{n,m})(u_{n,m} - u_{n,m-1})}{u_{n,m+1} - u_{n,m-1}}, \tag{54}$$

where c_m is an arbitrary function, such that $c_m \neq 0$ for any m. This example is Darboux integrable, as it possesses the following first integrals, see (20), (21):

$$\begin{aligned}V_{n,m} &= \chi_n u_{n+1,m}u_{n,m} + \chi_{n+1}(u_{n+1,m} + u_{n,m}), \\ W_{n,m} &= \frac{(u_{n,m+3} - u_{n,m})(u_{n,m+2} - u_{n,m+1})}{(u_{n,m+3} - u_{n,m+2})(u_{n,m+1} - u_{n,m})}.\end{aligned} \tag{55}$$

In the next section, a general solution of equation (53) will be constructed.

There exist examples with degenerate generalized symmetries, too. Such examples can be taken from the following statement: equation (54) is the symmetry of equation (32) iff

$$c_m(\omega_m - \omega_{m-1}) = c_m k_m = c_m k_{m-1} = 0 \quad \text{for all } m.$$

There may be very few points of degeneration of a symmetry, for instance, if

$$k_m \equiv 0, \quad \omega_m = -1, \ m \leq -1, \quad \omega_m = 1, \ m \geq 0,$$

then $c_0 = 0$ and c_m may be a nonzero constant in all the other points m. Such examples probably have not been considered in the literature up to now. Equations of this kind seem to be very close to the integrable ones and, in our opinion, deserve further study.

5.2. General solution

In this section we present, by an example, a scheme of construction of the general solution for the Darboux integrable discrete equations. We use an observation of [6] for polylinear equations (1) that, in many cases, first integrals of the first order (i.e. such that $k_1 = 1$ or $k_2 = 1$ in (21)) can be rewritten as a linear equation. Using this fact, the corresponding discrete equation can be equivalently rewritten as a nonautonomous linear equation.

We see that $V_{n,m}$ of (55) is a first integral of the first order, and the relation $(T_2 - 1)V_{n,m} = 0$ is equivalent to the discrete equation (53). We can solve this relation and obtain

$$V_{n,m} = \chi_n u_{n+1,m}u_{n,m} + \chi_{n+1}(u_{n+1,m} + u_{n,m}) = \eta_n, \tag{56}$$

where η_n is an arbitrary n-dependent function of integration. We see that this is the discrete Riccati equation. As it is known for such type equations, if we know a particular solution, then we can rewrite the equation as a linear one and find its general solution.

Let ρ_n be a particular solution of equation (56):

$$\chi_n \rho_{n+1} \rho_n + \chi_{n+1}(\rho_{n+1} + \rho_n) = \eta_n.$$

Then the discrete Riccati equation (56) is linearized by the following point transformation:

$$u_{n,m} = \frac{1}{v_{n,m}} + \rho_n, \tag{57}$$

and we get

$$(\chi_{n+1} + \chi_n \rho_n) v_{n,m} + (\chi_{n+1} + \chi_n \rho_{n+1}) v_{n+1,m} + \chi_n = 0.$$

Using an integrating factor $\chi_n \nu_n - \chi_{n+1} \kappa_n$, we can represent this linear equation as a total difference:

$$(T_1 - 1) w_{n,m} = 0, \quad w_{n,m} = (\chi_n \kappa_{n-1} - \chi_{n+1} \kappa_n) v_{n,m} + \chi_n \kappa_n + \chi_{n+1} \kappa_{n+1},$$

if $\nu_n = \kappa_n - \kappa_{n+2}$ and κ_n satisfies the following linear equation:

$$(\chi_n \nu_n + \chi_{n+1} \nu_{n-1}) \rho_n = \chi_n \kappa_{n-1} + \chi_{n+1} \kappa_n.$$

We are led to the relation $w_{n,m} = \theta_m$ with a new function of integration θ_m. From this relation we find $v_{n,m}$ and, taking into account (57), we obtain

$$u_{n,m} = \chi_n \frac{\kappa_{n-1}(\theta_m + \kappa_{n+2})}{(\kappa_{n+2} - \kappa_n)(\theta_m + \kappa_n)} + \chi_{n+1} \frac{\kappa_n(\theta_m + \kappa_{n-1})}{(\kappa_{n+1} - \kappa_{n-1})(\theta_m + \kappa_{n+1})}. \tag{58}$$

This formula (58) defines the general solution of the discrete equation (53) in the sense that this solution depends on two arbitrary functions of one discrete variable, namely, κ_n and θ_m. Probably, the formula (58) does not describe all solutions of equation (53), as it corresponds to a special choice of the function η_n in equation (56).

5.3. Modified Volterra case

Here we consider the discrete equations (35), (36) corresponding to the modified Volterra equations (10), (11).

As $b_{n,m}$ of equation (35) is a 2-periodic function with respect to n, we can introduce $d_m = b_{n,m} b_{n+1,m}$ which does not depend on n. After rescaling $u_{n,m} \to -u_{n,m} b_{n,m}$, $a_{n,m} \to -a_{n,m} b_{n,m}$, we obtain instead of (10), (35) the following equations:

$$\frac{du_{n,m}}{dt_1} = d_m (u_{n,m}^2 - 1)(u_{n+1,m} - u_{n-1,m}), \tag{59}$$

$$d_{m+1}(u_{n+1,m+1} - 1)(u_{n,m+1} + 1) = d_m (u_{n+1,m} + a_{n+1,m})(u_{n,m} - a_{n,m}), \tag{60}$$

where $d_m^2 = a_{n,m}^2 = 1$, $a_{n+2,m} = a_{n,m}$ for all n, m.

There are here two cases with a nondegenerate second generalized symmetry. In both cases $d_m \equiv 1$ and $a_{n,m} = A_m$, i.e. it does not depend on n. In the first case $A_m \equiv 1$, and equation (60) is nothing but the known equation (49). In the other case $A_m \equiv -1$, and the second generalized symmetry reads:

$$\frac{du_{n,m}}{dt_2} = c_m \frac{(u_{n,m+1} - u_{n,m})(u_{n,m} - u_{n,m-1})}{u_{n,m+1} - u_{n,m-1}}, \tag{61}$$

where c_m is an arbitrary function, such that $c_m \neq 0$ for any m. This discrete equation coincides, up to an autonomous Möbius transformation, with an example found in [25]. This equation is Darboux integrable, its first integrals and a general solution are presented in [25], too.

There is a degenerate example of the form (60) with $d_m \equiv 1$, $a_{n,m} = A_m$:

$$(u_{n+1,m+1} - 1)(u_{n,m+1} + 1) = (u_{n+1,m} + A_m)(u_{n,m} - A_m), \tag{62}$$

where A_m is such that $A_m^2 = 1$ and it is not constant. Equation (62) has a second symmetry of the form

$$\begin{aligned} \frac{du_{n,m}}{dt_2} &= c_m(A_m - 1)(A_{m-1} - 1)\frac{(u_{n,m+1} - u_{n,m})(u_{n,m} - u_{n,m-1})}{u_{n,m+1} - u_{n,m-1}} \\ &+ \hat{c}_m(A_m - 1)(A_{m-1} + 1)\frac{(u_{n,m+1} - u_{n,m})(u_{n,m} + u_{n,m-1})}{u_{n,m+1} + u_{n,m-1}}, \end{aligned} \tag{63}$$

with two arbitrary functions c_m, $\hat{c}_m$, such that $c_m \neq 0$, $\hat{c}_m \neq 0$ for any m. If $A_m = 1$ for some $m = M$, then $\frac{du_{n,M}}{dt_2} = 0$. For example, if $A_0 = 1$ and $A_m = -1$ for $m \neq 0$, then $\frac{du_{n,m}}{dt_2} = 0$ iff $m = 0$.

In case of the discrete equation (36), we cannot enumerate all particular cases possessing the second symmetry of the form (3). We just discuss here two Darboux integrable examples.

The first of them is equation (36) with $A_m \equiv B_m \equiv -1$:

$$(u_{n+1,m} - \chi_{n+1})(u_{n,m} + \chi_n) = (u_{n+1,m+1} - \chi_{n+1})(u_{n,m+1} + \chi_n). \tag{64}$$

Its first integrals read:

$$\begin{aligned} V_{n,m} &= (u_{n+1,m} - \chi_{n+1})(u_{n,m} + \chi_n), \\ W_{n,m} &= \frac{(u_{n,m+1} - u_{n,m-1})(u_{n,m+1}\chi_n + u_{n,m}\chi_{n+1})}{u_{n,m+1}(u_{n,m} - u_{n,m-1})}. \end{aligned} \tag{65}$$

Following the scheme described in Section 5.2, we construct a general solution with two arbitrary functions κ_n, θ_m:

$$u_{n,m} = \chi_n \frac{(\kappa_n + \kappa_{n+2})\theta_m - 1}{(\kappa_{n+2} - \kappa_n)\theta_m} + \chi_{n+1}\frac{\kappa_n \theta_m}{2\kappa_{n+1}\theta_m - 1}. \tag{66}$$

The discrete equation (64) is a particular case of equation (43) corresponding to $a_{n,m} = -\chi_n$ and $c_m \equiv -1$. We have $q_{n,m}^{(i)} \equiv 0$ for all three conservation laws given by (45)–(48). So, all these conservation laws turn into first integrals, providing $V_{n,m}$ of (65), in particular. This is natural for the Darboux integrable equations.

The second case corresponds to $A_m \equiv 1$, $B_m \equiv -1$ in equation (36), and this equation takes the form

$$(u_{n+1,m} + \chi_{n+1})(u_{n,m} - \chi_n) = (u_{n+1,m+1} - \chi_{n+1})(u_{n,m+1} + \chi_n). \tag{67}$$

The first integrals in this case are more complicated:

$$V_{n,m} = u_{n+1,m}u_{n-1,m}(u_{n,m}^2 - 1)\chi_n + (u_{n+1,m} + u_{n-1,m})u_{n,m}\chi_{n+1},$$
$$W_{n,m} = (u_{n,m+1} + u_{n,m})(u_{n,m} + u_{n,m-1})\left(\frac{\chi_n}{u_{n,m}^2 - 1} + \frac{\chi_{n+1}}{u_{n,m+1}u_{n,m-1}}\right). \tag{68}$$

This discrete equation is a particular case of equation (43) corresponding to $a_{n,m} = \chi_n$ and $c_m \equiv 1$. It can be checked that the conservation law (48) becomes trivial in the sense that $p_{n,m}^{(2)}$ is expressed in terms of a first integral and a total difference:

$$p_{n,m}^{(2)} = T_1(2V_{n,m}\chi_n + V_{n,m}^2\chi_{n+1}) + (1 - T_1)(u_{n,m}^2(u_{n+1,m} + 1)^2\chi_{n+1}), \tag{69}$$

and conservation laws (45) with $i \geq 3$ are trivial in the same sense, too. As the first integrals (68) are not of the first order, we cannot apply the scheme of Section 5.2. A general solution for equation (67) will be constructed in the next section in a different way.

6. Transformations

In this section we construct a few new integrable discrete examples, using equations found in Section 5.1 and some special transformations invertible on solutions of discrete equations. Those transformations provide Miura type transformations for both discrete equations and their generalized symmetries. We use a transformation theory developed in [17,26,27]. A formulation more close to the discrete equations together with some applications can be found in [25].

6.1. Transformation 1

Let us start from equation (32) with $k_m \equiv 0$. Multiplying by the function $\frac{\Omega_{n,m}}{u_{n,m+1}u_{n+1,m}} + \Omega_{n+1,m}$, which is nonzero for any n, m, we can rewrite the equation as:

$$\Omega_{n,m}\frac{u_{n+1,m+1}}{u_{n+1,m}} + \Omega_{n+1,m}(u_{n+1,m+1} - u_{n+1,m})$$
$$= \Omega_{n,m}\frac{u_{n,m}}{u_{n,m+1}} + \Omega_{n+1,m}(u_{n,m} - u_{n,m+1}), \tag{70}$$
$$\Omega_{n,m} = \frac{1 + \omega_m(-1)^n}{2}, \quad \omega_m^2 \equiv 1.$$

Denoting the right hand side of equation (70) by $v_{n,m}$, we get two formulas

$$v_{n-1,m} = \Omega_{n-1,m}\frac{u_{n,m+1}}{u_{n,m}} + \Omega_{n,m}(u_{n,m+1} - u_{n,m}),$$
$$v_{n,m} = \Omega_{n,m}\frac{u_{n,m}}{u_{n,m+1}} + \Omega_{n-1,m}(u_{n,m} - u_{n,m+1}) \tag{71}$$

relating the functions $u_{n,m}$, $u_{n,m+1}$ and $v_{n,m}$, $v_{n-1,m}$. These formulas define a transformation of the discrete equation (70) invertible on its solutions. The inverse transformation has the form:

$$u_{n,m} = -\Omega_{n-1,m}\frac{v_{n,m}}{v_{n-1,m} - 1} - \Omega_{n,m}\frac{v_{n,m}v_{n-1,m}}{v_{n,m} - 1},$$
$$u_{n,m+1} = -\Omega_{n,m}\frac{v_{n-1,m}}{v_{n,m} - 1} - \Omega_{n-1,m}\frac{v_{n,m}v_{n-1,m}}{v_{n-1,m} - 1}. \tag{72}$$

After an additional point transformation:

$$w_{n,m} = 2\Omega_{n,m}v_{n,m} + \Omega_{n+1,m}\frac{1 + v_{n,m}}{1 - v_{n,m}}, \tag{73}$$

we obtain the following discrete equation:

$$(w_{n+1,m+1} - \Omega_{n+1,m+1})(w_{n,m+1} + \Omega_{n,m+1}) = (w_{n,m} - \Omega_{n,m})(w_{n+1,m} + \Omega_{n+1,m}). \tag{74}$$

The generalized symmetry (8) of equation (70) turns into

$$\frac{dw_{n,m}}{dt_1} = (w_{n,m}^2 - \Omega_{n,m})(w_{n+1,m} - w_{n-1,m}) \tag{75}$$

up to scaling t_1, and this is a symmetry of equation (74) in the n-direction.

It should be remarked that the symmetry (75) is a particular case of equations (13), (37). The discrete equation (74) is a particular case of equation (40) and of equation (43) with $c_m \equiv 1$ and has, for this reason, an $L-A$ pair of the form (44).

As it has been stated in Section 5.1, there are two subcases among equations (32) possessing the second generalized symmetry of the form (3). The first of them is given by $k_m \equiv 0$, $\Omega_{n,m} = \chi_n$ and is equation (53). In this case equation (74) coincides with equation (67) and is Darboux integrable. Using the composition of transformations (71), (73) and the solution (58) of equation (53), we obtain a general solution for equation (67):

$$w_{n,m} = \chi_n \left(1 - \frac{2(\theta_m + \kappa_{n+2})(\theta_{m+1} + \kappa_n)}{(\kappa_{n+2} - \kappa_n)(\theta_{m+1} - \theta_m)}\right) - \chi_{n+1}\frac{2\kappa_n(\theta_{m+1} - \theta_m)}{(\theta_{m+1} + \kappa_{n+1})(\theta_m + \kappa_{n+1})}. \tag{76}$$

The second case is given by $\Omega_{n,m} = \chi_{n+m}$, and equation (74) takes the form:

$$(w_{n+1,m+1} - \chi_{n+m})(w_{n,m+1} + \chi_{n+m+1}) = (w_{n,m} - \chi_{n+m})(w_{n+1,m} + \chi_{n+m+1}). \tag{77}$$

Its second symmetry corresponding to (51) reads:

$$\begin{aligned}\frac{dw_{n,m}}{dt_2} =&\chi_{n+m+1}(C_{m-1}w_{n,m-1} - C_{m+1}w_{n,m+1})\\ &-\chi_{n+m}\frac{(C_{m-1}w_{n,m+1} - C_{m+1}w_{n,m-1})(w_{n,m}^2 - 1)}{w_{n,m+1}w_{n,m-1}},\end{aligned} \tag{78}$$

where $C_m = \alpha m + \beta$. After the point transformation $\xi_{n,m} = \chi_{n+m}w_{n,m} + \chi_{n+m+1}/w_{n,m}$, this symmetry takes the form:

$$\frac{d\xi_{n,m}}{dt_2} = (\xi_{n,m}^2 - \chi_{n+m})(C_{m+1}\xi_{n,m+1} - C_{m-1}\xi_{n,m-1}). \tag{79}$$

equation (79) with $C_m \equiv 1$ and equation (75) with $\Omega_{n,m} = \chi_{n+m}$ are the same up to the transformation $n \leftrightarrow m$. Equation (79) with $C_m \equiv 1$ is, for any fixed value of n, a $1+1$-dimensional differential-difference equation equivalent to the well-known modified Volterra equation (4) with $P_n(u_n) = u_n^2 - \chi_n$. Equation (79) with $C_m \equiv m$ is the master symmetry of equation (79) with $C_m \equiv 1$. So, we can see that the discrete equation (77) possesses two hierarchies of generalized symmetries in both directions n and m.

Equation (77) is a particular case of (43) with $c_m \equiv 1$ and $a_{n,m} = \chi_{n+m}$ and possesses, for this reason, an $L-A$ pair of the form (44). In addition to (45) we can construct in this case the second hierarchy of conservation laws

$$(T_1 - 1)\hat{q}_{n,m}^{(i)} = (T_2 - 1)\hat{p}_{n,m}^{(i)}, \tag{80}$$

using this $L-A$ pair and the scheme of [19]. First three conservation laws (80) of equation (77) are defined by:

$$\begin{aligned}
\hat{q}^{(0)}_{n,m} &= \log\left(\frac{\chi_{n+m}}{1-w_{n,m}} + \chi_{n+m+1}w_{n,m}\right),\\
\hat{p}^{(0)}_{n,m} &= -\chi_{n+m+1}\log(w_{n,m}(1-w_{n+1,m}));\\
\hat{q}^{(1)}_{n,m} &= \frac{1-w_{n,m}}{1+w_{n,m}}\chi_{n+m} - \frac{1+w_{n,m+1}}{w_{n,m}}\chi_{n+m+1},\\
\hat{p}^{(1)}_{n,m} &= \frac{2}{w_{n+1,m}}\chi_{n+m};\\
\hat{q}^{(2)}_{n,m} &= \frac{(1+w_{n,m+2})(1-2w_{n,m}-w_{n,m+2})}{w^2_{n,m+1}}\chi_{n+m}\\
&+\frac{(1-w_{n,m+1})(2w_{n,m+2}w_{n,m+1}+2w_{n,m+2}+w_{n,m+1}w_{n,m}-w_{n,m})}{w^2_{n,m+2}w_{n,m}}\chi_{n+m+1},\\
\hat{p}^{(2)}_{n,m} &= \frac{4w_{n+1,m}(w_{n,m+1}+1)^2}{(1+w_{n+1,m})^2w^2_{n,m}}\chi_{n+m+1}.
\end{aligned}$$

6.2. Transformation 2

Let us start now from equation (50) and rewrite it in the form:

$$\begin{aligned}
&\chi_{n+m}u_{n+1,m+1}u_{n,m+1} + \chi_{n+m+1}(u_{n+1,m+1}+u_{n,m+1})\\
&=\chi_{n+m}u_{n+1,m}u_{n,m} + \chi_{n+m+1}(u_{n+1,m}+u_{n,m}).
\end{aligned} \tag{81}$$

Denoting the right hand side of (81) by $U_{n,m+1}$, we get two formulas

$$\begin{aligned}
U_{n,m} &= \chi_{n+m+1}u_{n+1,m}u_{n,m} + \chi_{n+m}(u_{n+1,m}+u_{n,m}),\\
U_{n,m+1} &= \chi_{n+m}u_{n+1,m}u_{n,m} + \chi_{n+m+1}(u_{n+1,m}+u_{n,m}),
\end{aligned} \tag{82}$$

which define a transformation invertible on the solutions of equation (81). The inverse transformation is given by:

$$\begin{aligned}
u_{n,m} &= -\frac{\chi_{n+m+1}}{2}\left(U_{n,m+1}+\sqrt{U^2_{n,m+1}-4U_{n,m}}\right) - \frac{\chi_{n+m}}{2}\left(U_{n,m}+\sqrt{U^2_{n,m}-4U_{n,m+1}}\right),\\
u_{n+1,m} &= \frac{\chi_{n+m+1}}{2}\left(\sqrt{U^2_{n,m+1}-4U_{n,m}}-U_{n,m+1}\right) + \frac{\chi_{n+m}}{2}\left(\sqrt{U^2_{n,m}-4U_{n,m+1}}-U_{n,m}\right).
\end{aligned} \tag{83}$$

In terms of new function $U_{n,m}$, we obtain the following discrete equation:

$$\begin{aligned}
&\chi_{n+m+1}\left(U_{n,m+1}-U_{n+1,m}-\sqrt{U^2_{n+1,m}-4U_{n+1,m+1}}-\sqrt{U^2_{n,m+1}-4U_{n,m}}\right)\\
&+\chi_{n+m}\left(U_{n,m}-U_{n+1,m+1}-\sqrt{U^2_{n+1,m+1}-4U_{n+1,m}}-\sqrt{U^2_{n,m}-4U_{n,m+1}}\right)=0.
\end{aligned} \tag{84}$$

Unlike all the other examples presented in this paper, this equation is not polylinear and even not polynomial. Equations of this kind, with the square roots, have been discussed, e.g., in [28, 29].

A generalized symmetry in the n-direction, obtained from (8) by the invertible transformation (82), reads:

$$\frac{dU_{n,m}}{dt_1} = (\chi_{n+m}-\chi_{n+m+1}U_{n,m})(U_{n+1,m}-U_{n-1,m}). \tag{85}$$

In any fixed point m this equation is equivalent, up to the shift and rescaling $U_{n,m}$, to equation (4) with $P_n(u_n) = \chi_n u_n + \chi_{n+1}$. So, equation (84) defines a chain of Bäcklund auto-transformations for the Toda lattice equation.

A symmetry in the m-direction obtained from (51) is of the form:

$$\begin{aligned}\frac{dU_{n,m}}{dt_2} &= \frac{C_m\chi_{n+m+1}\left(U_{n,m-1}U_{n,m+1} - 4U_{n,m} + \sqrt{U_{n,m-1}^2 - 4U_{n,m}}\sqrt{U_{n,m+1}^2 - 4U_{n,m}}\right)}{2(U_{n,m+1} - U_{n,m-1})} \\ &+ \frac{\chi_{n+m}C_m\left(U_{n,m}^2 - 4U_{n,m+1} + \sqrt{U_{n,m}^2 - 4U_{n,m+1}}\sqrt{U_{n,m}^2 - 4U_{n,m-1}}\right)}{2(U_{n,m+1} - U_{n,m-1})} + \chi_{n+m}C_{m+1},\end{aligned} \tag{86}$$

where $C_m = \alpha m + \beta$. Equation (86) with $C_m \equiv m$ provides the master symmetry for equation (86) with $C_m \equiv 1$.

If we choose $C_m \equiv 2$ and introduce $V_m = U_{n,m}$ for any fixed n, we are led to the following equation up to the shift of m:

$$\begin{aligned}\frac{dV_m}{dt_2} &= \frac{R_m(V_{m+1}, V_m, V_{m-1}) + \sqrt{R_m(V_{m+1}, V_m, V_{m+1})}\sqrt{R_m(V_{m-1}, V_m, V_{m-1})}}{V_{m+1} - V_{m-1}}, \\ R_m(x_1, x_2, x_3) &= \chi_m(x_2^2 - 2x_1 - 2x_3) + \chi_{m+1}(x_1x_3 - 4x_2).\end{aligned} \tag{87}$$

This equation is, probably, a new nonautonomous generalization of the well-known integrable lattice equation presented in [10, 11].

The formulas (82) define two Miura type transformations of the discrete equation (50) into (84) and of the symmetry (8) into (85). Each of these transformations converts any solution $u_{n,m}$ of (50) into a solution of equation (84). These formulas also define transformations of the symmetry (51) into (86), but only on the solutions of the discrete equation (81). The inverse transformations (83) define two Miura type transformations which relate equations (84) and (81) as well as their symmetries (86) and (51). The formulas (71) and (72) define Miura type transformations of discrete equations and their symmetries as well.

7. Conclusions

In this paper we have constructed a number of new examples of integrable nonautonomoues discrete equations of the form (1).

In Section 3 we describe all polylinear nonautonomous Bäcklund auto-transformations for the differential-discrete equations (4)–(6) of the Volterra and Toda type. In this way we obtain discrete equations possessing one generalized symmetry of the form (2). In Section 5 we find cases in which there is the second generalized symmetry (3) and obtain in this way new integrable examples. In Section 6 we obtain some additional examples, using transformations invertible on the solutions of discrete equations.

As a result we get discrete equations of the sine-Gordon (50), (77), (84) and Liouville (53), (64), (67) type. For Liouville type equations, which are Darboux integrable, we construct the general solutions (58), (66), (76) by using a special linearization, see for details Section 5.2. The most interesting example of the Liouville type is equation (67) which has more complicated first integrals (68). Among sine-Gordon type equations, the most interesting from our point of view example is equation (77). For this equation we find an L–A pair and construct two hierarchies of conservation laws and generalized symmetries. The discrete equation (84) is the only example which is not polylinear. It generates a chain of Bäcklund auto-transformations for the Toda lattice equation. Its second symmetry provides a new nonautonomous generalization of the well-known integrable differential-discrete equation.

In Section 4 a discrete analog of the well-known dressing chain has been constructed. We find for it an $L-A$ pair and one hierarchy of generalized symmetries and conservation laws. In particular, this example includes equation (77) and provides it by an $L-A$ pair.

We also have found two degenerate examples of discrete equations, see Sections 5.1 and 5.3. One of their symmetries is degenerate in the sense that $\psi_{n,m}$ of (3) may be zero in some points. Examples of this kind seem to be very close to integrable ones and, in our opinion, deserve further study too.

Acknowledgments

R.G. was partially supported by Russian Science Foundation (project no. 14-11-00078), R.Y. was partially supported by Russian Foundation for Basic Research (grant numbers: 13-01-00070, 14-01-97008-r-povolzhie-a).

References

[1] Levi D, Petrera M, Scimiterna C and Yamilov R 2008 On Miura transformations and Volterra-type equations associated with the Adler–Bobenko–Suris equations *SIGMA Symmetry Integrability Geom. Methods Appl.* **4** 077
[2] Levi D and Yamilov R I 2011 Generalized symmetry integrability test for discrete equations on the square lattice *J. Phys.* A **44** 145207
[3] Xenitidis P 2009 Integrability and symmetries of difference equations: the Adler–Bobenko–Suris case *Proc. 4th Int. Workshop on Group Analysis of Differential Equations and Integrable Systems (Protaras, 2008)* (Nicosia: University of Cyprus) p 226 (*Preprint* arXiv:0902.3954)
[4] Xenitidis P D and Papageorgiou V G 2009 Symmetries and integrability of discrete equations defined on a black-white lattice *J. Phys.* A **42** 454025
[5] Adler V E 2011 On a discrete analog of the Tzitzeica equation *Preprint* arXiv:1103.5139
[6] Garifullin R N and Yamilov R I 2012 Generalized symmetry classification of discrete equations of a class depending on twelve parameters *J. Phys.* A **45** 345205
[7] Mikhailov A V and Xenitidis P 2014 Second order integrability conditions for difference equations: an integrable equation *Lett. Math. Phys.* **104**:4 431
[8] Scimiterna C, Hay M and Levi D 2014 On the integrability of a new lattice equation found by multiple scale analysis *J. Phys.* A **47** 265204
[9] Levi D and Yamilov R I 2009 The generalized symmetry method for discrete equations *J. Phys.* A **42** 454012
[10] Yamilov R I 1983 Classification of discrete evolution equations *Uspekhi Mat. Nauk* **38**:6 155 [in Russian]
[11] Yamilov R 2006 Symmetries as integrability criteria for differential difference equations *J. Phys.* A **39** R541
[12] Levi D and Yamilov R 1997 Conditions for the existence of higher symmetries of evolutionary equations on the lattice *J. Math. Phys.* **38** 6648
[13] Shabat A 1992 The infinite-dimensional dressing dynamical system *Inverse Problems* **8** 303
[14] Shabat A B and Yamilov R I 1990 Symmetries of nonlinear chains *Algebra i Analiz* **2**:2 183 [in Russian]; English transl. in *Leningrad Math. J.* **2**:2 377
[15] Veselov A P and Shabat A B 1993 Dressing chains and spectral theory of the Schrödinger operator *Funkts. Anal. Prilozh.* **27**:2 1 [in Russian]; English transl. in *Funct. Anal. Appl.* **27**:2 81
[16] Garifullin R N and Yamilov R I 2012 Examples of Darboux integrable discrete equations possessing first integrals of an arbitrarily high minimal order *Ufa Math. J.* **4**:3 174
[17] Yamilov R I 1994 Construction scheme for discrete Miura transformations *J. Phys.* A **27** 6839
[18] Levi D and Yamilov R I 2009 On a nonlinear integrable difference equation on the square *Ufa Math. J.* **1**:2 101
[19] Habibullin I T and Yangubaeva M V 2013 Formal diagonalization of the discrete Lax operator, and conservation laws and symmetries of dynamical systems *Teoret. Mat. Fiz.* **177**:3 441 [in Russian]; English transl. in *Theor. Math. Phys.* **177**:3 1655
[20] Nijhoff F and Capel H 1995 The discrete Korteweg–de Vries equation *Acta Appl. Math.* **39**:1-3 133
[21] Hirota R and Tsujimoto S 1995 Conserved quantities of a class of nonlinear difference-difference equations *J. Phys. Soc. Japan* **64** 3125
[22] Garifullin R N, Gudkova E V and Habibullin I T 2011 Method for searching higher symmetries for quad-graph equations *J. Phys.* A **44** 325202
[23] Habibullin I T 2005 Characteristic algebras of fully discrete hyperbolic type equations *SIGMA Symmetry Integrability Geom. Methods Appl.* **1** 023

[24] Cherdantsev I and Yamilov R 1996 Local master symmetries of differential-difference equations *Proc. Int. Conf. on Symmetries and Integrability of Difference Equations (Esterel, PQ, 1994)* (CRM Proceedings and Lecture Notes vol 9) (Providence: Amer. Math. Soc.) p 51

[25] Startsev S Ya 2010 On non-point invertible transformations of difference and differential-difference equations *SIGMA Symmetry Integrability Geom. Methods Appl.* **6** 092

[26] Yamilov R I 1990 Invertible changes of variables generated by Bäcklund transformations *Teoret. Mat. Fiz.* **85**:3 368 [in Russian]; English transl. in 1991 *Theor. Math. Phys.* **85**:3 1269

[27] Yamilov R I 1993 On the construction of Miura type transformations by others of this kind *Phys. Lett.* A **173** 53

[28] Atkinson J and Nieszporski M 2014 Multi-quadratic quad equations: integrable cases from a factorised-discriminant hypothesis *Int. Math. Res. Notices* **2014**:15 4215

[29] Kassotakis P and Nieszporski M 2012 On non-multiaffine consistent around the cube lattice equations *J. Phys.* A **376**:45 3135

Journal of Physics: Conference Series **621** (2015) 012006 doi:10.1088/1742-6596/621/1/012006

Application of group analysis to the spatially homogeneous and isotropic Boltzmann equation with source using its Fourier image

Yurii N Grigoriev[1], **Sergey V Meleshko**[2] **and Amornrat Suriyawichitseranee**[2]

[1] Institute of Computational Technology, Novosibirsk, 630090, Russia
[2] School of Mathematics, Institute of Science, Suranaree University of Technology, Nakhon Ratchasima, 30000, Thailand

E-mail: grigor@ict.nsc.ru, sergey@math.sut.ac.th, amornrat@math.sut.ac.th

Abstract. Group analysis of the spatially homogeneous and molecular energy dependent Boltzmann equations with source term is carried out. The Fourier transform of the Boltzmann equation with respect to the molecular velocity variable is considered. The correspondent determining equation of the admitted Lie group is reduced to a partial differential equation for the admitted source. The latter equation is analyzed by an algebraic method. A complete group classification of the Fourier transform of the Boltzmann equation with respect to a source function is given. The representation of invariant solutions and corresponding reduced equations for all obtained source functions are also presented.

1. Introduction

The classical Boltzmann equation which plays a central role in kinetic gas theory with mathematical point of view is a uniform integro-differential equation. At the same time, there is frequently the need to include additional source terms into it. In particular, the Boltzmann equation with autonomous sources, independent of the distribution function, are of interest for the kinetic description of initiation of high threshold processes by "hot" particles, for reacting gas flows over catalytic surfaces, and some others. A first attempt of application of group analysis for studying the Boltzmann equation with such a source term was done in [1], where by using the approach of [2], the spatially homogeneous and isotropic in velocity space Boltzmann equation with the Maxwell molecular model was reduced to the equation for a generating function of power moments. The complete Lie symmetry group of the equation for the moment generating function was recently found in [3]. However, the transformation of the invariant solutions thus obtained into the corresponding solutions of the original Boltzmann equation is a very complicated task. An example of typical difficulties can be found in [2], where this transition obstacle was overcome for obtaining the BKW-solution.

A more promising approach is to consider the Fourier image of the homogeneous and isotropic Boltzmann equation [4, 5]. The extensive group analysis of this equation without a source term was carried out in [6]. In the present paper these results are generalized for the non-uniform case.

The paper is organized as follows. After introducing the equation studied, the determining equation of an admitted Lie group is derived. The solution of the determining equation is reduced to the study of a partial differential equation for an admitted source function with some undefined constants.

For further study of the reduced equation an equivalence Lie group of considered integro-differential equation is obtained. It is shown that actions of the obtained equivalence transformations are similar to the automorphisms of the Lie algebra obtained in [6]. This allows us to use a method of constructing an optimal system of subalgebras for the group classification of the source function. These studies are followed by consideration of all possible inequivalent representations of invariant solutions and corresponding reduced equations of the Fourier image equation.

2. The equation studied

The Fourier image of the spatially homogeneous and isotropic Boltzmann equation with a source term has the form [4]:

$$\varphi_t(x,t) + \varphi(x,t)\varphi(0,t) = \int_0^1 \varphi(xs,t)\varphi(x(1-s),t)\,\mathrm{d}s + \hat{q}(x,t). \tag{1}$$

Here the function $\varphi(x,t)$ is related with the Fourier transform $\tilde{\varphi}(k,t)$ of the distribution function $f(v,t)$, isotropic in the 3D-space of molecular velocities by the formula

$$\varphi(x,t) \equiv \varphi(k^2/2,t) = \tilde{\varphi}(k,t) = \frac{4\pi}{k}\int_0^\infty v\sin(kv)f(v,t)\,\mathrm{d}v.$$

Similarly, the transform of the isotropic source function $q(v,t)$ is

$$\hat{q}(x,t) \equiv \hat{q}(k^2/2,t) = \tilde{\hat{q}}(k,t) = \frac{4\pi}{k}\int_0^\infty v\sin(kv)q(v,t)\,\mathrm{d}v.$$

The inverse Fourier transform of $\tilde{\varphi}(k,t)$ gives the distribution function

$$f(v,t) = \frac{4\pi}{v}\int_0^\infty k\sin(kv)\tilde{\varphi}(k,t)\,\mathrm{d}k.$$

In the process of solving the determining equation, we use the property that for any smooth function $\varphi_0(x)$ there exists a solution of the Cauchy problem of equation (1) with the initial data

$$\varphi(x,t_0) = \varphi_0(x).$$

3. Determining equation

The classical group analysis method cannot be applied to the integro-differential equation (1). One needs to use the method for equations with nonlocal, in particular, integral terms developed in [7–9]. A generator of the admitted Lie group is sought in the form[1]

$$X = \xi(x,t,\varphi)\partial_x + \eta(x,t,\varphi)\partial_t + \zeta(x,t,\varphi)\partial_\varphi.$$

According to the algorithm, the determining equation for equation (1) is

$$D_t\psi(x,t) + \psi(0,t)\varphi(x,t) + \psi(x,t)\varphi(0,t) - 2\int_0^1 \varphi(x(1-s),t)\psi(xs,t)\,\mathrm{d}s = 0, \tag{2}$$

[1] We follow here notations accepted in [10].

where D_t is the total derivative with respect to t, and the function $\psi(x,t)$ is

$$\psi(x,t)=\zeta(x,t,\varphi(x,t))-\xi(x,t,\varphi(x,t))\varphi_x(x,t)-\eta(x,t,\varphi(x,t))\varphi_t(x,t).$$

The determining equation (2) has to be satisfied for any solution of equation (1). This allows us to exclude the derivatives φ_t, φ_{xt} and φ_{tt} from the determining equation. In fact, differentiating (1) and substituting φ_t found from (1), we obtain

$$\begin{aligned}\varphi_{xt}(x,t)= &-\varphi_x(x,t)\varphi(0,t)+2\int_0^1 s\varphi_x(xs,t)\varphi(x(1-s),t)\,\mathrm{d}s+\hat{q}_x(x,t),\\ \varphi_{tt}(x,t)= &\ \varphi(x,t)\varphi^2(0,t)-3\varphi(0,t)\int_0^1 \varphi(xs,t)\varphi(x(1-s),t)\,\mathrm{d}s\\ &+2\int_0^1\int_0^1 \varphi(x(1-s),t)\varphi(xss',t)\varphi(xs(1-s'),t)\,\mathrm{d}s'\mathrm{d}s\\ &-\hat{q}(0,t)\varphi(x,t)-\hat{q}(x,t)\varphi(0,t)+2\int_0^1\varphi(x(1-s),t)\hat{q}(xs,t)\,\mathrm{d}s+\hat{q}_t(x,t).\end{aligned}$$

Analysing the determining equation (2) with a method similar to that developed in [9] for the case $\hat{q}=0$, we get that admitted generators have the form

$$X=c_0X_0+c_1X_1+c_2X_2+c_3X_3,$$

where

$$X_0=x\partial_x,\quad X_1=x\varphi\partial_\varphi,\quad X_2=\varphi\partial_\varphi-t\partial_t,\quad X_3=\partial_t, \tag{3}$$

the constants c_0, c_1, c_2, c_3 and the function $\hat{q}$ satisfy the classifying equation

$$(c_2t-c_3)\hat{q}_t-c_0x\hat{q}_x+(c_1x+2c_2)\hat{q}=0. \tag{4}$$

In order to find proper values of the constants c_0, c_1, c_2, c_3 and the function $\hat{q}$ we apply an algebraic method [11].

Note that the case with the function $\hat{q}=0$ was completely studied in [6], where it was shown that the admitted Lie algebra is four-dimensional and which is spanned by the generators X_0, X_1, X_2 and X_3.

3.1. Equivalence transformations

To carry out a group classification one needs to know equivalence transformations of equation (1). For convenience we introduce the operator L defined by

$$L\varphi=\varphi_t(x,t)+\varphi(x,t)\varphi(0,t)-\int_0^1\varphi(xs,t)\varphi(x(1-s),t)\,\mathrm{d}s. \tag{5}$$

Considering the transformations of $L\varphi$ corresponding to the generators X_0, X_1, X_2 and X_3, some equivalence transformations can be obtained.

The transformations corresponding to the generator $X_0=x\partial_x$ map a function $\varphi(x,t)$ into the function $\bar{\varphi}(\bar{x},\bar{t})=\varphi(\bar{x}e^{-a},\bar{t})$, where a is the group parameter. The transformed expression of (5) becomes

$$\begin{aligned}\bar{L}\bar{\varphi}&=\bar{\varphi}_{\bar{t}}(\bar{x},\bar{t})+\bar{\varphi}(\bar{x},\bar{t})\bar{\varphi}(0,\bar{t})-\int_0^1\bar{\varphi}(\bar{x}s,\bar{t})\bar{\varphi}(\bar{x}(1-s),\bar{t})\,\mathrm{d}s\\ &=\varphi_{\bar{t}}(\bar{x}e^{-a},\bar{t})+\varphi(\bar{x}e^{-a},\bar{t})\varphi(0,\bar{t})-\int_0^1\varphi(\bar{x}e^{-a}s,\bar{t})\varphi(\bar{x}e^{-a}(1-s),\bar{t})\,\mathrm{d}s\\ &=\varphi_t(x,t)+\varphi(x,t)\varphi(0,t)-\int_0^1\varphi(xs,t)\varphi(x(1-s),t)\,\mathrm{d}s\\ &=L\varphi.\end{aligned}$$

This means that

$$T_a^0: \quad \bar{x} = xe^a, \quad \bar{t} = t, \quad \bar{\varphi} = \varphi, \quad \bar{\hat{q}} = \hat{q}$$

are the equivalence transformations of equation (1). Similarly, the transformations corresponding to the generator $X_3 = \partial_t$ induce the equivalence Lie group

$$T_a^3: \quad \bar{x} = x, \quad \bar{t} = t + a, \quad \bar{\varphi} = \varphi, \quad \bar{\hat{q}} = \hat{q}.$$

The transformations corresponding to the generator $X_2 = \varphi\partial_\varphi - t\partial_t$ map a function $\varphi(x,t)$ to the function $\bar{\varphi}(\bar{x},\bar{t}) = e^a\varphi(\bar{x},\bar{t}e^a)$, which gives

$$\begin{aligned}
\bar{L}\bar{\varphi} &= \bar{\varphi}_{\bar{t}}(\bar{x},\bar{t}) + \bar{\varphi}(\bar{x},\bar{t})\bar{\varphi}(0,\bar{t}) - \int_0^1 \bar{\varphi}(\bar{x}s,\bar{t})\bar{\varphi}(\bar{x}(1-s),\bar{t})\,\mathrm{d}s \\
&= e^a\varphi_{\bar{t}}(\bar{x},\bar{t}e^a) + e^a\varphi(\bar{x},\bar{t}e^a)e^a\varphi(0,\bar{t}e^a) - \int_0^1 e^{2a}\varphi(\bar{x}s,\bar{t}e^a)\varphi(\bar{x}(1-s),\bar{t}e^a)\,\mathrm{d}s \\
&= e^{2a}\varphi_t(\bar{x},\bar{t}e^a) + e^{2a}\varphi(\bar{x},\bar{t}e^a)\varphi(0,\bar{t}e^a) - e^{2a}\int_0^1 \varphi(\bar{x}s,\bar{t}e^a)\varphi(\bar{x}(1-s),\bar{t}e^a)\,\mathrm{d}s \\
&= e^{2a}\left(\varphi_t(x,t) + \varphi(x,t)\varphi(0,t) - \int_0^1 \varphi(xs,t)\varphi(x(1-s),t)\,\mathrm{d}s\right) \\
&= e^{2a}L\varphi.
\end{aligned}$$

Hence we conclude that the transformations

$$T_a^2: \quad \bar{x} = x, \quad \bar{t} = te^{-a}, \quad \bar{\varphi} = \varphi e^a, \quad \bar{\hat{q}} = \hat{q}e^{2a}$$

compose another equivalence Lie group of equation (1).

The transformations corresponding to the generator $X_1 = x\varphi\partial_\varphi$ map a function $\varphi(x,t)$ to the function $\bar{\varphi}(\bar{x},\bar{t}) = e^{\bar{x}a}\varphi(\bar{x},\bar{t})$. Thus,

$$\begin{aligned}
\bar{L}\bar{\varphi} &= \bar{\varphi}_{\bar{t}}(\bar{x},\bar{t}) + \bar{\varphi}(\bar{x},\bar{t})\bar{\varphi}(0,\bar{t}) - \int_0^1 \bar{\varphi}(\bar{x}s,\bar{t})\bar{\varphi}(\bar{x}(1-s),\bar{t})\,\mathrm{d}s \\
&= e^{\bar{x}a}\varphi_{\bar{t}}(\bar{x},\bar{t}) + e^{\bar{x}a}\varphi(\bar{x},\bar{t})\varphi(0,\bar{t}) - \int_0^1 e^{\bar{x}as}\varphi(\bar{x}s,\bar{t})e^{\bar{x}(1-s)a}\varphi(\bar{x}(1-s),\bar{t})\,\mathrm{d}s \\
&= e^{\bar{x}a}\varphi_t(\bar{x},\bar{t}) + e^{\bar{x}a}\varphi(\bar{x},\bar{t})\varphi(0,\bar{t}) - e^{\bar{x}a}\int_0^1 \varphi(\bar{x}s,\bar{t})\varphi(\bar{x}(1-s),\bar{t})\,\mathrm{d}s \\
&= e^{xa}\left(\varphi_t(x,t) + \varphi(x,t)\varphi(0,t) - \int_0^1 \varphi(xs,t)\varphi(x(1-s),t)\,\mathrm{d}s\right) \\
&= e^{xa}L\varphi,
\end{aligned}$$

which gives the equivalence Lie group constituted by the transformations

$$T_a^1: \quad \bar{x} = x, \quad \bar{t} = t, \quad \bar{\varphi} = \varphi e^{xa}, \quad \bar{\hat{q}} = \hat{q}e^{xa}.$$

Summarizing these calculations one can conclude that the generators

$$X_0^{\mathrm{e}} = x\partial_x, \quad X_1^{\mathrm{e}} = x\varphi\partial_\varphi + x\hat{q}\partial_{\hat{q}}, \quad X_2^{\mathrm{e}} = \varphi\partial_\varphi - t\partial_t + x\hat{q}\partial_{\hat{q}}, \quad X_3^{\mathrm{e}} = \partial_t \tag{6}$$

define an equivalence Lie algebra of equation (1).

Note also that the transformation

$$E: \quad \bar{t} = -t, \quad \bar{\varphi} = -\varphi \tag{7}$$

does not change equation (1), i.e., this is a discrete equivalence transformation being an involution.

Let us study the change of a generator $X = x_0X_0 + x_1X_1 + x_2X_2 + x_3X_3$ under the transformations corresponding to these equivalence transformations. After the change defined by an equivalence transformation one gets the generator

$$X = \hat{x}_0\hat{X}_0 + \hat{x}_1\hat{X}_1 + \hat{x}_2\hat{X}_2 + \hat{x}_3\hat{X}_3, \tag{8}$$

where

$$\hat{X}_0 = \bar{x}\partial_{\bar{x}}, \quad \hat{X}_1 = \bar{x}\bar{\varphi}\partial_{\bar{\varphi}}, \quad \hat{X}_2 = \bar{\varphi}\partial_{\bar{\varphi}} - \bar{t}\partial_{\bar{t}}, \quad \hat{X}_3 = \partial_{\bar{t}}.$$

The corresponding transformations of the basis generators are

$$\begin{array}{lllll} T_a^0: & X_0 = \hat{X}_0, & X_1 = e^{-a}\hat{X}_1, & X_2 = \hat{X}_2, & X_3 = \hat{X}_3; \\ T_a^1: & X_0 = \hat{X}_0 + a\hat{X}_1, & X_1 = \hat{X}_1, & X_2 = \hat{X}_2, & X_3 = \hat{X}_3; \\ T_a^2: & X_0 = \hat{X}_0, & X_1 = \hat{X}_1, & X_2 = \hat{X}_2, & X_3 = e^{-a}\hat{X}_3; \\ T_a^3: & X_0 = \hat{X}_0, & X_1 = \hat{X}_1, & X_2 = \hat{X}_2 + a\hat{X}_3, & X_3 = \hat{X}_3. \end{array}$$

or coordinates of the generator X are changed as follows

$$\begin{array}{ll} T_a^0: & \hat{x}_1 = x_1e^{-a}, \\ T_a^1: & \hat{x}_1 = x_1 + ax_0, \\ T_a^2: & \hat{x}_3 = x_3e^{-a}, \\ T_a^3: & \hat{x}_3 = x_3 + ax_2, \end{array}$$

where only changeable coordinates are presented.

3.2. Algebraic approach for analyzing equation (4)

Group classification of equation (1) is carried out up to the equivalence transformations considered in the previous section. The method for classifying the source function $\hat{q}$ is similar to the method which was used for classifying equations for moment generating function in [3]. First of all we note that actions of the equivalence transformations T_a^i, $i = 0, \ldots, 3$, are equivalent to inner automorphisms of the Lie algebra L_4 spanned by the generators X_0, X_1, X_2 and X_3.

The table of commutators of these generators is

	X_0	X_1	X_2	X_3
X_0	0	X_1	0	0
X_1	$-X_1$	0	0	0
X_2	0	0	0	$-X_3$
X_3	0	0	X_3	0

Using the table of commutators, the inner automorphisms are obtained as follows

$$\begin{array}{ll} A_0: & \hat{x}_1 = x_1e^{a}, \\ A_1: & \hat{x}_1 = x_1 + ax_0, \\ A_2: & \hat{x}_3 = x_3e^{a}, \\ A_3: & \hat{x}_3 = x_3 + ax_2, \end{array}$$

where only changeable coordinates are presented.

Thus we conclude that the actions of equivalence transformations coincides with the actions of inner automorphisms. Because of this property we use an optimal system of subalgebras of the algebra L_4 for classifying equation (1).

The commutator table of the Lie algebra L_4 coincides with the commutator table considered in [3], where group classification of the equation for a moment generating function was studied. The difference in constructing an optimal system here consists of the set of involutions: in the present case the involution corresponding to $\hat{x}_1 = -x_1$ is absent comparing with [3]. The optimal system of subalgebras of the Lie algebra L_4 is presented in Table 1, where γ is an arbitrary constant.

Table 1. Optimal system of subalgebras of L_4

No.	Basis	No.	Basis
1.	$X_0,\ X_1,\ X_2,\ X_3$	13.	$X_0+X_3,\ X_1$
2.	$\gamma X_0+X_2,\ X_1,\ X_3$	14.	$X_1,\ X_3$
3.	$X_0,\ X_1,\ X_3$	15.	$X_2,\ X_3$
4.	$X_0,\ X_1,\ X_2$	16.	$X_0,\ X_1$
5.	$X_0,\ X_2,\ X_3$	17.	γX_0+X_2
6.	$X_2,\ X_3$	18.	X_1+X_2
7.	$X_2+X_0,\ X_1+\ X_3$	19.	X_1-X_2
8.	$\gamma X_2+2X_0,\ X_3$	20.	X_0+X_3
9.	$X_1+X_2,\ X_3$	21.	X_1+X_3
10.	$X_1-X_2,\ X_3$	22.	X_0
11.	$X_0,\ X_2$	23.	X_1
12.	$\gamma X_0+X_2,\ X_1$	24.	X_3

Using the optimal system of subalgebras for group classification of equation (1), the function $\hat{q}(t,x)$ is obtained by substituting, into equation (4), the constants c_i corresponding to the basis generators of a given subalgebra of the optimal system of subalgebras, and solving the obtained system of equations. For example, one can consider one-dimensional subalgebra $\{X_1 - X_2\}$ (see Case 19 of Table 1). For this case there is a single equation for the function $\hat{q}(t,x)$,

$$t\hat{q}_t - x\hat{q}_x + 2\hat{q} = 0.$$

The general solution of this equation is $\hat{q} = t^{x-2}\Phi(x)$, where $\Phi(x)$ is an arbitrary function.

As another example, the two-dimensional Lie subalgebra $\{\gamma X_2 + 2X_0,\ X_3\}$ (see Case 8 of Table 1) is considered. For this case there are two sets of the coefficients c_i, $i = 0,\ldots,3$:

$$\begin{aligned} \gamma X_2 + 2X_0: &\quad c_0 = 2, \quad c_1 = 0, \quad c_2 = \gamma, \quad c_3 = 0; \\ X_3: &\quad c_0 = 0, \quad c_1 = 0, \quad c_2 = 0, \quad c_3 = 1. \end{aligned}$$

These sets define a system of two equations for $\hat{q}(t,x)$. Substituting both sets of c_i into equation (4), we obtain

$$\gamma\left(\frac{1}{2}t\hat{q}_t + \hat{q}\right) - x\hat{q}_x = 0, \qquad \hat{q}_t = 0.$$

The general solution of these equations is $\hat{q} = \beta x^{\gamma}$, where β is a constant.

The complete group classification obtained by calculations of this kind is presented in Table 2.

Table 2. Group classification of equation (1)

No.	$\hat{q}(t,x)$	Generators
1.	0	$X_0,\ X_1,\ X_2\ X_3$
2.	$\beta x^2 e^{tx}$	$X_2+X_0,\ X_1+X_3$
3.	βx^γ	$\gamma X_2+2X_0,\ X_3$
4.	βt^{-2}	$X_0,\ X_2$
5.	$t^{-2}\Phi(xt^\gamma)$	γX_0+X_2
6.	$t^{-(x+2)}\Phi(x)$	X_1+X_2
7.	$t^{x-2}\Phi(x)$	X_1-X_2
8.	$\Phi(xe^{-t})$	X_0+X_3
9.	$e^{xt}\Phi(x)$	X_1+X_3
10.	$\Phi(t)$	X_0
11.	$\Phi(x)$	X_3

Here β and γ are arbitrary constants, and the function Φ is an arbitrary function of its argument. The other subalgebras of Table 1, which are omitted in Table 2, give inconsistent systems for $\hat{q}(t,x)$.

4. Reduced equations and invariant solutions

In this section, for each obtained function $\hat{q}$ we consider the admitted Lie algebra. Using an optimal system of subalgebras of these Lie algebras, we derive a form of invariant solutions and corresponding reduced equations. Similarly to differential equations, the reduced equations for finding invariant solutions have fewer number of the independent variables than equation (1). There are some trivial cases where invariant solutions are obtained in an explicit form.

4.1. The function $\hat{q}=\beta x^2 e^{xt}$

For the source function $\hat{q}=\beta x^2 e^{tx}$ the admitted Lie algebra of equation (1) is $\{X_2+X_0,\ X_1+X_3\}$. An optimal system of subalgebras of this Lie algebra consists of the subalgebras

$$\{X_2+X_0\},\quad \{X_1+X_3\},\quad \{X_2+X_0,\ X_1+X_3\}.$$

A representation of invariant solutions corresponding to the subalgebra $\{X_2+X_0\}$ is $\varphi=t^{-1}r(z)$, where $z=xt$. The substitution of this representation into equation (1) gives

$$zr'(z)-r(z)+r(z)r(0)-\int_0^1 r(zs)r(z(1-s))\,ds=\beta z^2e^z. \tag{9}$$

Equation (9) is an equation with the single independent variable z.

A representation of invariant solutions corresponding to the subalgebra $\{X_1+X_3\}$ is $\varphi=e^{xt}r(x)$. Substituting this representation into equation (1), we obtain the reduced equation

$$r(x)(x+r(0))-\int_0^1 r(xs)r(x(1-s))\,ds=\beta x^2.$$

The subalgebra $\{X_2+X_0,\ X_1+X_3\}$ gives invariant solutions in the form $\varphi=Cxe^{xt}$, where C is a constant. After substituting this representation into equation (1), we get an algebraic equation for the constant C,

$$C^2-6C+6\beta=0.$$

If $\beta\le\frac{3}{2}$, then $C=3\pm\sqrt{9-6\beta}$.

4.2. The source function $\hat{q} = \beta x^\gamma$
An optimal system of subalgebras is constituted by the subalgebras $\{\gamma X_2 + 2X_0, X_3\}$, $\{X_3\}$ and either $\{\gamma X_2 + 2X_0\}$ for $\gamma \neq 0$ or $\{X_3 + \alpha X_0\}$ for $\gamma = 0$.

A representation of invariant solutions corresponding to the subalgebra $\{\gamma X_2 + 2X_0, X_3\}$ is $\varphi = Cx^{\frac{\gamma}{2}}$, where the constant C satisfies the algebraic reduced equation

$$C^2 B + \beta = 0 \qquad \text{with} \qquad B := \mathrm{B}\left(\frac{\gamma}{2}, \frac{\gamma}{2}\right) = \int_0^1 s^{\frac{\gamma}{2}}(1-s)^{\frac{\gamma}{2}}\,\mathrm{d}s,$$

and B is the beta function [12].

A representation of invariant solutions corresponding to the subalgebra $\{X_3\}$ is $\varphi = r(x)$, and the associated reduced equation is

$$r(x)r(0) - \int_0^1 r(xs)r(x(1-s))\,\mathrm{d}s = \beta x^\gamma.$$

For $\gamma \neq 0$ and the subalgebra $\{\gamma X_2 + 2X_0\}$, invariant solutions are represented as $\varphi = t^{-1}r(z)$, where $z = t^2 x^\gamma$. Substituting this representation into equation (1), we have the reduced equation

$$2zr'(z) - r(z) + r(z)r(0) - \int_0^1 r(zs)r(z(1-s))\,\mathrm{d}s = \beta z$$

For $\gamma = 0$ and the subalgebra $\{X_3 + \alpha X_0\}$, invariant solutions takes in the form $\varphi = r(z)$, where $z = xe^{-\alpha t}$, and the function r is a solution of the reduced equation

$$-\alpha zr'(z) + r(z)r(0) - \int_0^1 r(zs)r(z(1-s))\,\mathrm{d}s = \beta.$$

4.3. The source function $\hat{q} = \beta t^{-2}$
The admitted Lie algebra of equation (1) with the source function $\hat{q} = \beta t^{-2}$ is $\{X_0, X_2\}$, whose optimal system of subalgebras consists of the subalgebras $\{X_0, X_2\}$, $X_2 + \alpha X_0\}$ and $\{X_0\}$.

The single invariant solution corresponding to the subalgebra $\{X_0, X_2\}$ is $\varphi = -\beta t^{-1}$.

For the subalgebra $\{X_2 + \alpha X_0\}$, a representation of invariant solutions is $\varphi = t^{-1}r(z)$, where $z = xt^\alpha$. Substituting this representation into equation (1), we get the reduced equation

$$\alpha zr'(z) - r(z) + r(z)r(0) - \int_0^1 r(zs)r(z(1-s))\,\mathrm{d}s = \beta.$$

A representation of invariant solutions corresponding to the subalgebra $\{X_0\}$ is $\varphi = r(t)$, which gives the the solution $\varphi = -\beta t^{-1} + C$, where C is a constant.

4.4. The source function $\hat{q} = t^{-2}\Phi(xt^\gamma)$
In this case the admitted Lie algebra is $\{\gamma X_0 + X_2\}$. Invariant solutions have the representation $\varphi = t^{-1}r(z)$, where $z = xt^\gamma$. Substituting this representation into equation (1), we obtain the reduced equation

$$\gamma zr'(z) - r(z) + r(z)r(0) - \int_0^1 r(zs)r(z(1-s))\,\mathrm{d}s = \Phi(z).$$

For the other one-dimensional algebras of Table 2 we just present the final results including representations of invariant solutions and reduced equations.

Case 6: $\varphi = t^{-(x+1)}r(x), \quad -(x+1)r(x) + r(x)r(0) - \int_0^1 r(xs)r(x(1-s))\,\mathrm{d}s = \Phi(x).$

Case 7: $\varphi = t^{x-1}r(x), \quad (x-1)r(x) + r(x)r(0) - \int_0^1 r(xs)r(x(1-s))\,\mathrm{d}s = \Phi(x).$

Case 8: $\varphi = r(z)$, where $z = xe^{-t}, \quad -zr'(z) + r(z)r(0) - \int_0^1 r(zs)r(z(1-s))\,\mathrm{d}s = \Phi(z).$

In particular, for BKW-solution $r = 6e^z(1-z)$ which gives that $\Phi = 0$.

Case 9: $\varphi = e^{xt}r(x), \quad xr(x) + r(x)r(0) - \int_0^1 r(xs)r(x(1-s))\,\mathrm{d}s = \Phi(x).$

Case 10: $\varphi = \int \Phi(t)\,dt.$

Case 11: $\varphi = r(x), \quad r(x)r(0) - \int_0^1 r(xs)r(x(1-s))\,\mathrm{d}s = \Phi(x).$

Acknowledgments

The authors are thankful to R O Popovych for useful discussion on applications of the algebraic method of group classification and to E Schulz for valuable remarks.

References

[1] Nonnenmacher T F 1984 Application of the similarity method to the nonlinear Boltzmann equation *Z. Angew. Math. Phys.* **35** 680
[2] Krook M and Wu T T 1976 Formation of Maxwellian tails *Phys. Rev. Lett.* **36** 1107
[3] Grigoriev Y N, Meleshko S V and Suriyawichitseranee A 2014 On group classification of the spatially homogeneous and isotropic Boltzmann equation with sources II *Int. J. Nonlinear Mech.* **61** 15
[4] Bobylev A V 1975 The method of the Fourier transform in the theory of the Boltzmann equation for Maxwell molecules *Dokl. Akad. Nauk SSSR* **225** 1041 [in Russian]
[5] Bobylev A V 1976 A class of invariant solutions of the Boltzmann equation *Dokl. Akad. Nauk SSSR* **231** 571 [in Russian]
[6] Grigoriev Y N and Meleshko S V 1987 Group analysis of the Boltzmann integro-differential equation *Dokl. Akad. Nauk SSSR* **297** 323 [in Russian]
[7] Grigoriev Y N and Meleshko S V 1986 Investigation of Invariant solutions of the Boltzmann Kinetic Euation and its Models *Preprint* of Institute of Theoretical and Applied Mechanics of Akad. Nauk SSSR. Sib. Otdelenie, no. 18 [in Russian]
[8] Meleshko S V 2005 *Methods for Constructing Exact Solutions of Partial Differential Equations. Mathematical and Analytical Techniques with Applications to Engineering* (New York: Springer)
[9] Grigoriev Y N, Ibragimov N H, Kovalev V F and Meleshko S V 2010 *Symmetries of Integro-Differential Equations and their Applications in Mechanics and Plasma Physics* (Lecture Notes in Physics vol 806) (Berlin / Heidelberg: Springer)
[10] Ovsiannikov L V 1978 *Group Analysis of Differential Equations* (Moscow: Nauka) [in Russian]; English transl. edited by Ames W F: Ovsiannikov L V 1982 *Group Analysis of Differential Equations* (New York-London: Academic Press)
[11] Popovych R O, Kunzinger M and Eshraghi H 2010 Admissible transformations and normalized classes of nonlinear Schrödinger equations *Acta Appl. Math.* **109** 315
[12] Abramowitz M and Stegun I A 1964 *Handbook of Mathematical Functions with Formulas, Graphs, and Mathematical Tables* (Washington: National Bureau of Standards)

Gardner's deformation of the Krasil'shchik–Kersten system

Arthemy V Kiselev[1] **and Andrey O Krutov**[2]

[1] Johann Bernoulli Institute for Mathematics and Computer Science, University of Groningen, P.O. Box 407, 9700 AK Groningen, The Netherlands

[2] Department of Higher Mathematics, Ivanovo State Power University, Rabfakovskaya str. 34, Ivanovo, 153003 Russia

E-mail: A.V.Kiselev@rug.nl, krutov@math.ispu.ru

Abstract. The classical problem of construction of Gardner's deformations for infinite-dimensional completely integrable systems of evolutionary partial differential equations (PDE) amounts essentially to finding the recurrence relations between the integrals of motion. Using the correspondence between the zero-curvature representations and Gardner deformations for PDE, we construct a Gardner's deformation for the Krasil'shchik–Kersten system. For this, we introduce the new nonlocal variables in such a way that the rules to differentiate them are consistent by virtue of the equations at hand and second, the full system of Krasil'shchik–Kersten's equations and the new rules contains the Korteweg–de Vries equation and classical Gardner's deformation for it.

Keywords: Integrable hierarchies, Krasil'shchik–Kersten system, conservation laws, Gardner's deformations, zero-curvature representations.

1. Introduction

The search for conservation laws and, in particular, the search for regular methods of construction of conservation laws [1, 2] are the classical problems in the theory of infinite-dimensional completely integrable systems. The existence of infinitely many integrals of motion allows one to detect relevant bi-Hamiltonian hierarchies [3]; we refer to [4] for an almost detective story how the quest for integrability took place some fifty years ago. Suppose now that "all" or "sufficiently many" nontrivial conserved currents are known for a PDE system under study. Then one can use them to balance solutions at the adjacent faces of a shock wave. Second, each conserved current for a system $\mathcal{E}$ with two independent variables (e.g., the time t and spatial coordinate x) determines an *Abelian* nonlocality[1] over $\mathcal{E}$; resolving the analytic obstructions, several layers of the new variables introduced in such a way are often enough for finding a recursion operator or symplectic structures [5]. Third, the knowledge of integrals of motion helps one to increase the precision of numerical simulations. Finally, we recall that advanced analytic methods for solution of Cauchy problems for nonlinear PDEs refer explicitly or tacitly to various types of conservation in the model at hand. How can the infinite chains of conservation laws be sought for in a systematic way?

[1] The new variables which we introduce in Section 3 of this paper over the Krasil'shchik–Kersten system do not belong to the class of Abelian nonlocalities for that model: the new structures stem from a zero-curvature representation, which corresponds to the non-Abelian picture.

Let us specify at once that in this paper we address the problem of finding recurrence relations between the conserved quantities. In other words, we discard the trivial idea of try-and-fault search for *isolated* solutions of the determining equations for generating sections of conservation laws [6]; software for symbolic calculations [7,8] can be used here. That approach would rely on the standard techniques in the local geometry of differential equation [6,9,10]; it often involves various case-dependent tricks such as the use of weight-homogeneity calculus for scaling-invariant systems.

Such an algorithm put aside, there remain three strategies to mention. The Lax-pair approach and its generalisation by Zakharov, Manakov, and Shabat prescribes the calculation of residues for fractional powers of the spectral operator [11,12]. Let us note that in the frames of this approach, each conserved density ρ_i is obtained without any reference to the already known ones; it is then hard to detect any relation between elements of the infinite sequence with $i \in \mathbb{N}$.

Another option would be to take a "good" recursion operator $R\colon \varphi_i \mapsto \varphi_{i+1}$ for local symmetries of the evolutionary system at hand; the adjoint operator $R^\dagger\colon \psi_i \mapsto \psi_{i+1}$ is known to propagate the *cosymmetries* (this is true for the class of evolutionary models, see [5]). With a bit of luck, one has that $\psi_i = \mathbf{E}(\rho_i)$, meaning that those cosymmetries ψ_i are true generating sections of nontrivial conserved currents whose conserved densities are ρ_i (we denote by $\mathbf{E}$ the Euler variational derivative). The densities ρ_i can be reconstructed by the homotopy formula (see [13] and references therein). It is readily seen that for bi-Hamiltonian hierarchies this algorithm can be simplified. Namely, by taking the tower of commuting flows given by $\varphi_i = J_1(\psi_{i+1}) = J_2(\psi_i)$, one inverts the senior Hamiltonian operator J_2 and then reconstructs the next Hamiltonian functional with density ρ_{i+1}. (The formalism of Hamiltonian structures for non-evolutionary systems, which exceeds the frames of this text, was developed by Krasil'shchik *et al.* in [5,14,15].) However, we conclude that again, the second approach yields the hoard of conserved densities ρ_i but not an explicit relation between them.

Let us recall finally that the renowned seminal paper [1] led not only to our understanding of the geometry of zero-curvature representations and Miura transformations, but also to an important class of deformation techniques for integrable models [16–18]. Gardner's deformations are the powerful instrument that produces explicit recurrence relations between the integrals of motion (and moreover, *only* between them, which means that neither the fractional powers of any pseudodifferential operators are involved nor the fragile homotopy formula is used to invert the Euler variational derivative).

It is the Gardner deformation problem for the Krasil'shchik–Kersten system which we solve in this paper, yet it should be emphasised that the solution method which we implement here can be applied to a wide class of deformation problems for nonlinear models of mathematical physics.

The Krasil'shchik–Kersten system of two evolution equations,[2]

$$\underline{u}_{12;t} = -\underline{u}_{12;xxx} + 6\underline{u}_{12}\underline{u}_{12;x} - 3u_0u_{0;xxx} - 3u_{0;x}u_{0;xx} + 3\underline{u}_{12;x}u_0^2 + 6\underline{u}_{12}u_0u_{0;x}, \tag{1a}$$

$$u_{0;t} = -u_{0;xxx} + 3u_0^2u_{0;x} + 3\underline{u}_{12}u_{0;x} + 3\underline{u}_{12;x}u_0. \tag{1b}$$

is the bosonic limit of the $N{=}2$ supersymmetric $a{=}1$ Korteweg–de Vries equation [19]: to obtain system (1) from the super-equation, one sets equal to zero the two fermionic components u_1, u_2 of the superfield $\boldsymbol{u}$ that contains $N{=}2$ Grassmann variables θ_1 and θ_2. The parent $N{=}2$ SKdV equation with $a{=}1$ belongs to the triplet $a \in \{-2, 1, 4\}$ of completely integrable cases, see [19,20]. The issue of integrability at $a{=}1$ is difficult, compared to the well-studied cases $a{=}-2$ and $a{=}4$ (e.g., see [21] and [22], also [23]). The three super-systems share the second

[2] The choice of sign for the coefficient $u_{12} = -\underline{u}_{12}$ reflects the freedom of order, $\theta_1\theta_2 \otimes u_{12} = \theta_2\theta_1 \otimes \underline{u}_{12}$, in the superfield expansion $\boldsymbol{u} = 1 \otimes u_0 + \theta_1 \otimes u_1 + \theta_2 \otimes u_2 + \theta_1\theta_2 \otimes u_{12}$, where $\theta_1\theta_2 = -\theta_2\theta_1$ is the product of Grassmann variables.

Journal of Physics: Conference Series **621** (2015) 012007 doi:10.1088/1742-6596/621/1/012007

Hamiltonian operator $J_2^{a=-2,1,4}$, whereas the first Hamiltonian structure $J_1^{a=1}$ for the case a=1 is highly non-local. Kersten and Sorin obtained it in [24] by factorising the recursion super-operator $R = J_2^{a=-2,1,4} \circ \left(J_1^{a=1}\right)^{-1}$, thus solving P. Mathieu's Open problem 5 from [21] at least formally. (It still remains to verify that the non-local super-operator $J_1^{a=1}$ is skew-adjoint and endows the tower of super-Hamiltonians with the Poisson bracket; it remains also to inspect whether the symmetries of the a=1 SKdV equation produced by such mapping R remain local.) It is clear however that the bosonic-limit system (1) inherits the bi-Hamiltonian structure $\left(J_1^{a=1}\big|_{\text{Fermi}:=0}, J_2^{a=-2,1,4}\big|_{\text{Fermi}:=0}\right)$ from the parent super-equation.

To the best of our knowledge, the study of standard geometric structures related to (1) was initiated by Krasil'shchik and Kersten in [25] and continued in [5], where the proper reduction of the second Hamiltonian structure for P. Mathieu's N=2 SKdV super-equation was re-discovered; a recursion operator for symmetries of (1) was obtained via the introduction of suitable nonlocalities, c.f. [26] in this context. (Let us note that the coefficients of that recursion operator depend on the new nonlocal variables so that the locality of such operator's output is arguable; yet it could well be that system (1) is but a precursor to the larger model with physical applications.) Around the same time, Karasu-Kalkanlı *et al.* [27] approached system (1) with the Painlevé test, performing the singularity analysis, and constructed an $\mathfrak{sl}_3(\mathbb{C})$-valued zero-curvature representation α_1^{KK} for (1). We shall use this Lie algebra-valued one-form for solving the Gardner's deformation problem of recursive production of Hamiltonians for the hierarchy of the Krasil'shchik–Kersten system.

Our approach is based on the earlier work [28, 29], see also [16, 17, 30] and [18]. By understanding that zero-curvature representations and Gardner's deformations are structures of the same nature within the nonlocal geometry of PDE, we reformulate the Gardner deformation problem for Krasil'shchik–Kersten's system in terms of construction of parameter-dependent families of new 'nonlocal' variables. We require that these nonlocalities reproduce the classical Gardner deformation from [1] under the shrinking of extension (1) for the Korteweg–de Vries equation back to

$$u_{12;t} = -u_{12;xxx} - 6u_{12}u_{12;x} \iff \underline{u}_{12;t} = -\underline{u}_{12;xxx} + 6\underline{u}_{12}\underline{u}_{12;x} \quad \text{for } \underline{u}_{12} = -u_{12}. \qquad (2)$$

We discover that the nonlocalities which encode the gauge class of parameter-dependent zero-curvature representation from [27] are a key to solution of the problem: they yield the recurrence relation between the hierarchy of integrals of motion for the Krasil'shchik–Kersten system.

2. Basic concept

Let us recall the definitions (see [6, 13, 31] and [32] for detail); this material is standard so we now fix some notation and review the concept.

2.1. *The geometry of infinite jet space $J^\infty(\pi)$*

Let M^n be a smooth real n-dimensional orientable manifold. Consider a smooth vector bundle $\pi\colon E^{n+m} \to M^n$ with m-dimensional fibres; let us construct the space $J^\infty(\pi)$ of infinite jets of sections for π. Let $\mu_{\boldsymbol{x}_0}^k\Gamma(\pi)$ be the space of (local) sections $s \in \Gamma(\pi)$ such that all partial derivatives of s up to and including order $k \geqslant 0$ vanish at a point $\boldsymbol{x}_0 \in M^n$:

$$\mu_{\boldsymbol{x}_0}^k\Gamma(\pi) = \left\{ s \in \Gamma(\pi) \;\middle|\; \frac{\partial^{|\sigma|}s}{\partial \boldsymbol{x}^\sigma}(\boldsymbol{x}_0) = 0 \text{ for all } \sigma \text{ such that } 0 \leqslant |\sigma| \leqslant k \right\};$$

by convention, the zeroth-order derivative of any function is the function itself. The rules for transformation of first- and higher-order derivatives under local reparametrisations of the independent variables $\boldsymbol{x} = (x^1, \ldots, x^n)$ in a chart $U \subseteq M^n$ imply that the space $\mu_{\boldsymbol{x}_0}^k\Gamma(\pi)$ is well defined – in a coordinate-free way. Consider the quotient space of equivalence classes of (local) sections near a point $\boldsymbol{x}_0$,

$$J^k_{\boldsymbol{x}_0}(\pi) = \Gamma(\pi)/\mu^k_{\boldsymbol{x}_0}\Gamma(\pi).$$

The space $J^k(\pi)$ of kth jets of sections for the vector bundle π is the union

$$J^k(\pi) = \bigcup_{\boldsymbol{x}_0\in M^n} J^k_{\boldsymbol{x}_0}(\pi),$$

naturally equipped with the smooth manifold structure. The *infinite jet space* $J^\infty(\pi)$ is the projective limit

$$J^\infty(\pi) = \varprojlim_{k\to+\infty} J^k(\pi).$$

A convenient organisation of local coordinates on $J^\infty(\pi)$ is as follows: let x^i be some coordinate system on a chart in the base M^n and denote by u^j the fibre coordinates in the bundle π so that the variables u^j play the rôle of unknowns; one obtains the collection u^j_σ of jet variables along fibres of the vector bundle $J^\infty(\pi) \to M^n$ (here $|\sigma| \geqslant 0$ and $u^j_\varnothing \equiv u^j$). In particular, we have that $n = 2$, $m = 1$, $x^1 = x$, $x^2 = t$, $u^1 = u_{12}$ for KdV equation (2) and $n = 2$, $m = 2$, $x^1 = x$, $x^2 = t$, $u^1 = \underline{u}_{12}$, $u^2 = u_0$ for Krasil'shchik–Kersten's system (1).

Define the ring of smooth function on $J^\infty(\pi)$ as the inductive limit

$$C^\infty(J^\infty(\pi)) = \Bigl\{f\colon J^\infty(\pi) \to \mathbb{R} \mid f \in C^\infty(M^n) \text{ or } \exists k \in \mathbb{N}_{\geqslant 0} \text{ such that } f \in C^\infty(J^k(\pi))\Bigr\}.$$

For a function from $C^\infty(J^\infty(\pi))$ we denote by $[u]$ its differential dependence on finitely many coordinates along the fibre of the infinite jet bundle over M^n: such differential order can be arbitrarily large but it is always finite. In this setup, the *total derivatives* D_{x^i} are the commuting vector fields

$$D_{x^i} = \frac{\partial}{\partial x^i} + \sum_{j=1}^{m}\sum_{|\sigma|\geqslant 0} u^j_{\sigma\cup\{i\}}\,\frac{\partial}{\partial u^j_\sigma}$$

on $J^\infty(\pi)$. Let us denote the total derivatives also by $\frac{\mathrm{d}}{\mathrm{d}x^i}$, making no distinction between the two ways of notation.

Let us impose some mild restrictions on the class of PDEs which we deal with and which are given in local coordinates by using the relations[3]

$$\mathcal{E} = \Bigl\{F^\ell(x^i, u^j, \ldots, u^j_{\sigma_\ell}) = 0, \quad \ell = 1, \ldots, r < \infty,\ 0 < |\sigma_\ell| \leqslant K < \infty\Bigr\}. \tag{3}$$

[3] It is very well known that the *differential order* of a (system of partial) differential equation(s) can depend on a choice of the jet space in which the equation(s) is – or are – realised by using the jet fibre coordinates. For example, the *second*-order equation $\mathcal{E}' = \{u_{xx} = 0\} \subset J^2(\pi'\colon \mathbb{R}\times\mathbb{R} \to \mathbb{R})$ is obviously equivalent to the *first*-order system $\mathcal{E}'' = \{u_x = v,\ v_x = 0\} \subset J^1(\pi''\colon \mathbb{R}\times\mathbb{R}^2 \to \mathbb{R})$. Therefore, one can safely think that $K = 1$ in (3) whenever the surjectivity $\mathcal{E}^\infty \to J^{K-1}(\pi)$ is required to outline the class of "good" differential equations (see below). At the same time, the realisation of Krasil'shchik–Kersten's system and Korteweg–de Vries' equation by using (1) and (2) of differential order three also makes no harm.

Let us emphasize that in this paper, we operate with the *covering* structures over PDEs $\mathcal{E}$ by viewing them not just as larger systems $\tilde{\mathcal{E}}$ of equations satisfying some properties (see Definition 3 below) but as the structures indeed; the coverings are built over the underlying systems $\mathcal{E}$ that are given in advance. In other words, the formal integrability and other requirements which we describe on pp. 4–5 refer to the systems $\mathcal{E}$ and their prolongations (specifically, to system (1) and equation (2)). But the key idea of the reasoning that follows is the use of several algebraic realisations for the covering structures over $\mathcal{E}$. By this we avoid a necessity to re-write – in terms of a larger system of differential order one – the covering equation $\tilde{\mathcal{E}}$ formed by third-order Korteweg–de Vries' equation (2) and by rules (10) to differentiate the nonlocality (here, of order one for (10a) and order two for (10b)). Leaving the details of that particular example to a curious reader, we thank the referee for addressing this issue.

Namely, we study only (systems of) partial differential equations[4] which are *formally integrable* (cf. [6, 13] and [33] by Goldschmidt); this class of PDEs $\mathcal{E}$ is defined as follows. By definition, put $\mathcal{E}^{(0)} = \mathcal{E}$. For every given PDE system (3) consider its differential consequences $\mathcal{E}^{(k)} = \{D_\tau(F^\ell) = 0 \mid |\tau| \leqslant k > 0\}$ for all admissible ℓ. Let us assume that at each $k > 0$, all these differential-algebraic relations determine a smooth submanifold in $J^{K+k}(\pi)$ such that the projection $\mathcal{E}^{(k)} \to \mathcal{E}^{(k-1)}$ yields a fibre bundle. Suppose further that the inverse limit $\operatorname{proj}\lim_{k\to\infty} \mathcal{E}^{(k)} = \mathcal{E}^\infty$ is a smooth submanifold in $J^\infty(\pi)$; the object $\mathcal{E}^\infty$ is called the *infinite prolongation*[5] of the underlying PDE system $\mathcal{E} \subseteq J^K(\pi)$. Without loss of generality, let us assume that the mapping $\mathcal{E}^\infty \to M^n$ under $\pi_{\infty,-\infty}\colon J^\infty(\pi) \to M^n$ is onto (otherwise, shrink the base manifold M^n); assume also that the projection $\mathcal{E}^\infty \to J^{K-1}(\pi)$ under $\pi_{\infty,K-1}\colon J^\infty(\pi) \to J^{K-1}(\pi)$ is a surjection as well.[6] By imposing some geometric conditions on the symbols of such PDEs, Goldschmidt proves in [33, Theorem 9.1] that formally integrable PDE systems with (locally-) analytic left-hand sides do possess (local-) analytic solutions for all Cauchy's data.[7]

For *evolutionary* PDE systems (e.g., for equation (2) and system (1)) the spatial derivatives $u^i_{;\varnothing} \equiv u^i$, $u^i_{;\boldsymbol{x}}$, $u^i_{;\boldsymbol{xx}}$, ... of all orders together with the independent variable(s) $\boldsymbol{x}$ and time t constitute the collection of convenient internal coordinates on the set $\mathcal{E}^\infty$, which is readily seen to be a smooth submanifold in $J^\infty(\pi)$. Let us denote by $\bar{D}_{x^i}$ the restrictions of total derivatives D_{x^i} to the infinite prolongation $\mathcal{E}^\infty$. Thanks to the assumptions which were made in the preceding paragraph, these vector fields are tangent to the nonsingular submanifold $\mathcal{E}^\infty \subset J^\infty(\pi)$, spanning the Cartan distribution $\mathcal{C} \subset T\mathcal{E}^\infty$ on it. At every point $\theta^\infty \in \mathcal{E}^\infty$ the tangent space $T_{\theta^\infty}\mathcal{E}^\infty$ splits in a direct sum of two subspaces. The one which is spanned by the Cartan distribution on $\mathcal{E}^\infty$ is *horizontal* and the other is *vertical*, forming the kernel of the differential of the projection $\mathcal{E}^\infty \to M^n$; we have that $T_{\theta^\infty}\mathcal{E}^\infty = \mathcal{C}_{\theta^\infty} \oplus V_{\theta^\infty}\mathcal{E}^\infty$. We denote by $\Lambda^{1,0}(\mathcal{E}^\infty) = \operatorname{Ann}\mathcal{C}$ and $\Lambda^{0,1}(\mathcal{E}^\infty) = \operatorname{Ann} V\mathcal{E}^\infty$ the $C^\infty(\mathcal{E}^\infty)$-modules of contact and horizontal one-forms which vanish on $\mathcal{C}$ and $V\mathcal{E}^\infty$, respectively. Denote further by $\Lambda^r(\mathcal{E}^\infty)$ the $C^\infty(\mathcal{E}^\infty)$-module of r-forms on $\mathcal{E}^\infty$. There is a natural decomposition $\Lambda^r(\mathcal{E}^\infty) = \bigoplus_{q+p=r} \Lambda^{p,q}(\mathcal{E}^\infty)$, where $\Lambda^{p,q}(\mathcal{E}^\infty) = \bigwedge^p \Lambda^{1,0}(\mathcal{E}^\infty) \wedge \bigwedge^q \Lambda^{0,1}(\mathcal{E}^\infty)$. This implies that the de Rham differential $\bar{\mathrm{d}}$ on $\mathcal{E}^\infty$ is subjected to the decomposition $\bar{\mathrm{d}} = \bar{\mathrm{d}}_h + \bar{\mathrm{d}}_{\mathcal{C}}$, where $\bar{\mathrm{d}}_h\colon \Lambda^{p,q}(\mathcal{E}^\infty) \to \Lambda^{p,q+1}(\mathcal{E}^\infty)$ is the horizontal differential and $\bar{\mathrm{d}}_{\mathcal{C}}\colon \Lambda^{p,q}(\mathcal{E}^\infty) \to \Lambda^{p+1,q}(\mathcal{E}^\infty)$ is the vertical differential. Let $f(x^j, [u])$ be a function (of finite differential order) on the infinite prolongation $\mathcal{E}^\infty$. The horizontal differential $\bar{\mathrm{d}}_h$ acts on it by the rule $f \mapsto \sum_{i=1}^n \bar{D}_{x^i}(f)\,\mathrm{d}x^i$. This formula's extension to the spaces $\Lambda^{0,q}(\mathcal{E}^\infty)$ of horizontal q-forms is immediate: for any $\eta = f(x^j, [u])\,\mathrm{d}x^{i_1} \wedge \cdots \wedge \mathrm{d}x^{i_q}$ we have that $\bar{\mathrm{d}}_h(\eta) = \sum_{i=1}^n \bar{D}_{x^i}(f)\,\mathrm{d}x^i \wedge \mathrm{d}x^{i_1} \wedge \cdots \wedge \mathrm{d}x^{i_q} \in \Lambda^{0,q+1}(\mathcal{E}^\infty)$. For $p > 0$, the action of horizontal differential $\bar{\mathrm{d}}_h$ on the space $\Lambda^{p,q}(\mathcal{E}^\infty)$ of differential forms containing p Cartan's differentials is highly nontrivial[8] (see [9, 10], also [15]). However, in this paper we deal with the forms that

[4] We shall primarily deal with the *evolutionary* systems of KdV-type, equipping them further with the geometric structures such as the nonlocalities, or *coverings* (see Definition 3).

[5] A geometric distinction between the smooth submanifold $\mathcal{E}^\infty \subseteq J^\infty(\pi)$ and its description by using the smooth left-hand sides in the infinite system $D_\tau(F^\ell) = 0$ is that the latter is always defined yet it can describe the *empty set*. For instance, consider the overdetermined equation $\mathcal{E} = \{u_{xx} = 1,\ u_y = x^2\}$ for which $(u_{xx})_y = 0 \neq 2 = (u_y)_{xx}$. Admitting the slightest abuse of language, we shall make no distinction between the geometric objects $\mathcal{E}^\infty$ and their algebraic descriptions.

[6] For example, the equation $\mathcal{E} = \{v_x = u,\ v_y = u\}$ can be solved only if the compatibility condition $v_{xy} = v_{yx}$ is satisfied, thus $u_x = u_y$ is the constraint due to which the projection from $\mathcal{E}^{(1)}$ down to $\mathcal{E}$ is not onto, hence not a vector bundle.

[7] The books [6, 13] contain an extensive study of the properties which partial differential equations – not only evolutionary – must have in order to admit formal solutions and possess infinitesimal symmetries.

[8] The horizontal differential $\bar{\mathrm{d}}_h$ acts on the spaces $\Lambda^{p,q}(\mathcal{E}^\infty)$ of differential forms via the graded Leibniz rule; its application to Cartan's forms $\bar{\mathrm{d}}_{\mathcal{C}}(u^j_\sigma)$ is deduced from the identity $\bar{\mathrm{d}}^2 = 0$ for the de Rham differential $\bar{\mathrm{d}} = \bar{\mathrm{d}}_h + \bar{\mathrm{d}}_{\mathcal{C}}$ on $\mathcal{E}^\infty$. Specifically, from $\bar{\mathrm{d}}_h^2 = \bar{\mathrm{d}}_h \circ \bar{\mathrm{d}}_{\mathcal{C}} + \bar{\mathrm{d}}_{\mathcal{C}} \circ \bar{\mathrm{d}}_h = \bar{\mathrm{d}}_{\mathcal{C}}^2 = 0$ one infers that $\bar{\mathrm{d}}_h \circ \bar{\mathrm{d}}_{\mathcal{C}} = -\bar{\mathrm{d}}_{\mathcal{C}} \circ \bar{\mathrm{d}}_h$, thus reducing the

contain no Cartan's differentials. By definition, we put $\bar{\Lambda}(\mathcal{E}^\infty) = \bigoplus_{q\geqslant 0} \Lambda^{0,q}(\mathcal{E}^\infty)$ and we denote by $\overline{H}^n(\mathcal{E}^\infty)$ the senior $\bar{\mathrm{d}}_h$-cohomology group (also called senior *horizontal cohomology*).

A conserved current η for the system $\mathcal{E}$ is the continuity equation

$$\sum_{i=1}^{n} \bar{D}_{x^i}(\eta_i) \doteq 0 \text{ on } \mathcal{E}^\infty,$$

where the symbol $\doteq$ denotes the equality by virtue of the system $\mathcal{E}$ and its differential consequences. The quantities $\eta_i(x^j, [u^k])$ convene to the $\bar{\mathrm{d}}_h$-closed horizontal $(n-1)$-form

$$\eta = \sum_{i=1}^{n} (-1)^{i+1} \eta_i \cdot \mathrm{d}x^1 \wedge \cdots \wedge \widehat{\mathrm{d}x^i} \wedge \cdots \wedge \mathrm{d}x^n \in \bar{\Lambda}^{n-1}(\pi),$$

in which the wedge factors $\widehat{\mathrm{d}x^i}$ are omitted; the conservation of the current η is the equality $\bar{\mathrm{d}}_h(\eta) \doteq 0$ on $\mathcal{E}^\infty$. The coefficient η^n is called the *conserved density* and coefficients $\eta^1, \ldots, \eta^{n-1}$ are the *flux* components. By definition, a current η is *trivial* if it is $\bar{\mathrm{d}}_h$-exact: $\eta = \bar{\mathrm{d}}_h(\xi)$ for some $(n-2)$-form ξ on $\mathcal{E}^\infty$ (here $n \geqslant 2$). A *conservation law* $\int \eta \in \overline{H}^{n-1}(\mathcal{E}^\infty)$ for an equation $\mathcal{E}$ is the equivalence class of conserved currents η taken modulo globally defined[9] exact forms $\bar{\mathrm{d}}_h\xi \in \int 0$. In other words, two conserved currents η_1 and η_2 are equivalent if they differ by a trivial current: $\eta_1 - \eta_2 = \bar{\mathrm{d}}_h\xi$. We denote by $\overline{H}^{n-1}(\mathcal{E}^\infty)$ the $(n-1)$th horizontal cohomology group for $\mathcal{E}^\infty$, that is, the set of equivalence classes of conserved currents which is equipped with the structure of Abelian group.

2.2. Gardner's deformations

Definition 1 ([1, 17, 18]). Let $\mathcal{E} = \{u_t = f(x, [u])\}$ be a system of evolution equations (in particular, a completely integrable system). Suppose $\mathcal{E}(\varepsilon) = \{\tilde{u}_t = f_\varepsilon(x, [\tilde{u}], \varepsilon) \mid f_\varepsilon \in \operatorname{im} \frac{\mathrm{d}}{\mathrm{d}x}\}$ is a deformation of $\mathcal{E}$ such that at each point $\varepsilon \in \mathcal{I}$ of an interval $\mathcal{I} \subseteq \mathbb{R}$ there is the *Miura contraction* $\mathfrak{m}_\varepsilon = \{u = u([\tilde{u}], \varepsilon)\} \colon \mathcal{E}(\varepsilon) \to \mathcal{E}$. Then the pair $(\mathcal{E}(\varepsilon), \mathfrak{m}_\varepsilon)$ is the (*classical*) *Gardner deformation* for the system $\mathcal{E}$.

Under the assumption that $\mathcal{E}(\varepsilon)$ be in the form of a conserved current, the Taylor coefficients $\tilde{u}^{(k)}$ of the formal power series $\tilde{u} = \sum_{k=0}^{+\infty} \tilde{u}^{(k)} \cdot \varepsilon^k$ are termwise conserved on $\mathcal{E}(\varepsilon)$ and hence

action of $\bar{\mathrm{d}}_h$ to the case when it has already been defined. In brief, the formula $\bar{\mathrm{d}}_h = \sum_i \mathrm{d}x^i \wedge \bar{D}_{x^i}$ means that the vector fields $\bar{D}_{x^i}$ proceed by the Leibniz rule over the argument's wedge factors, acting on each factor – pushed leftmost – via the Lie derivative.

[9] Let us exclude – from the future consideration – the *topological* conservation laws (which arise from the geometry of the bundle π or from the topology $H^{n-1}(\mathcal{E}^\infty) \neq 0$ of the PDE system at hand, cf. [6] or [31]). This is legitimate for the geometry of Korteweg–de Vries' equation and Krasil'shchik–Kersten's system under study; we have that the vector bundles $\pi \colon \mathbb{R}^n \times \mathbb{R}^m \to \mathbb{R}^n =: M^n$ are topologically trivial in the both cases.

The referee recalls that the requirement for a trivial current $\bar{\mathrm{d}}_h(\xi)$ to be globally defined does not exclude a possibility for existence of the topological conservation laws. For example, let the n-dimensional base manifold M^n be such that its $(n-1)$th de Rham cohomology group is nonzero; now pick any closed differential $(n-1)$-form ω on M^n such that the de Rham cohomology class of ω is nonzero. Next, construct the trivial bundle $\pi \colon M^n \times \mathbb{R} \to M^n$ with a coordinate u in the fibre, and postulate the PDE $\mathcal{E} = \{u = 0\}$ so that the section $u = 0$ is its only solution. Then $\mathcal{E}^\infty \cong M^n$, and one can regard ω as a horizontal $(n-1)$-form on $\mathcal{E}^\infty$. Note that, specifically to this example, the horizontal differential $\bar{\mathrm{d}}_h$ on $\mathcal{E}^\infty \cong M^n$ is equal to the usual de Rham differential, so that $\bar{\mathrm{d}}_h(\omega) = 0$. Since the de Rham cohomology class of ω is nonzero, there does not exist a globally defined form ξ on $M^n \cong \mathcal{E}^\infty$ satisfying $\bar{\mathrm{d}}_h(\xi) = \omega$. (Let us note that the globally defined topological conserved current ω on $\mathcal{E}^\infty$ is *not trivial* by the definition of cohomology.) The referee concludes that the conservation law $\int \omega \in \overline{H}^{n-1}(\mathcal{E}^\infty)$ is topological because it depends only on the topology of $\mathcal{E}^\infty \cong M^n$. However, all of this is irrelevant to the problem that we deal with in this paper.

on $\mathcal{E}$. Therefore, the contraction $\mathfrak{m}_\varepsilon$ yields the recurrence relations, ordered by the powers of ε, between these densities $\tilde{u}^{(k)}$, while the equality $\mathcal{E}(0)=\mathcal{E}$ specifies the initial condition for those relations.

Example 1 ([2]). The contraction

$$\mathfrak{m}_\varepsilon = \{u_{12} = \tilde{u}_{12} \pm \varepsilon \tilde{u}_{12;x} - \varepsilon^2 \tilde{u}_{12}^2\} \tag{4a}$$

maps solutions $\tilde{u}_{12}(x,t;\varepsilon)$ of the extended equation

$$\mathcal{E}(\varepsilon) = \{\tilde{u}_{12;t} + (\tilde{u}_{12;xx} + 3\tilde{u}_{12}^2 - 2\varepsilon^2 \cdot \tilde{u}_{12}^3)_x = 0\}, \tag{4b}$$

to solutions $u_{12}(x,t)$ of the Korteweg–de Vries equation

$$\mathcal{E} = \{u_{12;t} = -u_{12;xxx} - 6u_{12}u_{12;x}\}. \tag{2}$$

Plugging the series $\tilde{u}_{12} = \sum_{k=0}^{+\infty} u_{12}^{(k)} \cdot \varepsilon^k$ into expression (4a) for $\tilde{u}_{12}$, we obtain the chain of equations ordered by powers of ε,

$$u_{12} = \sum_{k=0}^{+\infty} \tilde{u}_{12}^{(k)} \cdot \varepsilon^k \pm \tilde{u}_{12;x}^{(k)} \cdot \varepsilon^{k+1} - \sum_{\substack{i+j=k\\ i,j\geq 0}} \tilde{u}_{12}^{(i)}\tilde{u}_{12}^{(j)} \cdot \varepsilon^{k+2}.$$

Let us fix the plus sign in (4a) by reversing $\varepsilon \to -\varepsilon$ if necessary. Equating the coefficients of ε^k, we obtain the relations

$$u = \tilde{u}_{12}^{(0)}, \quad 0 = \tilde{u}_{12}^{(1)} + \tilde{u}_{12;x}^{(0)}, \quad 0 = \tilde{u}_{12}^{(k)} + \tilde{u}_{12;x}^{(k-1)} - \sum_{\substack{i+j=k-2\\ i,j\geq 0}} \tilde{u}_{12}^{(i)}\tilde{u}_{12}^{(j)}, \quad k \geq 2.$$

Hence, from the initial condition $\tilde{u}_{12}^{(0)} = u_{12}$ we recursively generate the densities

$$\begin{aligned}
&\tilde{u}_{12}^{(1)} = -u_{12;x}, \quad \tilde{u}_{12}^{(2)} = u_{12;xx} - u_{12}^2, \quad \tilde{u}_{12}^{(3)} = -u_{12;xxx} + 4u_{12;x}u_{12},\\
&\tilde{u}_{12}^{(4)} = u_{12;4x} - 6u_{12;xx}u_{12} - 5u_{12;x}^2 + 2u_{12}^3,\\
&\tilde{u}_{12}^{(5)} = -u_{12;5x} + 8u_{12;xxx}u_{12} + 18u_{12;xx}u_{12;x} - 16u_{12;x}u_{12}^2,\\
&\tilde{u}_{12}^{(6)} = u_{12;6x} - 10u_{12;4x}u_{12} - 28u_{12;xxx}u_{12;x} - 19u_{12;xx}^2 + 30u_{12;xx}u_{12}^2 + 50u_{12;x}^2u_{12} - 5u_{12}^4,\\
&\tilde{u}_{12}^{(7)} = -u_{12;7x} + 12u_{12;5x}u_{12} + 40u_{12;4x}u_{12;x} + 68u_{12;xxx}u_{12;xx} - 48u_{12;xxx}u_{12}^2\\
&\qquad - 216u_{12;xx}u_{12;x}u_{12} - 60u_{12;x}^3 + 64u_{12;x}u_{12}^3, \quad \text{etc.}
\end{aligned}$$

The conservation $\tilde{u}_{12;t} = \frac{\mathrm{d}}{\mathrm{d}x}(\cdot)$ implies that each coefficient $\tilde{u}_{12}^{(k)}$ is conserved on (2), and one proves easily that the densities $\tilde{u}_{12}^{(2k)}$ with even indexes $2k \in 2\mathbb{N}_{\geq 0}$ determine the hierarchy of nontrivial conservation laws for the Korteweg–de Vries equation ([1, 2] vs [3]).

2.3. Zero-curvature representations

Let G be a finite-dimensional matrix complex Lie group and $\mathfrak{g}$ be its Lie algebra. Consider the tensor product $\mathfrak{g} \otimes_{\mathbb{R}} \bar{\Lambda}(\mathcal{E}^\infty)$ of $\mathfrak{g}$ with the exterior algebra $\bar{\Lambda}(\mathcal{E}^\infty) = \bigoplus_i \Lambda^{0,i}(\mathcal{E}^\infty)$. The product is endowed with the bracket

$$[A \otimes \mu, B \otimes \nu] = [A, B] \otimes \mu \wedge \nu$$

for $A, B \in \mathfrak{g}$ and $\mu, \nu \in \bar{\Lambda}(\mathcal{E}^\infty)$. Define the operator $\bar{\mathrm{d}}_h$ that acts on elements of $\mathfrak{g} \otimes \bar{\Lambda}(\mathcal{E}^\infty)$ by the rule

$$\bar{\mathrm{d}}_h(A \otimes \mu) = A \otimes \bar{\mathrm{d}}_h \mu,$$

where the horizontal differential $\bar{\mathrm{d}}_h$ in the right-hand side is already defined (see p. 5). Elements of $\mathfrak{g} \otimes C^\infty(\mathcal{E}^\infty)$ are called $\mathfrak{g}$-*matrices* [32].

Definition 2 ([32,34]). A horizontal 1-form $\alpha \in \mathfrak{g} \otimes \bar{\Lambda}^1(\mathcal{E}^\infty)$ is called a $\mathfrak{g}$-valued *zero-curvature representation* (ZCR) for the equation $\mathcal{E}$ if the Maurer–Cartan condition,

$$\bar{\mathrm{d}}_h \alpha \doteq \tfrac{1}{2}[\alpha, \alpha], \tag{5}$$

holds by virtue of $\mathcal{E}$ and its differential consequences.

Example 2. The $\mathfrak{sl}_2(\mathbb{C})$-valued zero-curvature representation $\alpha_1^{\mathrm{KdV}} = A\,\mathrm{d}x + B\,\mathrm{d}t$ for KdV equation (2),

$$\alpha_1^{\mathrm{KdV}} = \begin{pmatrix} 0 & \lambda - u_{12} \\ 1 & 0 \end{pmatrix} \mathrm{d}x + \begin{pmatrix} -u_{12;x} & -4\lambda^2 + 2\lambda u_{12} + 2u_{12}^2 + u_{12;xx} \\ -4\lambda - 2u_{12} & u_{12;x} \end{pmatrix} \mathrm{d}t, \tag{6}$$

is known from the paper [35].

Recall that $\mathfrak{g}$ is the Lie algebra of a given Lie group G, see above. Elements of $C^\infty(\mathcal{E}^\infty, G)$, i.e., G-valued functions on $\mathcal{E}^\infty$, are called *G-matrices*. Let α and α' be $\mathfrak{g}$-valued zero-curvature representations, then α and α' are called *gauge-equivalent* if there exists a G-matrix $S \in C^\infty(\mathcal{E}^\infty, G)$ such that

$$\alpha' = \bar{\mathrm{d}}_h S \cdot S^{-1} + S \cdot \alpha \cdot S^{-1} =: \alpha^S. \tag{7}$$

Not only that the notions of Gardner's deformations and zero-curvature representations are intimately related but moreover, the idea of gauge equivalence allows us to revise, simplify, and solve the deformation problem for (1).

2.4. Differential coverings

We now recall the definition of differential covering over (the infinite prolongation of) a given PDE system $\mathcal{E}$. This notion brings together the procedure of Gardner's deformations and the construction of zero-curvature representations, allowing – in principle – to interpret both of them in terms of new, larger set of differential equations that contains $\mathcal{E}$ as sub-system. (By construction, the new differential equations will not necessarily be evolutionary even if the system $\mathcal{E}$ is.) Let us remark however that in this text, the coverings over evolution equations are viewed and operated with as differential-geometric structures over the underlying systems (1), (2) of KdV-type.[10] More specifically, the covering structures will be described by using matrix Lie groups and algebras and by using Lie subalgebras in the algebras of vector fields on some fibre bundles over the manifolds $\mathcal{E}^\infty$.

Definition 3 ([6,36]). A *covering* (or *differential covering*) over a formally integrable equation $\mathcal{E}$ is another (usually, larger) system of partial differential equations $\tilde{\mathcal{E}}$ endowed with the n-dimensional Cartan distribution $\tilde{\mathcal{C}}$ and such that there is a mapping $\tau\colon \tilde{\mathcal{E}} \to \mathcal{E}^\infty$ for which at each point θ of the manifold $\tilde{\mathcal{E}}$, the tangent map $\tau_{*,\theta}$ is an isomorphism of the plane $\tilde{\mathcal{C}}_\theta$ to the Cartan plane $\mathcal{C}_{\tau(\theta)}$ at the point $\tau(\theta)$ in $\mathcal{E}^\infty$.

[10] This approach is highlighted by using the notation: nominally realised as differential equations, the coverings' total spaces $\tilde{\mathcal{E}}$ fibre over the infinite prolongations $\mathcal{E}^\infty$ of evolutionary systems.

The construction of a covering over $\mathcal{E}$ means the introduction of new variables in such a way that the compatibility of their mixed derivatives is valid by virtue of the underlying $\mathcal{E}^\infty$. In practice (see [31]), it is the rules to differentiate the new variable(s) which are specified in a consistent way; this implies that those new variables acquire the nature of nonlocalities if their derivatives are local but the variables themselves are not (e.g., consider the potential $\mathfrak{v} = \int u_{12}\,dx$ satisfying $\mathfrak{v}_x = u_{12}$ and $\mathfrak{v}_t = -u_{12;xx} - 3u_{12}^2$ for the KdV equation $u_{12;t} + u_{12;xxx} + 6u_{12}u_{12;x} = 0$). Whenever the covering is indeed realised as the fibre bundle $\tau\colon \tilde{\mathcal{E}} \to \mathcal{E}$, the forgetful map τ discards the nonlocalities.

In these terms, zero-curvature representations and Gardner's deformations are coverings of special kinds[11] (see Examples 3 and 6 below). Each zero-curvature representation with coefficients belonging to a matrix Lie algebra determines a (linear) covering, whereas each covering with fibre W can be regarded as a zero-curvature representation whose coefficients take values in the Lie algebra of vector fields on W. Indeed, let $x^1, \ldots, x^n$ be the independent variables in a given PDE and $\bar{D}_{x^i}$ be the corresponding total derivative operators. Then the zero-curvature representations and coverings are described by the same equation (5),

$$[\bar{D}_{x^i} + A_i, \bar{D}_{x^j} + A_j] \doteq 0, \quad 1 \leqslant i < j \leqslant n.$$

In the case of zero-curvature representations, the coefficients A_i and A_j are functions on $\mathcal{E}^\infty$ taking values in a Lie algebra. In the case of coverings, the objects A_i and A_j are vertical vector fields on the covering manifold. This correspondence between zero-curvature representations and coverings very often allows one to transfer results on ZCRs to results on coverings and *vice versa*, see [29, 38] and [39, 40] for detail.

The use of geometric similarity of the two notions allows us to construct new Gardner's deformations from known zero-curvature representations that take values in finite-dimensional complex Lie algebras.[12]

Example 3 (zero-curvature representations as coverings). Let $\mathfrak{g} := \mathfrak{sl}_2(\mathbb{C})$ as in Example 2 and introduce the standard basis e, h, f in $\mathfrak{g}$ so that

$$[e, h] = -2e, \quad [e, f] = h, \quad [f, h] = 2f.$$

Let us consider the matrix representation

$$\rho\colon\ \mathfrak{sl}_2(\mathbb{C}) \to \{A \in \mathrm{Mat}(2,2) \mid \mathrm{tr}\, A = 0\}$$

of $\mathfrak{g}$ and, simultaneously, its representation ϱ in the space of vector fields with polynomial coefficients on the complex line with the coordinate w:

$$\begin{aligned} \rho(e) &= \begin{pmatrix} 0 & 1 \\ 0 & 0 \end{pmatrix}, & \rho(h) &= \begin{pmatrix} 1 & 0 \\ 0 & -1 \end{pmatrix}, & \rho(f) &= \begin{pmatrix} 0 & 0 \\ 1 & 0 \end{pmatrix}, \\ \varrho(e) &= 1 \cdot \partial/\partial w, & \varrho(h) &= -2w \cdot \partial/\partial w, & \varrho(f) &= -w^2 \cdot \partial/\partial w. \end{aligned}$$

[11] The link between zero-curvature representations and recursion operators is discussed in the papers [26, 37].

[12] Whenever the vector field realisation of a covering structure over $\mathcal{E}^\infty$ is given *a priori*, the problem of reconstruction and recognition of a Lie algebra that could determine that covering is nontrivial; e.g., take the vector fields that encode a Gardner deformation for $\mathcal{E}$ in terms of a covering over its infinite prolongation. Suppose for definition that the coefficients of these vector fields are polynomial in the nonlocal variables. Then, as soon as one starts taking the fields' iterated commutators, either they close to a manifestly finite-dimensional Lie subalgebra in the Lie algebra of vector fields on the covering's fibre – or the degrees of such polynomials grow infinitely. In that situation, the polynomials of different degrees determine linearly independent elements within a basis of the generated Lie subalgebra in the Lie algebra of all vector fields on the fibres. For example, such is the case of Gardner's deformation for the Kaup–Boussinesq equation [41]. However, there still remain two possible options: either the Lie algebra realised by using all the commutator-generated vector fields is truly infinite-dimensional or there is a *finite*-dimensional Lie algebra such that the vector fields on the covering's fibres provide its infinite-dimensional representation.

Let us decompose the matrices $A_i \in C^\infty(\mathcal{E}^\infty)\otimes\mathfrak{g}$, occurring in the zero-curvature representation $\alpha = \sum_i A_i\,dx^i$, with respect to the basis in the space $\rho(\mathfrak{g})$:

$$A_i = a_e^{(i)} \otimes \rho(e) + a_h^{(i)} \otimes \rho(h) + a_f^{(i)} \otimes \rho(f) \quad \text{for } a_j^{(i)} \in C^\infty(\mathcal{E}^\infty).$$

To construct the covering $\tilde{\mathcal{E}}$ over $\mathcal{E}^\infty$ with a new fibre variable w (the 'nonlocality'), we switch from the representation ρ to ϱ. We thus obtain the map

$$A_i \mapsto V_{A_i} \tag{8}$$

that takes the $\mathfrak{g}$-matrices A_i to the vector fields

$$V_{A_i} = a_e^{(i)} \otimes \varrho(e) + a_h^{(i)} \otimes \varrho(h) + a_f^{(i)} \otimes \varrho(f);$$

the prolongations of total derivatives D_{x^i} to $\tilde{\mathcal{E}}$ are now defined by the formula

$$\tilde{D}_{x^i} = D_{x^i} - V_{A_i}. \tag{9}$$

The extended derivatives act on the nonlocal variable w as follows,

$$\tilde{D}_{x^i} w = dw \lrcorner (-V_{A_i}).$$

We shall use this approach in the construction of the covering in Example 4, see below.

Remark 1. The commutativity $[\tilde{D}_{x^i}, \tilde{D}_{x^j}] = 0$ of prolonged total derivatives for all $i \neq j$ is equivalent to Maurer–Cartan's equation (5). Indeed, we have that

$$\begin{aligned} 0 = [\tilde{D}_{x^i}, \tilde{D}_{x^j}] &= [D_{x^i} - V_{A_i}, D_{x^j} - V_{A_j}] = [D_{x^i}, D_{x^j}] - [D_{x^i}, V_{A_j}] - [V_{A_j}, D_{x^j}] + [V_{A_i}, V_{A_j}] = \\ &= -V_{D_{x^i}A_i} + V_{D_{x^j}A_i} + V_{[A_i,A_j]} = V_{D_{x^j}A_i - D_{x^i}A_j + [A_i,A_j]} \Leftrightarrow D_{x^j}A_i - D_{x^i}A_j + [A_i, A_j] = 0. \end{aligned}$$

This motivates our choice of the minus sign in (9).

Example 4 (one-dimensional covering over the KdV equation). One obtains the covering over the KdV equation from the zero-curvature representation α_1^{KdV} (see Example 2) by using the realisation of Lie algebra $\mathfrak{sl}_2(\mathbb{C})$ in the space of vector fields. Applying (8), we construct the following vector fields with the nonlocal variable w:

$$\begin{aligned} V_A &= -(u_{12} + w^2 - \lambda)\,\tfrac{\partial}{\partial w}, \\ V_B &= -\big(-u_{12;xx} - 2u_{12}^2 - 2\lambda u_{12} + 4\lambda^2 - 2u_{12;x}w - (2u_{12} + 4\lambda)w^2\big)\,\tfrac{\partial}{\partial w}. \end{aligned}$$

The prolongations of total derivatives act on the nonlocality w by the rules

$$w_x = u_{12} + w^2 - \lambda, \tag{10a}$$

$$w_t = -u_{12;xx} - 2u_{12}^2 - 2\lambda u_{12} + 4\lambda^2 - 2u_{12;x}w - (2u_{12} + 4\lambda)w^2. \tag{10b}$$

This yields the one-dimensional covering over KdV equation (2).

Example 5 (The projective substitution and nonlinear realisations of Lie algebras in the spaces of vector fields [30, 39, 40]). Let N be a $(k_0 + 1)$-dimensional manifold.[13] Because the reasoning

[13] The realisation scheme which we outline here can be translated verbatim into supergeometry of supermanifolds N of superdimension $(k_0 + 1|k_1)$ and zero-curvature representations with values in Lie superalgebras $\mathfrak{g} \subseteq \mathfrak{gl}(k_0 + 1|k_1)$.

is local, consider a chart $\mathcal{U} \subseteq N$ equipped with a (k_0+1)-tuple of rectifying coordinates $\boldsymbol{v} = (v^0, \ldots, v^{k_0})$ that establish a one-to-one correspondence between the points of $\mathcal{U}$ in N and a domain in the vector space $\mathbb{R}^{k_0+1}$. By definition, put

$$\partial_{\boldsymbol{v}} = (\partial_{v^0}, \ldots, \partial_{v^{k_0}})^{\mathrm{t}}.$$

Second, let $\mathfrak{g} \subseteq \mathfrak{gl}(k_0+1)$ be a Lie algebra (see footnote 13 again). Take any matrix $g \in \mathfrak{g}$ and represent it in the space of *linear* vector fields on the domain in $\mathbb{R}^{k_0+1}$ by using the formula

$$g \longmapsto V_g = \boldsymbol{v} g \partial_{\boldsymbol{v}}.$$

By construction, the linear vector field representation $g \mapsto V_g$ of the matrix Lie algebra $\mathfrak{g}$ preserves all the commutation relations in it,

$$[V_g, V_h] = [\boldsymbol{v}g\partial_{\boldsymbol{v}}, \boldsymbol{v}h\partial_{\boldsymbol{v}}] = \boldsymbol{v}[g,h]\partial_{\boldsymbol{v}} = V_{[g,h]}, \quad h, g \in \mathfrak{g}.$$

The problem we are solving now is the realisation of matrix Lie algebra $\mathfrak{g}$ by using vector fields with (non)linear coefficients. To begin with, fix a nonzero constant $\mu \in \mathbb{R}$; without loss of generality suppose $v^0 \neq 0$. Consider the locally defined mapping $p\colon \mathbb{R}^{k_0+1} \to \mathbb{R}^{k_0}$ that takes every point $\boldsymbol{v} = (v^0, \ldots, v^{k_0})$ from the domain at hand to the point $(w^1, \ldots, w^{k_0}) \in \mathbb{R}^{k_0}$, where

$$w^i = \frac{\mu v^i}{v^0}, \quad 1 \leqslant i \leqslant k_0.$$

The differential $\mathrm{d}p_{\boldsymbol{v}}\colon T_{\boldsymbol{v}}\mathbb{R}^{k_0+1} \to T_{p(\boldsymbol{v})}\mathbb{R}^{k_0}$ at the point $\boldsymbol{v} \in \mathbb{R}^{k_0+1}$ acts on the basic vectors from the (k_0+1)-tuple $\partial_{\boldsymbol{v}}$ as follows,

$$\mathrm{d}p_{\boldsymbol{v}}\left(\frac{\partial}{\partial v^0}\right) = \sum_{j=1}^{k_0} \frac{\partial w^j}{\partial v^0}\frac{\partial}{\partial w^j} = \sum_{j=1}^{k_0} -\frac{\mu v^j}{(v^0)^2}\frac{\partial}{\partial w^j},$$

$$\mathrm{d}p_{\boldsymbol{v}}\left(\frac{\partial}{\partial v^i}\right) = \sum_{j=1}^{k_0} \frac{\partial w^j}{\partial v^i}\frac{\partial}{\partial w^j} = \frac{\mu}{v^0}\frac{\partial}{\partial w^i}, \quad 1 \leqslant i \leqslant k_0.$$

Using these formulae, let us calculate the action of differential $\mathrm{d}p_{\boldsymbol{v}}$ on the linear vector field V_g at the point $\boldsymbol{v}$ in the domain:

$$\begin{aligned}
\mathrm{d}p_{\boldsymbol{v}}(V_g) &= \mathrm{d}p_{\boldsymbol{v}}\left(\sum_{i,j=0}^{k_0} v^i g_{ij}\partial_{\boldsymbol{v}}^j\right) = \sum_{i=0}^{k_0} v^i g_{i0} \sum_{j=1}^{k_0}\left(-\frac{\mu v^j}{(v^0)^2}\frac{\partial}{\partial w^j}\right) + \sum_{i=0}^{k_0}\sum_{j=1}^{k_0} v^i g_{ij}\frac{\mu}{v^0}\frac{\partial}{\partial w^j} \\
&= \sum_{i=0}^{k_0} \frac{\mu v^i}{v^0} g_{i0}\left(-\frac{1}{\mu}\sum_{j=1}^{k_0}\frac{\mu v^j}{v^0}\frac{\partial}{\partial w^j}\right) + \sum_{i=0}^{k_0}\sum_{j=1}^{k_0}\frac{\mu v^i}{v^0} g_{ij}\frac{\partial}{\partial w^j} \\
&= \mu g_{00}\left(-\frac{1}{\mu}\sum_{j=1}^{k_0} w^j\frac{\partial}{\partial w^j}\right) + \sum_{i=1}^{k_0} w^i g_{i0}\left(-\frac{1}{\mu}\sum_{j=1}^{k_0} w^j \frac{\partial}{\partial w^j}\right) + \sum_{j=1}^{k_0}\mu g_{0j}\frac{\partial}{\partial w^j} \\
&\quad + \sum_{i=1}^{k_0}\sum_{j=1}^{k_0} w^i g_{ij}\frac{\partial}{\partial w^j},
\end{aligned}$$

where $\partial_{\boldsymbol{v}}^j$ is the jth element of the tuple $\partial_{\boldsymbol{v}}$. By definition, put

$$\boldsymbol{w} = (\mu, w^1, \ldots, w^{k_0}), \quad \partial_{\boldsymbol{w}} = \left(-\frac{1}{\mu}\sum_{j=1}^{k_0} w^j\frac{\partial}{\partial w^j}, \frac{\partial}{\partial w^1}, \ldots, \frac{\partial}{\partial w^{k_0}}\right)^{\mathrm{t}}.$$

We conclude that the vector field $X_g = \mathrm{d}p\,(V_g)$ is expressed by the formula

$$X_g = \boldsymbol{w} g \partial_{\boldsymbol{w}}. \tag{11}$$

Generally speaking, the vector field X_g on the respective subset of the target space $\mathbb{R}^{k_0}$ is nonlinear with respect to the variables $w^0, \ldots, w^{k_0}$. Nevertheless, the commutation relations between vector fields of such type are inherited from the relations in Lie algebra $\mathfrak{g} \ni g, h$:

$$[X_g, X_h] = [\mathrm{d}p\,(V_g), \mathrm{d}p\,(V_h)] \overset{(*)}{=} \mathrm{d}p\,([V_g, V_h]) = \mathrm{d}p\,(V_{[g,h]}) = X_{[g,h]},$$

see Appendix A for an explicit proof of equality $(*)$. Take X_g for the representation $\varrho(g)$ of elements g of Lie algebra $\mathfrak{g}$; now that the representation ϱ is specified, either leave the parameter μ free or set it equal to any convenient nonzero constant. We refer to [39, 40] for other examples of realisations of Lie algebras by using vector fields.

For the sake of definition let us take $k_0 = 1$ so that $w^1 = w$ for $n = 2$ with $x^1 = x$ and $x^2 = t$; set $\mu = 1$. Using the representation ϱ, we construct the prolongations of total derivatives,

$$\tilde{D}_x = \bar{D}_x + w_x \frac{\partial}{\partial w}, \quad \tilde{D}_t = \bar{D}_t + w_t \frac{\partial}{\partial w},$$

and inspect the way in which they act on the nonlocal variable w along W:

$$w_x = \bar{D}_x \lrcorner\, \mathrm{d}w, \quad w_t = \bar{D}_t \lrcorner\, \mathrm{d}w.$$

We thus obtain a one-dimensional covering $\tau\colon \tilde{\mathcal{E}} = W \times \mathcal{E}^\infty \to \mathcal{E}^\infty$ with nonlocal variable w.

We claim that Gardner's deformation (4) and zero-curvature representation (2) for KdV equation (2) determine the coverings which are related by using an $\mathrm{SL}(2,\mathbb{C})$-valued gauge transformation.

Example 6 (the covering which is based on Gardner's deformation)**.** Consider the Gardner deformation of Korteweg–de Vries equation (2),

$$\mathfrak{m}_\varepsilon = \bigl\{u_{12} = \tilde{u}_{12} - \varepsilon \tilde{u}_{12;x} - \varepsilon^2 \tilde{u}_{12}^2\bigr\} : \mathcal{E}_\varepsilon \to \mathcal{E}_0, \tag{4a}$$
$$\mathcal{E}_\varepsilon = \bigl\{\tilde{u}_{12;t} = -(\tilde{u}_{12;xx} + 3\tilde{u}_{12}^2 - 2\varepsilon^2 \tilde{u}_{12}^3)_x\bigr\}. \tag{4b}$$

Expressing $\tilde{u}_{12;x}$ from (4a) and substituting it in (4b), we obtain the one-dimensional covering over the KdV equation,

$$\tilde{u}_{12;x} = \frac{1}{\varepsilon}(\tilde{u}_{12} - u_{12}) - \varepsilon \tilde{u}_{12}^2, \tag{12a}$$
$$\begin{aligned}\tilde{u}_{12;t} &= \frac{1}{\varepsilon}(u_{12;xx} + 2u_{12}^2) + \frac{1}{\varepsilon^2} u_{12;x} + \frac{1}{\varepsilon^3} u_{12} \\ &\quad + \left(-2u_{12;x} - \frac{2}{\varepsilon} u_{12} - \frac{1}{\varepsilon^3}\right) \tilde{u}_{12} + \left(2\varepsilon u_{12} + \frac{1}{\varepsilon}\right) \tilde{u}_{12}^2.\end{aligned} \tag{12b}$$

From this covering we derive the $\mathfrak{sl}_2(\mathbb{C})$-valued zero-curvature representation for (2):

$$\alpha_2^{\mathrm{KdV}} = \begin{pmatrix} \frac{1}{2\varepsilon} & \frac{u_{12}}{\varepsilon} \\ -\varepsilon & -\frac{1}{2\varepsilon} \end{pmatrix} \mathrm{d}x + \begin{pmatrix} u_{12;x} + \frac{1}{\varepsilon} u_{12} + \frac{1}{2\varepsilon^3} & -\frac{1}{\varepsilon}(u_{12;xx} + 2u_{12}^2) - \frac{1}{\varepsilon^2} u_{12;x} - \frac{1}{\varepsilon^3} u_{12} \\ 2\varepsilon u_{12} + \frac{1}{\varepsilon} & -u_{12;x} - \frac{1}{\varepsilon} u_{12} - \frac{1}{2\varepsilon^3} \end{pmatrix} \mathrm{d}t. \tag{13}$$

The gauge transformation between zero-curvature representations (6) and (13) is given by the group element

$$S = \begin{pmatrix} \boldsymbol{i}/\sqrt{\varepsilon} & \boldsymbol{i}/(2\varepsilon\sqrt{\varepsilon}) \\ 0 & -\boldsymbol{i}\sqrt{\varepsilon} \end{pmatrix}, \tag{14}$$

where we set $\lambda = \frac{1}{4\varepsilon^2}$ to match the spectral parameter λ in (6) and Gardner's deformation parameter ε.

Let us apply the same construction to Krasil'shchik–Kersten's system (1) and by this, derive the recurrence relation between the integrals of motion in its hierarchy.

3. The deformation of Krasil'shchik–Kersten's system

From the paper [27] we know that Krasil'shchik–Kersten's system (1) admits the $\mathfrak{sl}_3(\mathbb{C})$-valued zero-curvature representation $\alpha_1^{\mathrm{KK}} = A_1^{\mathrm{KK}}\,dx + B_1^{\mathrm{KK}}\,dt$, where

$$A_1^{\mathrm{KK}} = \begin{pmatrix} \eta & \underline{u}_{12} - u_0^2 + 9\eta^2 & u_0 \\ 1 & \eta & 0 \\ 0 & 6\eta u_0 & -2\eta \end{pmatrix},$$

$$B_1^{\mathrm{KK}} = \begin{pmatrix} b_{11} & b_{12} & -18\eta^2 u_0 - 3\eta u_{0;x} - u_{0;xx} + u_0^3 + 2u_0\underline{u}_{12} \\ -36\eta^2 + u_0^2 + 2\underline{u}_{12} & -b_{11} - 72\eta^3 - 6\eta u_0^2 & -6\eta u_0 - u_{0;x} \\ -36\eta^2 u_0 + 6\eta u_{0;x} & b_{32} & 72\eta^3 - 6\eta u_0^2 \end{pmatrix};$$

the elements b_{11}, b_{12}, and b_{32} of the matrix B_1^{KK} are as follows:

$$\begin{aligned} b_{11} &= -36\eta^3 + 3\eta u_0^2 + u_{0;x}u_0 + \underline{u}_{12;x}, \\ b_{12} &= -324\eta^4 + 9\eta^2(u_0^2 - 2\underline{u}_{12}) - u_{0;xx}u_0 - u_{0;x}^2 - \underline{u}_{12;xx} - u_0^4 - u_0^2\underline{u}_{12} + 2\underline{u}_{12}^2, \\ b_{32} &= -108\eta^3 u_0 + 18\eta^2 u_{0;x} + 6\eta(-u_{0;xx} + u_0^3 + 2u_0\underline{u}_{12}). \end{aligned}$$

Let us find the matrix $S^{\mathrm{KK}} \in \mathrm{SL}_3(\mathbb{C}) \hookrightarrow C^\infty(\mathcal{E}^\infty, \mathrm{SL}_3(\mathbb{C}))$ of the gauge transformation that makes the classical formulae by Gardner a part of the covering over system (1). By definition, we put $\varepsilon = \eta^2$ in (14). Next, let us enlarge the old group $\mathrm{SL}(2,\mathbb{C})$ for equation (2) to the gauge group $\mathrm{SL}(3,\mathbb{C})$ of zero-curvature representation α_1^{KK} for system (1). We set

$$S^{\mathrm{KK}} = \begin{pmatrix} i\eta^{-1} & \frac{1}{2}i\eta^{-3} & 0 \\ 0 & -i\eta & 0 \\ 0 & 0 & 1 \end{pmatrix}.$$

Applying the gauge transformation S^{KK} to the zero-curvature representation α_1^{KK}, we obtain the gauge-equivalent zero-curvature representation $\alpha_2^{\mathrm{KK}} := (\alpha_1^{\mathrm{KK}})^{S^{\mathrm{KK}}} = A_2^{\mathrm{KK}}\,dx + B_2^{\mathrm{KK}}\,dt$ for Krasil'shchik–Kersten's system (1):

$$A_2^{\mathrm{KK}} = \begin{pmatrix} \frac{2}{3}\eta^{-2} & u_0^2 - \underline{u}_{12} & i\eta^{-1}u_0 \\ -1 & -\frac{1}{3}\eta^{-2} & 0 \\ 0 & i\eta^{-1}u_0 & -\frac{1}{3}\eta^{-2} \end{pmatrix},$$

$$B_2^{\mathrm{KK}} = \begin{pmatrix} b_{11} & b_{12} & b_{13} \\ -u_0^2 - 2\underline{u}_{12} + \eta^{-4} & -u_{0;x}u_0 - \underline{u}_{12;x} - \eta^{-2}\underline{u}_{12} + \frac{1}{3}\eta^{-6} & i\eta^{-1}u_{0;x} + i\eta^{-3}u_0 \\ -i\eta^{-1}u_{0;x} + i\eta^{-3}u_0 & \eta^{-1}(-iu_{0;xx} + iu_0^3 + 2iu_0\underline{u}_{12}) & -\eta^{-2}u_0^2 + \frac{1}{3}\eta^{-6} \end{pmatrix},$$

where the elements b_{11}, b_{12}, and b_{13} of the matrix B_2^{KK} are as follows:

$$\begin{aligned} b_{11} &= u_{0;x}u_0 + \underline{u}_{12;x} + \eta^{-2}(u_0^2 + \underline{u}_{12}) - \tfrac{2}{3}\eta^{-6}, \\ b_{12} &= u_{0;xx}u_0 + u_{0;x}^2 + \underline{u}_{12;xx} + u_0^4 + u_0^2\underline{u}_{12} - 2\underline{u}_{12}^2 + \eta^{-2}(u_{0;x}u_0 + \underline{u}_{12;x}) + \eta^{-4}\underline{u}_{12}, \\ b_{13} &= \eta^{-1}(-iu_{0;xx} + iu_0^3 + 2iu_0\underline{u}_{12}) - i\eta^{-3}u_{0;x} - i\eta^{-5}u_0. \end{aligned}$$

Recalling that formula (11) yields the representation of matrices A_2^{KK} and B_2^{KK} in terms of vector fields, from the zero-curvature representation α_2^{KK} we obtain the two-dimensional covering over Krasil'shchik–Kersten's system (1). Denoting the new nonlocal variables by $\tilde{u}_0$ and $\tilde{\underline{u}}_{12}$, we have that their derivatives with respect to the spatial variable x and time t are equal to

$$\tilde{u}_{0;x} = -\tilde{u}_0\tilde{\underline{u}}_{12} - i\eta^{-1}u_0 + \eta^{-2}\tilde{u}_0, \tag{15a}$$

$$\tilde{\underline{u}}_{12;x} = -\tilde{\underline{u}}_{12}^2 - u_0^2 + \underline{u}_{12} - i\eta^{-1}\tilde{u}_0 u_0 + \eta^{-2}\tilde{\underline{u}}_{12}, \tag{15b}$$

and

$$
\begin{aligned}
\tilde{u}_{0;t} &= \tilde{u}_0(u_{0;x}u_0 + \underline{u}_{12;x} - \tilde{\underline{u}}_{12}u_0^2 - 2\tilde{\underline{u}}_{12}\underline{u}_{12}) \\
&\quad + \eta^{-1}(\boldsymbol{i}u_{0;xx} - \boldsymbol{i}u_{0;x}\tilde{u}_0^2 - \boldsymbol{i}u_{0;x}\tilde{\underline{u}}_{12} - \boldsymbol{i}u_0^3 - 2\boldsymbol{i}u_0\underline{u}_{12}) + \eta^{-2}\tilde{u}_0(2u_0^2 + \underline{u}_{12}) \\
&\quad + \eta^{-3}(\boldsymbol{i}u_{0;x} + \boldsymbol{i}\tilde{u}_0^2u_0 - \boldsymbol{i}\tilde{\underline{u}}_{12}u_0) + \eta^{-4}\tilde{u}_0\tilde{\underline{u}}_{12} + \boldsymbol{i}\eta^{-5}u_0 - \eta^{-6}\tilde{u}_0, \qquad (15c)\\
\tilde{\underline{u}}_{12;t} &= -u_{0;xx}u_0 - u_{0;x}^2 + 2u_{0;x}\tilde{\underline{u}}_{12}u_0 - \underline{u}_{12;xx} + 2\underline{u}_{12;x}\tilde{\underline{u}}_{12} - \tilde{\underline{u}}_{12}^2u_0^2 - 2\tilde{\underline{u}}_{12}^2\underline{u}_{12} - u_0^4 - u_0^2\underline{u}_{12} \\
&\quad + 2\underline{u}_{12}^2 + \eta^{-1}\tilde{u}_0(\boldsymbol{i}u_{0;xx} - \boldsymbol{i}u_{0;x}\tilde{\underline{u}}_{12} - \boldsymbol{i}u_0^3 - 2\boldsymbol{i}u_0\underline{u}_{12}) \\
&\quad + \eta^{-2}(-u_{0;x}u_0 - \underline{u}_{12;x} + \tilde{\underline{u}}_{12}u_0^2 + 2\tilde{\underline{u}}_{12}\underline{u}_{12}) \\
&\quad + \boldsymbol{i}\eta^{-3}\tilde{u}_0\tilde{\underline{u}}_{12}u_0 + \eta^{-4}(\tilde{\underline{u}}_{12}^2 - \underline{u}_{12}) - \eta^{-6}\tilde{\underline{u}}_{12}. \qquad (15d)
\end{aligned}
$$

We note that under the reduction $u_0 = 0$ and by virtue of the relation $u_{12} = -\underline{u}_{12}$, this covering retracts to Gardner's deformation (12) for KdV equation (2).

Theorem 1 (Gardner's deformation of Krasil'shchik–Kersten's system (1))**.** The extension $\mathcal{E}(\varepsilon)$ of (1) consists of, first, the evolution equation which is not in the form of a conserved current,

$$
\begin{aligned}
\tilde{u}_{0;t} &= 3\varepsilon^4\tilde{u}_0^2\tilde{\underline{u}}_{12}(2\tilde{u}_{0;x}\tilde{\underline{u}}_{12} + \tilde{\underline{u}}_{12;x}\tilde{u}_0) + 3\varepsilon^3\tilde{u}_0(-\tilde{u}_{0;xx}\tilde{u}_0\tilde{\underline{u}}_{12} - 3\tilde{u}_{0;x}^2\tilde{\underline{u}}_{12} - \tilde{u}_{0;x}\tilde{\underline{u}}_{12;x}\tilde{u}_0) \\
&\quad + 3\varepsilon^2(\tilde{u}_{0;xx}\tilde{u}_{0;x}\tilde{u}_0 + \tilde{u}_{0;x}^3 + 3\tilde{u}_{0;x}\tilde{u}_0^2\tilde{\underline{u}}_{12} + \tilde{u}_{0;x}\tilde{\underline{u}}_{12}^2 + \tilde{\underline{u}}_{12;x}\tilde{u}_0^3 + \tilde{\underline{u}}_{12;x}\tilde{u}_0\tilde{\underline{u}}_{12}) \\
&\quad + 3\varepsilon\tilde{u}_0(-\tilde{u}_{0;xx}\tilde{u}_0 - 2\tilde{u}_{0;x}^2) - \tilde{u}_{0;xxx} + 3\tilde{u}_{0;x}\tilde{u}_0^2 + 3\tilde{u}_{0;x}\tilde{\underline{u}}_{12} + 3\tilde{\underline{u}}_{12;x}\tilde{u}_0, \qquad (16a)
\end{aligned}
$$

and second, the continuity relation

$$
\begin{aligned}
\tilde{\underline{u}}_{12;t} &= \frac{\mathrm{d}}{\mathrm{d}x}\Big(3\varepsilon^4\tilde{u}_0^2\tilde{\underline{u}}_{12}^3 + 3\varepsilon^3\tilde{u}_0\tilde{\underline{u}}_{12}(\tilde{u}_{0;x}\tilde{\underline{u}}_{12} - \tilde{\underline{u}}_{12;x}\tilde{u}_0) \\
&\quad + \varepsilon^2(-3\tilde{u}_{0;xx}\tilde{u}_0\tilde{\underline{u}}_{12} - 3\tilde{u}_{0;x}\tilde{\underline{u}}_{12;x}\tilde{u}_0 + 2\tilde{\underline{u}}_{12}^3 + 6\tilde{u}_0^2\tilde{\underline{u}}_{12}^2) \\
&\quad + 3\varepsilon(\tilde{u}_{0;xx}\tilde{u}_{0;x} - \tilde{u}_{0;x}\tilde{u}_0\tilde{\underline{u}}_{12} + \tilde{\underline{u}}_{12;x}\tilde{u}_0^2) - 3\tilde{u}_{0;xx}\tilde{u}_0 - \tilde{\underline{u}}_{12;xx} + 3\tilde{u}_0^2\tilde{\underline{u}}_{12} + 3\tilde{\underline{u}}_{12}^2\Big). \qquad (16b)
\end{aligned}
$$

The Miura contraction from (16) to (1) is

$$
\begin{aligned}
u_0 &= \tilde{u}_0 - \varepsilon\tilde{u}_{0;x} + \varepsilon^2\tilde{\underline{u}}_{12}\tilde{u}_0, \qquad (17a)\\
\underline{u}_{12} &= \tilde{\underline{u}}_{12} - \varepsilon(\tilde{\underline{u}}_{12;x} + \tilde{u}_{0;x}\tilde{u}_0) + \varepsilon^2(\tilde{u}_{0;x}^2 + \tilde{\underline{u}}_{12}^2 + \tilde{\underline{u}}_{12}\tilde{u}_0^2) - 2\varepsilon^3u_{0;x}\tilde{\underline{u}}_{12}\tilde{u}_0 + \varepsilon^4\tilde{\underline{u}}_{12}^2\tilde{u}_0^2. \qquad (17b)
\end{aligned}
$$

Under the reduction $u_0 = 0$, this deformation retracts to classical Gardner's formulas (4).

Proof. Let us express u_0 and $\underline{u}_{12}$ from (15a-15b) and plug them in (15c-15d). We get

$$
\begin{aligned}
u_0 &= \boldsymbol{i}\eta(\tilde{u}_{0;x} + \tilde{u}_0\tilde{\underline{u}}_{12}) - \boldsymbol{i}\eta^{-1}\tilde{u}_0, \\
\underline{u}_{12} &= \eta^2(-\tilde{u}_{0;x}^2 - 2\tilde{u}_{0;x}\tilde{u}_0\tilde{\underline{u}}_{12} - \tilde{u}_0^2\tilde{\underline{u}}_{12}^2) + \tilde{u}_{0;x}\tilde{u}_0 + \tilde{\underline{u}}_{12;x} + \tilde{u}_0^2\tilde{\underline{u}}_{12} + \tilde{\underline{u}}_{12}^2 - \eta^{-2}\tilde{\underline{u}}_{12}, \\
u_{0;t} &= 3\eta^2(-\tilde{u}_{0;xx}\tilde{u}_{0;x}\tilde{u}_0 - \tilde{u}_{0;xx}\tilde{u}_0^2\tilde{\underline{u}}_{12} - \tilde{u}_{0;x}^3 - 3\tilde{u}_{0;x}^2\tilde{u}_0\tilde{\underline{u}}_{12} - \tilde{u}_{0;x}\tilde{\underline{u}}_{12;x}\tilde{u}_0^2 - 2\tilde{u}_{0;x}\tilde{u}_0^2\tilde{\underline{u}}_{12}^2 \\
&\quad - \tilde{\underline{u}}_{12;x}\tilde{u}_0^3\tilde{\underline{u}}_{12}) - \tilde{u}_{0;xxx} + 3\tilde{u}_{0;xx}\tilde{u}_0^2 + 6\tilde{u}_{0;x}^2\tilde{u}_0 + 9\tilde{u}_{0;x}\tilde{u}_0^2\tilde{\underline{u}}_{12} + 3\tilde{u}_{0;x}\tilde{\underline{u}}_{12}^2 + 3\tilde{\underline{u}}_{12;x}\tilde{u}_0^3 \\
&\quad + 3\tilde{\underline{u}}_{12;x}\tilde{u}_0\tilde{\underline{u}}_{12} - 3\eta^{-2}(\tilde{u}_{0;x}\tilde{u}_0^2 + \tilde{u}_{0;x}\tilde{\underline{u}}_{12} + \tilde{\underline{u}}_{12;x}\tilde{u}_0), \\
\underline{u}_{12;t} &= 3\eta^2(\tilde{u}_{0;xxx}\tilde{u}_{0;x} + \tilde{u}_{0;xxx}\tilde{u}_0\tilde{\underline{u}}_{12} + \tilde{u}_{0;xx}^2 + \tilde{u}_{0;xx}\tilde{u}_{0;x}\tilde{\underline{u}}_{12} + 2\tilde{u}_{0;xx}\tilde{\underline{u}}_{12;x}\tilde{u}_0 - \tilde{u}_{0;xx}\tilde{u}_0\tilde{\underline{u}}_{12}^2 \\
&\quad + \tilde{u}_{0;x}^2\tilde{\underline{u}}_{12;x} - \tilde{u}_{0;x}^2\tilde{\underline{u}}_{12}^2 + \tilde{u}_{0;x}\tilde{\underline{u}}_{12;xx}\tilde{u}_0 - 2\tilde{u}_{0;x}\tilde{u}_0\tilde{\underline{u}}_{12}^3 + \tilde{\underline{u}}_{12;xx}\tilde{u}_0^2\tilde{\underline{u}}_{12} + \tilde{\underline{u}}_{12;x}^2\tilde{u}_0^2 - 3\tilde{\underline{u}}_{12;x}\tilde{u}_0^2\tilde{\underline{u}}_{12}^2) \\
&\quad - 3\tilde{u}_{0;xxx}\tilde{u}_0 - 3\tilde{u}_{0;xx}\tilde{u}_{0;x} + 3\tilde{u}_{0;xx}\tilde{u}_0\tilde{\underline{u}}_{12} + 3\tilde{u}_{0;x}^2\tilde{\underline{u}}_{12} - 3\tilde{u}_{0;x}\tilde{\underline{u}}_{12;x}\tilde{u}_0 + 12\tilde{u}_{0;x}\tilde{u}_0\tilde{\underline{u}}_{12}^2 \\
&\quad - \tilde{\underline{u}}_{12;xxx} - 3\tilde{\underline{u}}_{12;xx}\tilde{u}_0^2 + 12\tilde{\underline{u}}_{12;x}\tilde{u}_0^2\tilde{\underline{u}}_{12} + 6\tilde{\underline{u}}_{12;x}\tilde{\underline{u}}_{12}^2 \\
&\quad + 3\eta^{-2}(-2\tilde{u}_{0;x}\tilde{u}_0\tilde{\underline{u}}_{12} - \tilde{\underline{u}}_{12;x}\tilde{u}_0^2 - 2\tilde{\underline{u}}_{12;x}\tilde{\underline{u}}_{12}).
\end{aligned}
$$

Setting $\tilde{u}_{0;\text{new}} = \boldsymbol{i}\eta\,\tilde{u}_{0;\text{old}}$ and $\tilde{\underline{u}}_{12;\text{new}} = \eta^2\tilde{\underline{u}}_{12;\text{old}}$, and putting $\eta = \sqrt{\varepsilon}$, we derive formulas (16), (17). □

Remark 2. This change of variables removes the singularity at $\eta = \varepsilon = 0$ in formulas (15). (Let us recall that Gardner's deformations with a singularity at $\varepsilon = 0$ do yield the recurrence relations between the conserved densities, see [28, 42] for an example of such deformation for the N=2 supersymmetric a=4-Korteweg–de Vries equation.)

In Theorem 1 we obtained the deformation of Krasil'shchik–Kersten's system such that one of the extended equations is not a continuity relation, which makes the construction different from the classical concept of Gardner's deformation. Nevertheless, deformations of such unconventional type do yield the recurrence relations between the conserved densities.

Theorem 2. Gardner's deformation (16), (17) for Krasil'shchik–Kersten's system (1) yields the following recurrence relations between the conserved densities, which we denote by $\underline{w}_n$ for $n \in \mathbb{N}_{\geqslant 0}$:

$$\begin{aligned}
&\underline{w}_0 = \underline{u}_{12}, \quad \underline{w}_1 = \underline{u}_{12;x} + u_{0;x}u_0,\\
&\underline{w}_2 = D_x\underline{w}_1 + D_x(v_0v_1) - u_{0;x}^2 - \underline{u}_{12}^2 - \underline{u}_{12}u_0^2,\\
&\underline{w}_3 = D_x\underline{w}_2 + \frac{1}{2}\sum_{k=0}^{2} D_x(v_kv_{2-k}) - 2D_x(v_0)D_x(v_1) - 2\underline{w}_1\underline{w}_0 - \underline{w}_1v_0^2 - 2\underline{w}_0v_1v_0 + 2u_{0;x}u_0\underline{u}_{12},\\
&\underline{w}_n = D_x\underline{w}_{n-1} + \frac{1}{2}\sum_{k=0}^{n-1} D_x(v_kv_{n-1-k}) - \sum_{k=0}^{n-2}\left(D_x(v_k)D_x(v_{n-2-k}) - \underline{w}_k\underline{w}_{n-2-k}\right)\\
&\qquad - \sum_{k+l+j=n-2}\underline{w}_kv_lv_j + 2\sum_{k+l+j=n-3}\underline{w}_kv_lD_xv_j - \sum_{k+l+j+i=n-4}\underline{w}_k\underline{w}_lv_jv_i, \quad n \geqslant 4,
\end{aligned}$$

where the quantities v_i are given by the formulas

$$v_0 = u_0, \quad v_1 = u_{0;x}, \quad v_n = D_xv_{n-1} - \sum_{k=0}^{n-2}\underline{w}_kv_{n-2-k} \quad \text{for } n \geqslant 2.$$

The generating function $\breve{\underline{w}}(u_0, \underline{u}_{12}, \varepsilon)$ of the zero differential order component of the series $w([u_0, \underline{u}_{12}], \varepsilon)$ is given by the formula

$$\breve{\underline{w}} = \frac{12\varepsilon^2(-u_0^2 + \underline{u}_{12}) + q^2 - 4q + 4}{6\varepsilon^2 q}, \tag{18}$$

where we put

$$\begin{aligned}
q = 2^{2/3}\Big(&9\varepsilon^2(2u_0^2 + \underline{u}_{12}) + 2\\
&+ 3\sqrt{3}\varepsilon\sqrt{4\varepsilon^4(u_0^6 - 3u_0^4\underline{u}_{12} + 3u_0^2\underline{u}_{12}^2 - \underline{u}_{12}^3) + \varepsilon^2(8u_0^4 + 20u_0^2\underline{u}_{12} - \underline{u}_{12}^2) + 4u_0^2}\Big)^{1/3}.
\end{aligned}$$

Proof. Plugging the series $\tilde{u}_0 = \sum_{k=0}^{+\infty}\varepsilon^kv_k$ and $\tilde{\underline{u}}_{12} = \sum_{k=0}^{+\infty}\varepsilon^k\underline{w}_k$ into (16), we obtain the recurrence relations between v_k and $\underline{w}_k$. The coefficients $\underline{w}_k$ are conserved because $\tilde{\underline{u}}_{12;t}$ is in divergent form (i.e., the velocity of $\tilde{\underline{u}}_{12}$ is in the image of $\mathrm{d}/\mathrm{d}x$). The coefficients v_k are auxiliary quantities which, generally speaking, could be not conserved (note that the density v_0 is also conserved by virtue of (1) at $\varepsilon = 0$).

The respective zero-order components $\breve{v}(u_0, \underline{u}_{12}, \varepsilon)$ and $\breve{\underline{w}}(u_0, \underline{u}_{12}, \varepsilon)$ of $\tilde{u}_0$ and $\tilde{\underline{u}}_{12}$ satisfy the equations

$$u_0 = \breve{v} + \varepsilon^2\breve{v}\breve{\underline{w}}, \tag{19a}$$
$$\underline{u}_{12} = \breve{\underline{w}} + \varepsilon^2(\breve{v}^2\breve{\underline{w}} + \breve{\underline{w}}^2) + \varepsilon^4\breve{\underline{w}}^2\breve{v}^2. \tag{19b}$$

Expressing $\breve{v}$ from (19a), we obtain that

$$\breve{v} = \frac{u_0}{1+\varepsilon^2 \underline{\breve{w}}}.$$

Substituting this expression further in (19b), we obtain the cubic equation for $\underline{\breve{w}}$,

$$\varepsilon^4 \underline{\breve{w}}^3 + 2\varepsilon^2 \underline{\breve{w}}^2 + (\varepsilon^2 u_0^2 - \varepsilon^2 \underline{u}_{12} + 1)\underline{\breve{w}} - \underline{u}_{12} = 0.$$

The limit behaviour $\lim_{\varepsilon\to 0} \underline{\breve{w}} = \underline{u}_{12}$ of its solution at the origin of deformation parameter prescribes that we pick the root (18). □

Let us calculate several low-order conserved densities for system (1) by using the recurrence relations for $\underline{w}_n$:

$$\begin{aligned}
\underline{w}_0 &= \underline{u}_{12},\\
\underline{w}_1 &= u_{0;x}u_0 + \underline{u}_{12;x},\\
\underline{w}_2 &= -\underline{u}_{12}u_0^2 - \underline{u}_{12}^2 + 2u_0u_{0;xx} + u_{0;x}^2 + u_{0;xx},\\
\underline{w}_3 &= -u_0^3u_{0;x} - 6\underline{u}_{12}u_0u_{0;x} - 3\underline{u}_{12;x}u_0^2 - 4\underline{u}_{12}\underline{u}_{12;x} + 3u_0u_{0;xxx} + 4u_{0;x}u_{0;xx} + \underline{u}_{12;xxx},\\
\underline{w}_4 &= \underline{u}_{12}u_0^4 + 4\underline{u}_{12}^2u_0^2 - 4u_0^3u_{0;xx} - 8u_0^2u_{0;x}^2 + 2\underline{u}_{12}^3 - 13\underline{u}_{12}u_0u_{0;xx} - 8\underline{u}_{12}u_{0;x}^2 - 19\underline{u}_{12;x}u_0u_{0;x}\\
&\quad - 6\underline{u}_{12;xx}u_0^2 - 6\underline{u}_{12}\underline{u}_{12;xx} - 5\underline{u}_{12;x}^2 + 4u_0u_{0;4x} + 7u_{0;x}u_{0;xxx} + 4u_{0;xx}^2 + \underline{u}_{12;4x}, \quad \text{etc.}
\end{aligned}$$

Lemma 1. *The conserved densities $\underline{w}_{2k}$ with even indexes $2k \geqslant 0$ are non-trivial.*

This follows from the non-triviality of conserved densities $\tilde{u}_{12}^{(2k)}$ generated at all even indexes by Gardner's deformation (4) for KdV equation (2).

Proof. The reduction $u_0 = 0$ maps Krasil'shchik–Kersten's system (1) to the Korteweg–de Vries equation; likewise, the reduction $\tilde{u}_0 = 0$ takes the recurrence relations between the conserved densities $\underline{w}_k$ for (1) to Gardner's formulas for the densities $\tilde{u}_{12}^{(k)}$ which are conserved on equation (2). Therefore, the quantities $\underline{w}_{2k}$ have the form

$$\underline{w}_{2k}\bigl([\underline{u}_{12}],[u_0]\bigr) = \tilde{u}_{12}^{(2k)}\bigl([\underline{u}_{12}]\bigr) + f_{2k}\bigl([\underline{u}_{12}],[u_0]\bigr), \quad k \in \mathbb{N},$$

where the differential polynomials $f_{2k}\bigl([\underline{u}_{12}],[u_0]\bigr)$ are such that $f_{2k}|_{u_0=0} = 0$. On the one hand, the densities $\tilde{u}_{12}^{(2k)}$ are known to be equal to $\tilde{u}_{12}^{(2k)} = c_k \cdot \underline{u}_{12}^k + \cdots$, where c_k are nonzero constants [43]. On the other hand, every monomial in f_{2k} essentially depends on either u_0 or its derivatives with respect to x, hence f_{2k} cannot contain the monomial $\underline{u}_{12}^k$. Consequently, $\underline{w}_{2k} = c_k \cdot \underline{u}_{12}^k + \cdots$ with $c_k \neq 0$, so that these quantities may not belong to the image of total derivative $\mathrm{d}/\mathrm{d}x$, whence they are nontrivial. □

It must be expected that the densities $\underline{w}_{2k+1}$ with odd indexes, not contributing to the hierarchy of (1), are trivial; this is confirmed by a straightforward calculation of small-index terms in that auxiliary sequence. The triviality of irrelevant quantities $\underline{w}_{2k+1}$, $k \in \mathbb{N}$ can be approached, e.g., by using the technique from [43]; that method's idea is a realisation of the generating function for – in retrospect, trivial – conserved densities $\underline{w}_{2k+1}$ via the non-trivial quantities $\underline{w}_{2k}$ and the generating function for them.

In the meantime, the problem of finding recurrence relations for the hierarchy of integrals of motion for the Krasil'shchik–Kersten system is solved.

Remark 3. In the recent papers [18, 44] it was shown that Gardner's deformations provide the initial data for construction of new integrable systems. Applying the algorithm described in [18, 44] to Gardner's deformation (16-17) for Krasil'shchik–Kersten's system, we obtain the Kaup–Newell hierarchy [45].

Let us conclude this paper by recalling that the recurrence relation for the hierarchy of Hamiltonian functionals for the bosonic-limit system (1) is the initial datum for solution of Gardner's deformation problem for the full $N{=}2$ supersymmetric $a{=}1$ Korteweg–de Vries equation from [19, 21]. This technique of recursive construction of the Hamiltonian superfunctionals that depend on the $N{=}2$ superfield was developed in [41].

Appendix A.

Let $\mathfrak{g} \subseteq \mathfrak{gl}(k_0+1)$ be a finite-dimensional Lie algebra and $g \in \mathfrak{g}$. On the one hand, the element g is represented in the space of $(k_0+1)\times(k_0+1)$ matrices. On the other hand, the element g can be represented in the space of vector fields on some open domain in $\mathbb{R}^{k_0+1}$ by using the formula

$$g \longmapsto X_g = \boldsymbol{w} g \partial_{\boldsymbol{w}},$$

where

$$\boldsymbol{w} = \begin{pmatrix} \mu & w^1 & \dots & w^{k_0} \end{pmatrix}, \quad \partial_{\boldsymbol{w}} = \begin{pmatrix} -\frac{1}{\mu}\sum_{i=1}^{k_0} w^i \partial_{w^i} \\ \partial_{w^1} \\ \vdots \\ \partial_{w^{k_0}} \end{pmatrix}.$$

Let us show that the vector field representation preserves all the commutation relations in $\mathfrak{g}$, that is, let us verify the identity

$$[X_g, X_h] = X_g(X_h) - X_h(X_g) = X_{[g,h]}$$

for all $g, h \in \mathfrak{g}$.

By convention, summation over repeated indexes is performed (here, not necessarily over one *upper* and one *lower* index, which is due to the notation for matrix elements of g, h and for the components of $\partial_{\boldsymbol{w}}$).

Proof. We have that

$$\begin{aligned}
X_g(X_h) &= \boldsymbol{w}^i g_{ij} \partial^j_{\boldsymbol{w}} \left(\boldsymbol{w}^p h_{pq} \partial^q_{\boldsymbol{w}}\right) \\
&= \boldsymbol{w}^i g_{i0} \partial^0_{\boldsymbol{w}} (\boldsymbol{w}^p h_{pq} \partial^q_{\boldsymbol{w}}) + \sum_{j\neq 0} \boldsymbol{w}^i g_{ij} \partial^j_{\boldsymbol{w}} \left(\boldsymbol{w}^p h_{pq} \partial^q_{\boldsymbol{w}}\right) \\
&= \sum_{p\neq 0} \boldsymbol{w}^i g_{i0} \left(-\tfrac{1}{\mu}\boldsymbol{w}^p\right) h_{pq} \partial^q_{\boldsymbol{w}} + \boldsymbol{w}^i g_{i0} \boldsymbol{w}^p h_{p0} \left(-\tfrac{1}{\mu}\partial^0_{\boldsymbol{w}}\right) \\
&\qquad + \sum_{j\neq 0} \boldsymbol{w}^i g_{ij} h_{jq} \partial^q_{\boldsymbol{w}} + \sum_{j\neq 0} \boldsymbol{w}^i g_{ij} \boldsymbol{w}^p h_{p0} \left(-\tfrac{1}{\mu}\partial^j_{\boldsymbol{w}}\right) \\
&= \sum_{p\neq 0} \boldsymbol{w}^i g_{i0} \left(-\tfrac{1}{\mu}\boldsymbol{w}^p\right) h_{pq} \partial^q_{\boldsymbol{w}} - \boldsymbol{w}^p h_{p0} g_{00} \partial^0_{\boldsymbol{w}} + \sum_{i\neq 0} \boldsymbol{w}^i g_{i0} \boldsymbol{w}^p h_{p0} \left(-\tfrac{1}{\mu}\partial^0_{\boldsymbol{w}}\right) + \sum_{j\neq 0} \boldsymbol{w}^i g_{ij} h_{jq} \partial^q_{\boldsymbol{w}} \\
&\qquad - \sum_{j\neq 0} \boldsymbol{w}^p h_{p0} g_{0j} \partial^j_{\boldsymbol{w}} + \sum_{j\neq 0, i\neq 0} \boldsymbol{w}^i g_{ij} \boldsymbol{w}^p h_{p0} \left(-\tfrac{1}{\mu}\partial^j_{\boldsymbol{w}}\right) \\
&= \sum_{j\neq 0} \boldsymbol{w}^i g_{ij} h_{jq} \partial^q_{\boldsymbol{w}} - \boldsymbol{w}^p h_{p0} g_{0j} \partial^j_{\boldsymbol{w}} + \sum_{i\neq 0} \boldsymbol{w}^i g_{ij} \boldsymbol{w}^p h_{p0} \left(-\tfrac{1}{\mu}\partial^j_{\boldsymbol{w}}\right) + \sum_{p\neq 0} \boldsymbol{w}^i g_{i0} \left(-\tfrac{1}{\mu}\boldsymbol{w}^p\right) h_{pq} \partial^q_{\boldsymbol{w}}.
\end{aligned}$$

An almost the same calculation yields that

$$\begin{aligned}X_h(X_g) &= \boldsymbol{w}^p h_{pq}\partial^q_{\boldsymbol{w}}\left(\boldsymbol{w}^i g_{ij}\partial^j_{\boldsymbol{w}}\right)\\ &= \sum_{q\neq 0}\boldsymbol{w}^p h_{pq} g_{qj}\partial^j_{\boldsymbol{w}} - \boldsymbol{w}^i g_{i0}h_{0q}\partial^q_{\boldsymbol{w}} + \sum_{p\neq 0}\boldsymbol{w}^p h_{pq}\boldsymbol{w}^i g_{i0}\left(-\tfrac{1}{\mu}\partial^q_{\boldsymbol{w}}\right) + \sum_{i\neq 0}\boldsymbol{w}^p h_{p0}\left(-\tfrac{1}{\mu}\boldsymbol{w}^i\right) g_{ij}\partial^j_{\boldsymbol{w}}.\end{aligned}$$

Therefore, the commutator of vector fields X_g and X_h is equal to

$$\begin{aligned}[X_g, X_h] &= \boldsymbol{w}^i g_{ij}\partial^j_{\boldsymbol{w}}\left(\boldsymbol{w}^p h_{pq}\partial^q_{\boldsymbol{w}}\right) - \boldsymbol{w}^p h_{pq}\partial^q_{\boldsymbol{w}}\left(\boldsymbol{w}^i f_{ij}\partial^j_{\boldsymbol{w}}\right)\\ &= \sum_{j\neq 0}\boldsymbol{w}^i g_{ij}h_{jp}\partial^p_{\boldsymbol{w}} - \boldsymbol{w}^p h_{p0}g_{0j}\partial^j_{\boldsymbol{w}} - \sum_{q\neq 0}\boldsymbol{w}^p h_{pq}g_{qj}\partial^j_{\boldsymbol{w}} + \boldsymbol{w}^i g_{i0}h_{0q}\partial^q_{\boldsymbol{w}}\\ &= \boldsymbol{w}^i g_{ij}h_{jp}\partial^p_{\boldsymbol{w}} - \boldsymbol{w}^p h_{pq}g_{qj}\partial^j_{\boldsymbol{w}} = \boldsymbol{w}^i\left(g_{ij}h_{jq} - h_{ip}g_{pq}\right)\partial^q_{\boldsymbol{w}} = \boldsymbol{w}^i\left([g,h]\right)_{iq}\partial^q_{\boldsymbol{w}} = X_{[g,h]}.\end{aligned}$$

This proves the claim. □

Acknowledgments

The authors thank the Organising committee of the 7th International workshop 'Group Analysis of Differential Equations and Integrable Systems' (15–19 June 2014, Larnaca, Cyprus) for a warm atmosphere during the meeting. The authors thank M.A. Nesterenko, D.A. Leites, and A.V. Mikhailov for stimulating discussions and constructive criticism; the authors are grateful to the anonymous referee for remarks and advice. The research of A. V. K. was partially supported by JBI RUG project 103511 (Groningen); the research of A. O. K. was partially supported by JBI RUG project 135110 (Groningen).

References

[1] Miura R M 1968 Korteweg–de Vries equation and generalizations. I. A remarkable explicit nonlinear transformation *J. Math. Phys.* **9** 1202

[2] Miura R M, Gardner C S and Kruskal M D 1968 Korteweg–de Vries equation and generalizations. II. Existence of conservation laws and constants of motion *J. Math. Phys.* **9** 1204

[3] Magri F 1978 A simple model of the integrable Hamiltonian equation *J. Math. Phys.* **19** 1156

[4] Newell A C 1985 *Solitons in Mathematics and Physics* (CBMS-NSF Regional Conference Series in Applied Mathematics vol 48) (Philadelphia: Society for Industrial and Applied Mathematics (SIAM))

[5] Kersten P, Krasil'shchik I and Verbovetsky A 2004 Hamiltonian operators and ℓ^*-coverings *J. Geom. Phys.* **50** 273 (*Preprint* `math.DG/0304245`)

[6] Bocharov A V, Chetverikov V N, Duzhin S V, Khor'kova N G, Krasil'shchik I S, Samokhin A V, Torkhov Y N, Verbovetsky A M and Vinogradov A M 1999 *Symmetries and Conservation Laws for Differential Equations of Mathematical Physics* (Translations of Mathematical Monographs vol 182) (Providence: American Mathematical Society)

[7] Kiselev A V and Wolf T 2007 Classification of integrable super-systems using the SsTools environment *Comput. Phys. Comm.* **177** 315 (*Preprint* `nlin.SI/0609065`)

[8] Marvan M 2009 Sufficient set of integrability conditions of an orthonomic system *Found. Comput. Math.* **9** 651 (*Preprint* `nlin.SI/0605009`)

[9] Vinogradov A M 1984 The $\mathcal{C}$-spectral sequence, Lagrangian formalism, and conservation laws. I. The linear theory *J. Math. Anal. Appl.* **100** 1

[10] Vinogradov A M 1984 The $\mathcal{C}$-spectral sequence, Lagrangian formalism, and conservation laws. II. The nonlinear theory *J. Math. Anal. Appl.* **100** 41

[11] Lax P D 1968 Integrals of nonlinear equations of evolution and solitary waves *Comm. Pure Appl. Math.* **21** 467

[12] Faddeev L D and Takhtajan L A 1987 *Hamiltonian Methods in the Theory of Solitons* Springer Series in Soviet Mathematics (Berlin: Springer-Verlag)

[13] Olver P J 1993 *Applications of Lie Groups to Differential Equations* 2nd ed (Graduate Texts in Mathematics vol 107) (New York: Springer-Verlag)

[14] Kersten P, Krasil'shchik I S, Verbovetsky A M and Vitolo R 2009 *Differential Equations: Geometry, Symmetries and Integrability* (Abel Symp. vol 5) (Berlin: Springer) p 187 (*Preprint* `arXiv:0812.4895`)

[15] Krasil'shchik J and Verbovetsky A 2011 Geometry of jet spaces and integrable systems *J. Geom. Phys.* **61** 1633 (*Preprint* `arXiv:1002.0077`)
[16] Fordy A P 1983 Projective representations and deformations of integrable systems *Proc. Roy. Irish Acad. Sect.* A **83** 75
[17] Kupershmidt B A 1983 Deformations of integrable systems *Proc. Roy. Irish Acad. Sect.* A **83** 45
[18] Kiselev A V 2007 Algebraic properties of Gardner deformations of integrable systems *Theor. Math. Phys.* **152** 963 (*Preprint* `nlin.SI/0610072`)
[19] Labelle P and Mathieu P 1991 A new $N = 2$ supersymmetric Korteweg–de Vries equation *J. Math. Phys.* **32** 923
[20] Laberge C A and Mathieu P 1988 $N = 2$ superconformal algebra and integrable O(2) fermionic extensions of the Korteweg–de Vries equation *Phys. Lett.* B **215** 718
[21] Mathieu P 2001 *Bäcklund and Darboux Transformations. The Geometry of Solitons (Halifax, NS, 1999)* (CRM Proc. Lecture Notes vol 29) (Providence: Amer. Math. Soc.) p 325 (*Preprint* `math-ph/0005007`)
[22] Popowicz Z 1993 The Lax formulation of the "new" $N = 2$ SUSY KdV equation *Phys. Lett.* A **174** 411
[23] Kiselev A and Hussin V 2009 Hirota's virtual multisoliton solutions of $N = 2$ supersymmetric Korteweg–de Vries equations *Theor. Math. Phys.* **159** 833 (*Preprint* `arXiv:0810.0930`)
[24] Kersten P H M and Sorin A S 2002 Bi-Hamiltonian structure of the $N = 2$ supersymmetric $\alpha = 1$ KdV hierarchy *Phys. Lett.* A **300** 397 (*Preprint* `nlin.SI/0201061`)
[25] Kersten P and Krasil'shchik J 2002 *Lie Groups, Geometric Structures and Differential Equations – One Hundred Years after Sophus Lie (Kyoto/Nara, 1999)* (Adv. Stud. Pure Math. vol 37) (Tokyo: Math. Soc. Japan) p 151
[26] Sakovich S Y 2004 Cyclic bases of zero-curvature representations: five illustrations to one concept *Acta Appl. Math.* **83** 69 (*Preprint* `nlin/0212019`)
[27] Karasu-Kalkanlı A, Sakovich S Y and Yurduşen I 2003 Integrability of Kerstein–Krasil'shchik coupled KdV-mKdV equations: singularity analysis and Lax pair *J. Math. Phys.* **44** 1703
[28] Kiselev A V and Krutov A O 2012 Gardner's deformations of the graded Korteweg–de Vries equations revisited *J. Math. Phys.* **53** 103511 (*Preprint* `arXiv:1108.2211`)
[29] Kiselev A V and Krutov A O 2014 On the (non)removability of spectral parameters in $\mathbb{Z}_2$-graded zero-curvature representations and its applications *Preprint* `arXiv:1301.7143v3`
[30] Roelofs M 1993 Ph.D. thesis *Prolongation Structures of Supersymmetric Systems* (Enschede: University of Twente)
[31] Kiselev A V 2012 The twelve lectures in the (non)commutative geometry of differential equations *Preprint* IHÉS/M/12/13 (Bures-sur-Yvette, France) URL `http://preprints.ihes.fr/2012/M/M-12-13.pdf`
[32] Marvan M 2002 On the horizontal gauge cohomology and nonremovability of the spectral parameter *Acta Appl. Math.* **72** 51
[33] Goldschmidt H 1967 Integrability criteria for systems of nonlinear partial differential equations *J. Differential Geometry* **1** 269
[34] Marvan M 2010 On the spectral parameter problem *Acta Appl. Math.* **109** 239 (*Preprint* `nlin.SI/0804.2031`)
[35] Zakharov V and Shabat A 1979 Integration of nonlinear equations of mathematical physics by the method of inverse scattering. II *Functional Analysis and Its Applications* **13** 166
[36] Krasil'shchik I S and Vinogradov A M 1989 Nonlocal trends in the geometry of differential equations: symmetries, conservation laws, and Bäcklund transformations *Acta Appl. Math.* **15** 161
[37] Baran H and Marvan M 2008 A conjecture on nonlocal terms of recursion operators *Journal of Mathematical Sciences* **151** 3083
[38] Igonin S, Kersten P H M and Krasil'shchik I 2003 On symmetries and cohomological invariants of equations possessing flat representations *Diff. Geom. Appl.* **19** 319 (*Preprint* `math.DG/0301344`)
[39] Popovych R O, Boyko V M, Nesterenko M O and Lutfullin M W 2003 Realizations of real low-dimensional Lie algebras *J. Phys.* A **36** 7337 (*Preprint* `math-ph/0301029`)
[40] Shchepochkina I M 2006 How to realize a Lie algebra by vector fields *Theor. Math. Phys.* **147** 821
[41] Hussin V, Kiselev A V, Krutov A O and Wolf T 2010 $N = 2$ supersymmetric $a = 4$-Korteweg–de Vries hierarchy derived via Gardner's deformation of Kaup–Boussinesq equation *J. Math. Phys.* **51** 083507 (*Preprint* `arXiv:0911.2681`)
[42] Krutov A O 2014 Ph.D. thesis *Deformations of Equations and Structures in Nonlinear Problems of Mathematical Physics* (Groningen: University of Groningen, JBI)
[43] Kupershmidt B A 1981 On the nature of the Gardner transformation *J. Math. Phys.* **22** 449
[44] Kiselev A V and Krutov A O 2014 Gardner's deformations as generators of new integrable systems *J. Phys.: Conf. Ser.* **482** 012021 (*Preprint* `arXiv:1312.6941`)
[45] Kaup D J and Newell A C 1978 An exact solution for a derivative nonlinear Schrödinger equation *J. Math. Phys.* **19** 798

Equivalence groupoid for (1+2)-dimensional linear Schrödinger equations with complex potentials

Célestin Kurujyibwami

College of Science and Technology, University of Rwanda, P.O. Box: 3900, Kigali, Rwanda
Linköping University, 58183 Linköping, Sweden

E-mail: celeku@yahoo.fr

Abstract. We describe admissible point transformations in the class of (1+2)-dimensional linear Schrödinger equations with complex potentials. We prove that any point transformation connecting two equations from this class is the composition of a linear superposition transformation of the corresponding initial equation and an equivalence transformation of the class. This shows that the class under study is semi-normalized.

1. Introduction

In the theory of group classification of differential equations, the *equivalence groupoid* of a class of differential equations is the set of *admissible transformations* of this class together with the composition operation of admissible transformations [1]. More specifically, an admissible (point) transformation of the class consists of a pair of similar equations (an initial and a target ones) from the class and a point transformation between these equations. The composition of two admissible transformations is well defined whenever the target equation of the first admissible transformation coincides with the initial equation of the second admissible transformation. For each equation from the class, there exists the admissible identity transformation. Each admissible transformation is invertible. The associativity of the composition of admissible transformations directly follows from those of point transformations.

The study of admissible transformations was initiated by Kingston and Sophocleous [2], who called an unformalized version of such transformations *form-preserving* [3]. Form-preserving transformations under the name *allowed transformations* also arose in the course of symmetry analysis of variable-coefficient Korteweg–de Vries equations [4] and variable-coefficient (1+1)-dimensional cubic Schrödinger equations [5]. Later on, formalizing the framework of admissible transformations was initiated [6] and the range of applicability of admissible transformations was extended to various classes of differential equations being important for applications, including nonlinear Schrödinger equations [7], variable-coefficient reaction–diffusion equations [8, 9, 10], eddy vorticity flux parameterizations of the inviscid barotropic vorticity equation [11] and nonlinear wave equations from the theory of elasticity [1].

In order to rigorously pose the problem under consideration, we need to precisely define the notions of classes of differential equations and their equivalence groupoids. Other notions and definitions related to classes of differential equations can be found, e.g., in [1, 7]. Consider a system of differential equations $\mathcal{L}_\theta$: $L(x, u_{(p)}, \theta_{(q)}(x, u_{(p)})) = 0$, parameterized by the tuple of arbitrary elements $\theta(x, u_{(p)}) = (\theta^1(x, u_{(p)}), \dots, \theta^k(x, u_{(p)}))$, where $x = (x_1, \dots, x_n)$ is the tuple of independent variables and $u_{(p)}$ is the set of the dependent variables $u = (u^1, \dots, u^m)$ together with all derivatives of u with respect to x up to order p. The symbol $\theta_{(q)}$ stands for the set of derivatives of θ of order not greater than q with respect to the variables x and $u_{(p)}$. The tuple of arbitrary elements θ runs through the set $\mathcal{S}$ of solutions of an auxiliary system of differential equations $S(x, u_{(p)}, \theta_{q'}(x, u_{(p)})) = 0$ and differential inequalities, like $\Sigma(x, u_{(p)}, \theta_{(q')}(x, u_{(p)})) \neq 0$, in which both x and $u_{(p)}$ play the role of independent variables and S and Σ are tuples of smooth functions depending on x, $u_{(p)}$ and $\theta_{(q')}$. Then the set $\{\mathcal{L}_\theta \mid \theta \in \mathcal{S}\} := \mathcal{L}|_{\mathcal{S}}$ is called a *class of differential equations* defined by the parameterized form $\mathcal{L}_\theta$ and the set $\mathcal{S}$ run by the arbitrary elements θ. Denote by $\mathrm{T}(\theta, \tilde{\theta})$ with $\theta, \tilde{\theta} \in \mathcal{S}$ the set of point transformations in the space of the variables (x, u) that map the system $\mathcal{L}_\theta$ to the system $\mathcal{L}_{\tilde{\theta}}$. A triple constituted by two arbitrary elements $\theta, \tilde{\theta} \in \mathcal{S}$ with $\mathrm{T}(\theta, \tilde{\theta}) \neq \varnothing$ and a point transformation $\varphi \in \mathrm{T}(\theta, \tilde{\theta})$ is called an *admissible transformation* of the class $\mathcal{L}|_{\mathcal{S}}$. The set of admissible transformations of the class $\mathcal{L}|_{\mathcal{S}}$ equipped by the natural composition of such transformations as an algebraic operation is called the *equivalence groupoid* of this class.

The notion of equivalence group originated from the work of Ovsiannikov [12] and became a powerful tool which the group classification of differential equations relies on. Thus, the use of equivalence groups constitutes a basis for group classification. Here and in what follows, we are concerned with usual equivalence groups. This notion can be generalized in several ways [6, 7, 13]. The *(usual) equivalence (pseudo)group* $G^\sim = G^\sim(\mathcal{L}|_{\mathcal{S}})$ of the class $\mathcal{L}|_{\mathcal{S}}$ consists of point transformations in the space of independent and dependent variables and arbitrary elements each of which is projectable to the space of $(x, u_{(p')})$ for any $0 \leqslant p' \leqslant p$, is consistent with the contact structure on the space of $(x, u_{(p)})$, and maps every system from the class $\mathcal{L}|_{\mathcal{S}}$ to another system from the same class. Its elements are called *equivalence transformations* [7].

Knowing the equivalence groupoid of a class of differential equations simplifies the process of classifying Lie symmetry extensions within this class. In particular, this gives an easy way of finding the corresponding equivalence group. The presence of a nice relation between the equivalence groupoid and the equivalence group makes the entire procedure of symmetry classification less cumbersome and more harmonious and allows for presenting a final classification list in a compact explicit form. Moreover, properties of the equivalence groupoid influences the choice of methods for group classification.

The class $\mathcal{L}|_{\mathcal{S}}$ is *normalized (in the usual sense)* if its equivalence groupoid is generated by its (usual) equivalence group and is said to be *semi-normalized* if its equivalence groupoid is generated by the transformations from the equivalence group simultaneously with transformations from symmetry groups of initial or transformed systems [7].

The algebraic method is the best choice for group classification of normalized or specifically semi-normalized classes of differential equations. A number of classes of differential equations that are important for applications are normalized or specifically semi-normalized or can be partitioned into normalized classes and hence were classified within the framework of the algebraic method. See, e.g., [1, 7, 11, 14, 15, 16].

The equivalence groupoid of the class of linear Schrödinger equations with complex potentials has been computed in [16] in the (1+1)-dimensional case. Then properties of the groupoid (more precisely, semi-normalization of the class) was taken as a base for carrying out the group classification of these equations by the algebraic method similarly to [15], where Lie symmetries of nth order ($n \geqslant 2$) of linear ordinary differential equations were classified using semi-normalization of the class of such equations. The final classification result, which is a complete list of inequivalent potentials corresponding to equations with nontrivial Lie symmetries, may be used in quantum theory, quantum field theory, optics and other branches of physics, cf. [17, 18, 19, 20].

The aim of this paper is to study the equivalence groupoid, the equivalence group and normalization properties of the class $\mathcal{F}$ of (1+2)-dimensional linear Schrödinger equations with complex potentials,

$$i\psi_t + \psi_{aa} + V(t, x)\psi = 0. \tag{1}$$

Here and in what follows t, x_1 and x_2 are the real independent variables, $x = (x_1, x_2)$, ψ is the complex dependent variable and V is an arbitrary smooth complex-valued potential depending on t and x. Subscripts of functions denote differentiation with respect to the corresponding variables. In particular, $\psi_t = \partial\psi/\partial t$ and $\psi_{ab} = \partial^2\psi/\partial x_a\partial x_b$. The indices a, b, c, and d run from 1 to 2, and we use summation convention over repeated indices.

The structure of this paper is organized as follows: In Section 2 we compute the equivalence groupoid of the class $\mathcal{F}$. It appears that the transformational parts of admissible transformations of $\mathcal{F}$ are uniformly parameterized by the corresponding initial values of V. The expression for target values of the arbitrary element V in terms of its initial values and transformation parameters is also found. Then the equivalence group of the class $\mathcal{F}$ and normalization properties of this class are described in Section 3. The last section includes a short summary and suggests a direction for the future work.

2. Equivalence groupoid

We find the equivalence groupoid $\mathcal{G}^\sim$ of the class $\mathcal{F}$ by using the direct method. We seek for all invertible point transformations of the form

$$\tilde t = T(t,x,\psi,\psi^*),\quad \tilde x_a = X^a(t,x,\psi,\psi^*),\quad \tilde\psi = \Psi(t,x,\psi,\psi^*),\quad \tilde\psi^* = \Psi^*(t,x,\psi,\psi^*) \tag{2}$$

with $J = \partial(T,X^a,\Psi,\Psi^*)/\partial(t,x,\psi,\psi^*) \neq 0$ that map a fixed equation from the class $\mathcal{F}$ to another equation of the same class,

$$i\tilde\psi_{\tilde t} + \tilde\psi_{\tilde x_a\tilde x_a} + \tilde V(\tilde t,\tilde x)\tilde\psi = 0. \tag{3}$$

Hereafter in case of a complex value β we use the notation

$$\hat\beta = \beta \quad\text{if}\quad T_t > 0 \quad\text{and}\quad \hat\beta = \beta^* \quad\text{if}\quad T_t < 0.$$

Lemma 1. *Any point transformation $\mathcal{T}$ connecting two equations from the class $\mathcal{F}$ satisfies the conditions*

$$\begin{gathered} T_x = T_\psi = T_{\psi^*} = 0,\quad X^a_\psi = X^a_{\psi^*} = 0,\\ \Psi_\psi = 0 \quad \text{if}\quad T_t < 0 \quad\text{and}\quad \Psi_{\psi^*} = 0 \quad\text{if}\quad T_t > 0. \end{gathered} \tag{4}$$

Proof. The proof is similar to that given in [7]. □

Theorem 1. *The equivalence groupoid $\mathcal{G}^\sim$ of the class $\mathcal{F}$ consists of the triples of the form $(V,\tilde V,\mathcal{T})$, where $\mathcal{T}$ is a point transformation in the space of variables, whose components are*

$$\tilde t = T,\quad \tilde x_a = |T_t|^{1/2}O^{ab}x_b + \mathcal{X}^a, \tag{5a}$$

$$\tilde\psi = \exp\left(\frac{i}{8}\frac{T_{tt}}{|T_t|}x_ax_a + \frac{i}{2}\frac{\varepsilon'\mathcal{X}^b_t}{|T_t|^{1/2}}O^{ba}x_a + \Lambda + i\Sigma\right)(\hat\psi + \hat\Omega), \tag{5b}$$

and the transformed potential $\tilde V$ is expressed via V as

$$\tilde V = \frac{\hat V}{|T_t|} + \frac{2T_{ttt}T_t - 3T_{tt}{}^2}{16\varepsilon' T_t{}^3}x_ax_a + \frac{\varepsilon'}{2|T_t|^{1/2}}\left(\frac{\mathcal{X}^a_t}{T_t}\right)_t O^{ba}x_a + \frac{\Sigma_t - i\Lambda_t}{T_t} - \frac{\mathcal{X}^a_t\mathcal{X}^a_t + iT_{tt}}{4T_t{}^2}. \tag{5c}$$

Here T, $\mathcal{X}^a$, Λ and Σ are arbitrary smooth real-valued functions of t with $T_t \neq 0$, $\varepsilon' = \operatorname{sgn} T_t$, $\Omega = \Omega(t,x)$ is an arbitrary solution of the initial equation and $O = (O^{ab})$ is an arbitrary constant 2×2 orthogonal matrix.

Proof. Let a point transformation $\mathcal{T}$ connect two equations from the class $\mathcal{F}$. Lemma 1 implies that $T = T(t)$ with $T_t \neq 0$, $X^a = X^a(t,x)$ and $\Psi_{\hat\psi} \neq 0$. Applying the chain rule for total derivatives with respect to t and x to the equality $\tilde\psi(\tilde t,\tilde x) = \Psi(t,x,\psi)$, we derive

$$\begin{gathered} D_t\tilde\psi(\tilde t,\tilde x) = \tilde\psi_{\tilde t}T_t + \tilde\psi_{\tilde x_b}X^b_t = D_t\Psi,\quad D_a\tilde\psi(\tilde t,\tilde x) = \tilde\psi_{\tilde x_c}X^c_a = D_a\Psi,\\ D_bD_a\tilde\psi(\tilde t,\tilde x) = \tilde\psi_{\tilde x_c\tilde x_d}X^c_bX^d_a + \tilde\psi_{\tilde x_d}X^d_{ab} = D_bD_a\Psi, \end{gathered}$$

where D_t and D_a are the total derivative operators with respect to t and x_a, respectively. The above equations are equivalent to

$$\tilde{\psi}_{\tilde{t}} = \frac{1}{T_t}\left(D_t\Psi - Y^a_b X^b_t D_a\Psi\right), \quad \tilde{\psi}_{\tilde{x}_c} = Y^a_c D_a\Psi, \tag{6}$$

$$\tilde{\psi}_{\tilde{x}_c\tilde{x}_d} = Y^a_c Y^b_d (D_b D_a\Psi - Y^d_c X^c_{ab} D_d\Psi), \tag{7}$$

where $Y^a_c X^c_b = \delta^a_b$ and δ^a_b is the Kronecker delta. In fact, the vector-function (Y^1, Y^2) is the inverse of the vector-function (X^1, X^2) with respect to x and $Y^a_c = \partial Y^a/\partial\tilde{x}_c$. We substitute the values of $\tilde{\psi}_{\tilde{t}}$ and $\tilde{\psi}_{\tilde{x}_c\tilde{x}_d}$ defined in (6) and (7) into the equation (3) and take into account the expression for $\hat{\psi}_t$, $\hat{\psi}_t = i\varepsilon'(\hat{\psi}_{aa} + \hat{V}\hat{\psi})$, where $\varepsilon' = \operatorname{sgn} T_t$. As a result, we derive the equation

$$\frac{i}{T_t}\left(\Psi_t + \Psi_{\hat{\psi}}(i\varepsilon'\hat{\psi}_{aa} + i\varepsilon'\hat{V}\hat{\psi}) - Y^a_b(\Psi_a + \Psi_{\hat{\psi}}\hat{\psi}_a)X^b_t\right) + Y^a_c Y^b_c\left(\Psi_{ab} + \Psi_{a\hat{\psi}}\hat{\psi}_b + \Psi_{\hat{\psi}b}\hat{\psi}_a\right)$$
$$+Y^a_c Y^b_c\left(\Psi_{\hat{\psi}\hat{\psi}}\hat{\psi}_b\hat{\psi}_a + \Psi_{\hat{\psi}}\hat{\psi}_{ab} - Y^d_c(\Psi_d + \Psi_{\hat{\psi}}\hat{\psi}_d)X^c_{ab}\right) + \tilde{V}\Psi = 0.$$

Then splitting this equation with respect to various derivatives of $\hat{\psi}$ and additionally arranging leads to the system

$$Y^a_c Y^b_c = 0,\ a \neq b, \quad Y^a_c Y^a_c = \frac{1}{|T_t|}, \quad \Psi_{\hat{\psi}\hat{\psi}} = 0, \tag{8}$$

$$\frac{2}{|T_t|}\Psi_{a\hat{\psi}} - \frac{i}{T_t}Y^a_b\Psi_{\hat{\psi}}X^b_t - \frac{1}{|T_t|}Y^a_c\Psi_{\hat{\psi}}X^c_{aa} = 0, \tag{9}$$

$$\frac{i}{T_t}\Psi_t - \frac{1}{|T_t|}\hat{V}\Psi_{\hat{\psi}}\hat{\psi} - \frac{i}{T_t}Y^a_b\Psi_a X^b_t + \frac{1}{|T_t|}\Psi_{aa} - Y^d_c\Psi_d X^c_{aa} + \tilde{V}\Psi = 0. \tag{10}$$

We first show that $X^c_{ab} = 0$ for all a,b and c. The first two equations in (8) together with the condition $Y^a_c X^c_b = \delta^a_b$ imply that $X^b_a = |T_t|Y^a_b$. Therefore, $X^c_a X^c_b = |T_t|\delta^a_b$, i.e. $X^c_a = |T_t|^{1/2}O^{ca}$, where $O = (O^{ca})$ is a 2×2 orthogonal matrix-function of t and x. Suppose that O is a special orthogonal matrix, i.e. $\det O = 1$. Then the matrix (X^c_a) can be written as

$$(X^c_a) = |T_t|^{1/2}\begin{pmatrix}\cos\theta & -\sin\theta\\ \sin\theta & \cos\theta\end{pmatrix} \tag{11}$$

for a smooth function $\theta = \theta(t,x)$. The representation (11) implies that $X^1_1 = X^2_2 = |T_t|^{1/2}\cos\theta$ and $X^1_2 = -X^2_1 = -|T_t|^{1/2}\sin\theta$, which means that transformation components X^1 and X^1 satisfy the Cauchy–Riemann system

$$X^1_1 = X^2_2, \quad X^1_2 = -X^2_1.$$

Hence both the components X^c are solution of the Laplace equation, i.e. $X^c_{bb} = 0$. Applying the Laplace operator ∂_{bb} to both sides of the equation $X^c_a X^c_a = |T_t|$, where there is no summation with respect to a, we derive $X^c_{ab}X^c_{ab} = 0$, i.e. $X^c_{ab} = 0$ for all a, b and c. The same result is obtained when $\det O = -1$. Thus, X^c is a linear function in x, which implies that the matrix O may depend only on t. From the previous consideration, the equation (9) becomes

$$\Psi_{a\hat{\psi}} = \frac{i}{2T_t}X^b_a X^b_t \Psi_{\hat{\psi}}. \tag{12}$$

Differentiating this equation for $a = 1$ with respect to x_2 and for $a = 2$ with respect to x_1, subtracting the first result to the second and taking into account the coincidence of mixed derivatives, we find the compatibility condition $X^b_1 X^b_{t2} = X^b_2 X^b_{t1}$. The representation (11) together with the compatibility condition results in $\theta_t = 0$. Therefore, the matrix O is a constant orthogonal matrix.

The general solution of the third equation in (8) is $\Psi = \Psi^1\hat{\psi}+\Psi^0$, where Ψ^0 and Ψ^1 are smooth complex-valued functions of t and x, and $\Psi^1 = \Psi_{\hat{\psi}} \neq 0$. The integration of the equation (12) gives the expression of Ψ^1,

$$\Psi^1 = \exp\left(\frac{i}{8}\frac{T_{tt}}{|T_t|}x_ax_a + \frac{i}{2}\frac{\varepsilon'\mathcal{X}_t^b}{|T_t|^{1/2}}O^{ba}x_a + \Lambda + i\Sigma\right), \tag{13}$$

where Λ and Σ are arbitrary smooth real-valued function of t arising in the course of the integration. Finally, we consider the equation (10), which reduces, under the derived conditions, to

$$\frac{i}{T_t}\Psi_t - \frac{1}{|T_t|}\hat{V}\Psi_{\hat{\psi}}\hat{\psi} - \frac{i}{T_t}\frac{X_a^b}{|T_t|}X_t^b\Psi_a + \frac{1}{|T_t|}\Psi_{aa} + \tilde{V}\Psi = 0.$$

Splitting with respect to $\hat{\psi}$ in view of the representation for Ψ and rearranging, we obtain

$$\tilde{V} = \frac{\hat{V}}{|T_t|} - \frac{i}{T_t\Psi^1}\left(\Psi_t^1 - \frac{X_a^bX_t^b}{|T_t|}\Psi_a^1\right) - \frac{1}{|T_t|}\frac{\Psi_{aa}^1}{\Psi^1}, \tag{14}$$

$$i\varepsilon'\Psi_t^0 - \frac{i}{T_t}X_a^bX_t^b\Psi_a^0 + \Psi_{aa}^0 + |T_t|\tilde{V}\Psi^0 = 0. \tag{15}$$

Let us introduce the function $\Omega = \hat{\Psi}^0/\hat{\Psi}^1$, i.e. $\Psi^0 = \Psi^1\hat{\Omega}$. The equation (15) is equivalent to the initial linear Schrödinger equation in terms of Ω. After the substitution of Ψ^1 by its expression from (13) into (14) and then additionally collecting coefficients of x, we derive the final expression for $\tilde{V}$. □

3. Equivalence group and normalization properties

There are several ways of computing the equivalence group for a given class of differential equations. Some of them are the direct or the infinitesimal method. At the same time, it is unnecessary to do the computation of the equivalence group $G^\sim$ of the class $\mathcal{F}$ using one of the above alternatives since we can derive the equivalence group $G^\sim$ of the class $\mathcal{F}$ from the knowledge of its equivalence groupoid $\mathcal{G}^\sim$.

Corollary 1. *The (usual) equivalence group $G^\sim$ of the class $\mathcal{F}$ consists of point transformations in the space of independent and dependent variables and arbitrary element that are of the form* (5) *with $\Omega = 0$.*

Proof. Let $\mathcal{T}$ be a point transformation connecting two equations from the class $\mathcal{F}$. Then $\mathcal{T}$ is necessarily of the form (5a)–(5b), and potentials of the equations are related by (5c). Any transformation from the group $G^\sim$ generates a family of admissible transformations of the class $\mathcal{F}$ and hence it has the form (5). On the other hand, from the definition of equivalence group, the group $G^\sim$ contains only point transformations whose components corresponding to variables do not depend on the arbitrary element V. This condition is satisfied if and only if Ω is a common solution for all equations from the class $\mathcal{F}$. The only common solution is $\Omega = 0$. □

Remark 1. In fact, the whole equivalence group $G^\sim$ is generated by the continuous family of transformations of the form (5), where $\Omega = 0$, $T_t > 0$ and $\det O = 1$, and two discrete transformations: the space reflection I_a for a fixed a ($\tilde{t} = t$, $\tilde{x}_a = -x_a$, $\tilde{x}_b = x_b$, $b \neq a$, $\tilde{\psi} = \psi$, $\tilde{V} = V$) and the Wigner time reflection I_t ($\tilde{t} = -t$, $\tilde{x} = x$, $\tilde{\psi} = \psi^*$, $\tilde{V} = V^*$).

Summing up, we state the following

Corollary 2. *The class $\mathcal{F}$ is semi-normalized. More precisely, for each admissible transformation $(V, \tilde{V}, \mathcal{T})$ in the class $\mathcal{F}$, its transformational part $\mathcal{T}$ is the composition of a linear superposition transformation $\mathcal{T}_1$ of the initial equation with the potential V and the projection of an equivalence transformation $\mathcal{T}_2$ of the class $\mathcal{F}$ to the space variables with $\mathcal{T}_2V = \tilde{V}$.*

Proof. Consider two fixed similar equations from the class $\mathcal{F}$ with potentials V and $\tilde{V}$ and let $\mathcal{T}$ be a point transformation connecting these equations. From Theorem 1, the transformation $\mathcal{T}$ is of the form (5a)–(5b), and the potentials V and $\tilde{V}$ are related by (5c). We define two point transformations. The first transformation $\mathcal{T}^1$ is the point transformation in the variable space with the components $\tilde{t} = t$, $\tilde{x} = x$, $\tilde{\psi} = \psi + \Omega$ with the same Ω as in $\mathcal{T}$. It does not change the potential V and, therefore, is a point symmetry transformation of linear superposition for the initial equation. The second transformation $\mathcal{T}^2$ is the point transformation in the extended space of variables and the potential V that is of the form (5) with the same values of parameters as in $\mathcal{T}$ except $\Omega = 0$. Hence it is an equivalence transformation, which connects the equations with the potentials V and $\tilde{V}$, $\tilde{V} = \mathcal{T}_2 V$. As a result, the transformation $\mathcal{T}$ coincides with the composition of $\mathcal{T}^1$ and the projection $\mathcal{T}^2|_{(t,x,\psi)}$ of $\mathcal{T}^2$, $\mathcal{T} = \mathcal{T}^2|_{(t,x,\psi)} \circ \mathcal{T}^1$. □

4. Conclusion

The equivalence groupoid of the class $\mathcal{F}$ of (1+2)-dimensional linear Schrödinger equations with complex potentials, which has an interesting algebraic structure, is exhaustively described in Theorem 1 using the direct method. The method is standard for the study of point transformations between differential equations. At the same time, it is not quite algorithmic, especially when considering the entire set of admissible transformations of a class of differential equations. The versions of the direct method for finding point symmetry transformations of a single differential equation or equivalence transformations of a class of differential equations are much easier to be realized. Moreover, the computations become trickier and more cumbersome if we increase the dimension of equations. New features also appear in the form of admissible transformations in comparison with the (1+1)-dimensional case. They are not exhausted by the formal extension of the set of space variables. Coupling of space variables leads, in particular, to involving rotations in the corresponding equivalence groupoid. In total, the above makes the entire consideration much more complicated than in (1+1)-dimensions.

Knowing the equivalence groupoid of the class $\mathcal{F}$, we construct the (usual) equivalence group of $\mathcal{F}$ in an easy way, roughly speaking, by singling out families of admissible transformations that are pointwise parameterized by the arbitrary element, which is the potential V for the class $\mathcal{F}$, and having the same transformational part. We relate the equivalence groupoid, the equivalence group and normalization properties of the class $\mathcal{F}$, and this relation is similar to that for the (1+1)-dimensional counterpart of the class $\mathcal{F}$, which is studied in [16]. We show that, roughly speaking, any point transformation connecting two fixed equations from the class $\mathcal{F}$ is the composition of a linear superposition symmetry transformation of the initial equation and an equivalence transformation of this class. In other words, the class $\mathcal{F}$ is semi-normalized in a quite specific way, which guarantees, in view of our experience with the (1+1)-dimensional counterpart of the class $\mathcal{F}$ [16], the effective use of the algebraic method for the exhaustive group classification of the class $\mathcal{F}$. Therefore, the results of the present paper can be considered as a first step in our future work on this classification.

Acknowledgments

The author thanks Prof Peter Basarab-Horwath and Prof Roman Popovych for their help during the preparation of this paper. The research was supported by International Science Programme (ISP) in collaboration with East African Universities Mathematics Programme (EAUMP).

References

[1] Bihlo A, Dos Santos Cardoso-Bihlo E and Popovych R O 2012 Complete group classification of a class of nonlinear wave equations *J. Math. Phys.* **53** 123515 (*Preprint* arXiv:1106.4801)

[2] Kingston J G and Sophocleous C 1991 On point transformations of a generalized Burgers equation *Phys. Lett.* A **155** 15

[3] Kingston J G and Sophocleous C 1998 On form-preserving point transformations of partial differential equations *J. Phys.* A **31** 1597

[4] Winternitz P and Gazeau J P 1992 Allowed transformations and symmetry classes of variable coefficient Korteweg–de Vries equations *Phys. Lett.* A **167** 246

[5] Gagnon L and Winternitz P 1993 Symmetry classes of variable coefficient nonlinear Schrödinger equations *J. Phys.* A **26** 7061
[6] Popovych R O 2006 Classification of admissible transformations of differential equations *Collection of Works of Institute of Mathematics* vol 3 (Kyiv: Institute of Mathematics) p 239
[7] Popovych R O, Kunzinger M and Eshraghi H 2010 Admissible transformations and normalized classes of nonlinear Schrödinger equations *Acta Appl. Math.* **109** 315 (*Preprint* arXiv:math-ph/0611061)
[8] Vaneeva O O, Johnpillai A G, Popovych R O and Sophocleous C 2007 Enhanced group analysis and conservation laws of variable coefficient reaction–diffusion equations with power nonlinearities *J. Math. Anal. Appl.* **330** 1363 (*Preprint* arXiv:math-ph/0605081)
[9] Vaneeva O O, Popovych R O and Sophocleous C 2009 Enhanced group analysis and exact solutions of variable coefficient semilinear diffusion equations with a power source *Acta Appl. Math.* **106** 1 (*Preprint* arXiv:0708.3457)
[10] Vaneeva O O, Popovych R O and Sophocleous C 2012 Extended group analysis of variable coefficient reaction–diffusion equations with exponential nonlinearities *J. Math. Anal. Appl.* **396** 225 (*Preprint* arXiv:1111.5198)
[11] Popovych R O and Bihlo A 2012 Symmetry preserving parameterization schemes *J. Math. Phys.* **53** 073102 (*Preprint* arXiv:1010.3010)
[12] Ovsiannikov L V 1982 *Group Analysis of Differential Equations* (New York: Academic Press)
[13] Ivanova N M, Popovych R O and Sophocleous C 2010 Group analysis of variable coefficient diffusion-convection equations. I. Enhanced group classification *Lobachevskii J. Math.* **31** 100 (*Preprint* arXiv:0710.2731)
[14] Basarab-Horwath P, Lahno V and Zhadanov R 2001 The structure of Lie algebras and the classification problem for partial differential equations *Acta Appl. Math.* **69** 43
[15] Boyko V M, Popovych R O and Shapoval N M 2014 Equivalence groupoids of classes of linear ordinary differential equations and their group classification *Preprint* arXiv:1403.6062
[16] Kurujyibwami C, Basarab-Horwath P and Popovych R O 2014 Algebraic method for group classification of (1+1)-dimensional linear Schrödinger equations (in preparation)
[17] Bender C M, Brody D C and Jones H F 2002 Complex extension of quantum mechanics *Phys. Rev. Lett.* **89** 270401
[18] Bender C M 2007 Making sense of non-Hermitian Hamiltonians *Rep. Prog. Phys.* **70** 947
[19] Fernandez F M, Guardiola R, Ros J and Znojil M 1999 A family of complex potentials with real spectrum *J. Phys.* A **32** 3105
[20] Mostafazadeh A 2014 A dynamical formulation of one-dimensional scattering theory and its applications in optics *Ann. Phys.* **341** 77

The Poincaré algebras $p(1,1)$ and $p(1,2)$: realizations and deformations

Maryna Nesterenko
Institute of Mathematics of NAS of Ukraine, 3 Tereshchenkivska Str., Kyiv 01601, Ukraine
E-mail: maryna@imath.kiev.ua

Abstract. All inequivalent realizations of the Poincaré algebras $p(1,1)$ and $p(1,2)$ acting in spaces of not more than three variables are constructed. First-order deformations of $p(1,1)$ and $p(1,2)$ are proposed and the generic realizations for the initial and deformed Poincaré algebras are presented.

1. Introduction

Realizations of Lie algebras by vector fields are widely applicable in mathematics and physics, indeed, representations by first-order differential operators are effectively used in Lie group theory of differential equations: reduction, integration, calculation of differential invariants, etc.; in group classification of partial differential equations; in the theory of differential invariants; in general relativity and other physical problems such as classification of gravity fields of a general form under the motion groups and groups of conformal transformations, or quantization based on Noether symmetries, see also [1–5] and [6] for review.

In this work we apply an algebraic approach proposed in [7] and the classification of subalgebras obtained in [8] for the explicit construction of vector fields of the lowest Poincaré algebra $p(1,1)$, the six-dimensional Poincaré algebra $p(1,2)$ and their deformations. Realizations of different Poincaré algebras were widely studied (see, e.g., [9–13]), but only covariant realizations were considered and a number of realizations are not represented explicitly as far as some of the coefficients require the solution of partial differential equations, that can not be presented in quadratures. Along with the realizations we also propose three one-parametric deformations of $p(1,1)$ and $p(1,2)$ and construct their generic realizations by the same method. The results of this paper is a part of the forthcoming general classification of the realizations of Poincaré algebras $p(1,2)$ and $p(1,3)$ [14].

Let $\mathfrak{g}$ be an n-dimensional Lie algebra over a field $\mathbb{R}$ or $\mathbb{C}$. We denote an open subset of $\mathbb{R}^m$ as M and the Lie algebra of vector fields on it as $\mathrm{Vect}(M)$, the vector fields are taken in a form of linear first-order differential operators with analytical coefficients and the Lie product of vector fields is given by their commutator. The group of all automorphisms of $\mathfrak{g}$ is denoted by $\mathrm{Aut}(\mathfrak{g})$ and the notion of realization of a Lie algebra is defined as follows (see [6] for details).

A *realization* of a Lie algebra $\mathfrak{g}$ in vector fields on M is a homomorphism $R\colon \mathfrak{g} \to \mathrm{Vect}(M)$. The realization is *faithful* if $\ker R = \{0\}$ and *unfaithful* otherwise. If the action of the local Lie group that corresponds to a fixed realization is transitive on M, then this realization is called *generic*.

Realizations of Lie algebras are constructed up to Aut($\mathfrak{g}$)-equivalence and nondegenerate change of coordinates on M. Note that for any generic realization the dimension of M coincides with the dimension of $\mathfrak{g}$ that is $m = n$.

Below we propose the practical scheme for construction of left-invariant vector fields extracted from [7]. The coefficients $\xi_k^i(x)$ of the generic realization $\Xi_i = \sum_{k=1}^{n} \xi_k^i(x)\frac{\partial}{\partial x_k}$, $i = 1, 2, \ldots, n$, can be recovered from the left-invariant differential one-forms $\Omega^i = \sum_{l=1}^{n} \omega_i^l(x)\mathrm{d}x_l$ using the duality of the left-invariant vector fields and differential one-forms $\omega_i^l(x)\xi_k^i(x) = \delta_k^l$ and the coefficients $\omega_i^l(x)$ of the differential one-forms are constructed as follows:

$$\omega_i^l(x) = \left(A^{(1)}(x^1)A^{(2)}(x^2)\cdots A^{(i-1)}(x^{i-1})\right)_i^l, \quad i = 2, 3, \ldots, n, \ l = 1, 2, \ldots, n, \ \omega_1^l = \delta_1^l,$$

where the matrices $A^{(p)}$, $p = 1, 2, \ldots, n$, are the exponential solutions of the system

$$\dot{A}^{(p)}(t) = -\mathrm{ad}_{e_p} A^{(p)}(t), \quad A^{(p)}(0) = I.$$

All the rest of realizations of a fixed Lie algebra are constructed by means of projection of the generic realization using the known set of Aut($\mathfrak{g}$)-inequivalent subalgebras and the following rule.

Let $\mathfrak{h} = \langle e_{m+1}, \ldots, e_n\rangle$ be a subalgebra of $\mathfrak{g} = \langle e_1, \ldots, e_n\rangle$ with a complementary space $\{e_1, \ldots, e_m\}$, then, using the above approach and the shortcut $\partial_i = \frac{\partial}{\partial x_i}$, we will obtain the basis elements in the form

$$\begin{aligned} e_i = {} & \xi_i^1(x_1, x_2, \ldots, x_m)\partial_1 + \xi_i^2(x_1, x_2, \ldots, x_m)\partial_2 + \cdots + \xi_i^m(x_1, x_2, \ldots, x_m)\partial_m + \\ & + \xi_i^{m+1}(x_1, x_2, \ldots, x_n)\partial_{m+1} + \xi_i^{m+2}(x_1, x_2, \ldots, x_n)\partial_{m+2} + \cdots + \xi_i^n(x_1, x_2, \ldots, x_n)\partial_n. \end{aligned}$$

The realization projected on the coordinates $x_1, x_2, \ldots, x_m$ is well defined and has the form

$$\mathrm{pr}_{\mathfrak{h}}(e_i) = \xi_i^1(x_1, x_2, \ldots, x_m)\partial_1 + \xi_i^2(x_1, x_2, \ldots, x_m)\partial_2 + \cdots + \xi_i^m(x_1, x_2, \ldots, x_m)\partial_m.$$

The number of necessary variables $m = \dim(\mathfrak{g}) - \dim(\mathfrak{h})$ coincides with the codimension of $\mathfrak{h}$ and the projected realization $R(\mathfrak{h})$ is called *realization corresponding to the subalgebra* $\mathfrak{h}$. In particular this means that the generic realization corresponds to zero subalgebra $\mathfrak{h}_0 = \langle 0\rangle$ and $R(\mathfrak{h}_0)$ is always realized in $n = \dim(\mathfrak{g})$ variables.

The structure of realizations constructed by means of the algebraic method reminds a tree diagram, namely: a realization corresponding to a subalgebra $\mathfrak{h}_1$ can be constructed by means of projection from a realization corresponding to a subalgebra $\mathfrak{h}_2$ if $\mathfrak{h}_2 \subset \mathfrak{h}_1$.

Note that all inequivalent realizations of a fixed Lie algebra can be obtained by the above method, as far as any realization corresponds to a quotient group G/H that acts effectively on some subspace M, where H is a subgroup that corresponds to some subalgebra $\mathfrak{h}$ [7].

2. Realizations of the Poincaré algebra $p(1,1)$

The 'smallest' Poincaré algebra $p(1,1)$ is supposed to act in the two-dimensional space-time $\mathbb{R}^{1,1}$ and has dimension three: $p(1,1) = \langle P_0, P_1, J_{01}\rangle$. Its basis elements satisfy the commutation relations

$$[P_0, P_1] = 0, \quad [P_0, J_{01}] = P_1, \quad [P_1, J_{01}] = P_0. \tag{1}$$

Only covariant realizations of the Lie algebra $p(1,1)$ in space of three variables were constructed in [9]. All inequivalent faithful realizations of $p(1,1)$ were also obtained by the direct

Table 1. Realizations of $p(1,1)$

Complementary vector space	Subalgebra	Realization
$\{P_0, P_1, J_{01}\}$	$\langle 0\rangle$	$P_0=\partial_1,\ P_1=\partial_2,\ J_{01}=x_2\partial_1+x_1\partial_2+\partial_3$
$\{J_{01}, P_1\}$	$\langle P_0\rangle$	$P_0=-\sinh x_1\partial_2,\ P_1=\cosh x_1\partial_2,\ J_{01}=\partial_1$
$\{P_0, P_1\}$	$\langle J_{01}\rangle$	$P_0=\partial_1,\ P_1=\partial_2,\ J_{01}=x_2\partial_1+x_1\partial_2$
$\{P_0, J_{01}\}$	$\langle P_0+P_1\rangle$	$P_0=\partial_1,\ P_1=-\partial_1,\ J_{01}=-x_1\partial_1+\partial_2$
$\{P_0, J_{01}\}$	$\langle P_0-P_1\rangle$	$P_0=\partial_1,\ P_1=\partial_1,\ J_{01}=x_1\partial_1+\partial_2$
$\{J_{01}\}$	$\langle P_0, P_1\rangle$	$P_0=0,\ P_1=0,\ J_{01}=\partial_1$
$\{P_0\}$	$\langle P_0+P_1, J_{01}\rangle$	$P_0=\partial_1,\ P_1=-\partial_1,\ J_{01}=-x_1\partial_1$
$\{P_0\}$	$\langle P_0-P_1, J_{01}\rangle$	$P_0=\partial_1,\ P_1=\partial_1,\ J_{01}=x_1\partial_1$

method in [6] for the isomorphic Lie algebra $A_{3.4}^{-1}$ with the commutation relations ($e_1=P_0+P_1$, $e_2=P_0-P_1$, $e_3=J_{01}$)

$$[e_1,e_2]=0,\quad [e_1,e_3]=e_1,\quad [e_2,e_3]=-e_2. \tag{2}$$

We list realizations of $p(1,1)$ together with the corresponding subalgebras and bases of complementary vector spaces in Table 1.

The subalgebras that contain one of ideals $\langle P_0\pm P_1\rangle$ of $p(1,1)$ lead to unfaithful realizations which are new and were not constructed in [6]. The rest of realizations are equivalent to the faithful realizations obtained for $A_{3.4}^{-1}$.

We purposely do not fix independent (say t, x) and dependent (say u, w) variables in the presented realizations. One can choose them in any order convenient for the certain application. For example, the first realization from Table 1 can be considered as

$$\{P_0=\partial_t,\ P_1=\partial_x,\ J_{01}=x\partial_t+t\partial_x+\partial_u\},\quad \text{for}\quad u=u(t,x)\quad \text{or}$$
$$\{P_0=\partial_u,\ P_1=\partial_w,\ J_{01}=w\partial_u+u\partial_w+\partial_t\},\quad \text{where}\quad u=u(t)\ \text{and}\ w=w(t).$$

3. Realizations of the Poincaré algebra $p(1,2)$

The Poincaré algebra

$$p(1,2)=\langle P_0,P_1,P_2,J_{01},J_{02},J_{12}\rangle$$

is six-dimensional and usually considered in (1+2)-dimensional Minkowski space with the basis elements that satisfy the nonzero commutation relations

$$\begin{aligned}
&[P_0,J_{01}]=P_1, && [P_1,J_{01}]=P_0, && [P_2,J_{02}]=P_0,\\
&[P_0,J_{02}]=P_2, && [P_1,J_{12}]=-P_2, && [P_2,J_{12}]=P_1,\\
&[J_{01},J_{02}]=-J_{12}, && [J_{02},J_{12}]=J_{01}, && [J_{01},J_{12}]=-J_{02}.
\end{aligned}$$

Since we are interested in the generic realization and realizations in not more than three variables, we need the subalgebras of $p(1,2)$ of the dimensions three, four and five only. Using the subalgebras of the Poincaré algebra obtained in [8] and applying the algorithm described in the first section we obtained the desired realizations. The result is arranged in tables together with the subalgebras and bases of the complementary vector spaces used in calculations.

Table 2. Realizations of $p(1,2)$ in three variables

Complementary vector space	Subalgebra	Realization
$\{J_{01}, J_{02}, J_{12}\}$	$\langle P_0, P_1, P_2\rangle$	$P_0 = 0,\ P_1 = 0,\ P_2 = 0,$ $J_{01} = \partial_1,$ $J_{02} = -\sinh x_1 \tanh x_2 \partial_1 + \cosh x_1 \partial_2 - \frac{\sinh x_1}{\cosh x_2}\partial_3,$ $J_{12} = \cosh x_1 \tanh x_2 \partial_1 - \sinh x_1 \partial_2 + \frac{\cosh x_1}{\cosh x_2}\partial_3$
$\{P_0 - P_2, J_{01} + J_{12}, J_{02}\}$	$\langle P_0 + P_2, P_1,$ $\beta(P_0 - P_2) +$ $+J_{01} - J_{12}\rangle,$ $\beta \in \{0, 1\}$	$P_0 = \frac{1}{2}(1 + x_2^2)\partial_1,\ P_1 = -x_2\partial_1,$ $P_2 = \frac{1}{2}(-1 + x_2^2)\partial_1,$ $J_{01} = (-x_1x_2 - \frac{\beta}{2}e^{-2x_3})\partial_1 + \frac{1}{2}(1 - x_2^2)\partial_2 + x_2\partial_3,$ $J_{02} = -x_1\partial_1 - x_2\partial_2 + \partial_3,$ $J_{12} = (x_1x_2 + \frac{\beta}{2}e^{-2x_3})\partial_1 + \frac{1}{2}(1 + x_2^2)\partial_2 - x_2\partial_3$
$\{P_0, P_1, P_2\}$	$\langle J_{01}, J_{02}, J_{12}\rangle$	$P_0 = \partial_1,\ P_1 = \partial_2,\ P_2 = \partial_3,$ $J_{01} = x_2\partial_1 + x_1\partial_2,\ J_{02} = x_3\partial_1 + x_1\partial_3,$ $J_{12} = x_3\partial_2 - x_2\partial_3$
$\{P_0 - P_2, J_{01}, J_{12}\}$	$\langle P_0 + P_2,$ $P_1, J_{02}\rangle$	$P_0 = (\sinh x_1 \sin x_3 + \cosh x_1)\partial_2,$ $P_1 = (-\sinh x_1 - \cosh x_1 \sin x_3)\partial_2,$ $P_2 = -\cos x_3 \partial_2,\ J_{01} = \partial_1,$ $J_{02} = -\cosh x_1 \tan x_3 \partial_1 - \frac{\cosh x_1\ x_2}{\cos x_3}\partial_2 - \sinh x_1 \partial_3,$ $J_{12} = \sinh x_1 \tan x_3 \partial_1 + \frac{\sinh x_1\ x_2}{\cos x_3}\partial_2 + \cosh x_1 \partial_3$
$\{J_{02}, J_{01}, P_0\}$	$\langle \alpha P_0 + J_{12},$ $P_1, P_2\rangle,$ $\alpha \geq 0$	$P_0 = \cosh x_1 \cosh x_2 \partial_3,\ P_1 = -\sinh x_2 \partial_3,$ $P_2 = -\sinh x_1 \cosh x_2 \partial_3,$ $J_{01} = -\sinh x_1 \tanh x_2 \partial_1 + \cosh x_1 \partial_2 - \alpha\frac{\sinh x_1}{\cosh x_2}\partial_3,$ $J_{02} = \partial_1,$ $J_{12} = -\cosh x_1 \tanh x_2 \partial_1 + \sinh x_1 \partial_2 - \alpha\frac{\cosh x_1}{\cosh x_2}\partial_3$
$\{J_{12}, J_{01}, P_1\}$	$\langle \alpha P_1 + J_{02},$ $P_0, P_2\rangle,$ $\alpha \geq 0$	$P_0 = -\sinh x_2 \partial_3,\ P_1 = \cos x_1 \cosh x_2 \partial_3,$ $P_2 = -\sin x_1 \cosh x_2 \partial_3,$ $J_{01} = -\sin x_1 \tanh x_2 \partial_1 + \cos x_1 \partial_2 - \alpha\frac{\sin x_1 \cosh(2x_1)}{2\cosh x_2}\partial_3,$ $J_{02} = -\cos x_1 \tanh x_2 \partial_1 - \sin x_1 \partial_2 - \alpha\frac{\cos x_1 \cosh(2x_1)}{2\cosh x_2}\partial_3,$ $J_{12} = \partial_1$
$\{P_0 - P_2, P_1, J_{01} + J_{12}\}$	$\langle P_0 + P_2,$ $\alpha P_1 + J_{02},$ $J_{01} - J_{12}\rangle,$ $\alpha \geq 0$	$P_0 = (1 - x_3^2)\partial_1 - x_3\partial_2,\ P_1 = \partial_2,\ P_2 = -(1 + x_3^2)\partial_1 - x_3\partial_2,$ $J_{01} = (x_2 - x_3^2(x_2 + 2\alpha))\partial_1 + (x_1 - x_3(x_2 + \alpha))\partial_2 + (1 - x_3^2)\partial_3,$ $J_{02} = -2(x_1 + a x_3)\partial_1 - \alpha\partial_2 - 2x_3\partial_3,$ $J_{12} = (x_2 + x_3^2(x_2 + 2\alpha))\partial_1 - (x_1 - x_3(x_2 + \alpha))\partial_2 + (1 + x_3^2)\partial_3$

Generic realization of $p(1,2)$ has the form

$$P_0 = \partial_1, \quad P_1 = \partial_2, \quad P_2 = \partial_3, \quad J_{01} = x_2\partial_1 + x_1\partial_2 + \partial_4,$$

$$J_{02} = x_3\partial_1 + x_1\partial_3 - \sinh x_4 \tanh x_5 \partial_4 + \cosh x_4 \partial_5 - \frac{\sinh x_4}{\cosh x_5}\partial_6,$$

$$J_{12} = x_3\partial_2 - x_2\partial_3 + \cosh x_4 \tanh x_5 \partial_4 - \sinh x_4 \partial_5 + \frac{\cosh x_4}{\cosh x_5}\partial_6.$$

Comparing the realizations of $p(1,2)$ obtained in this section with the realizations obtained in [13] we conclude that the result of [13] is a partial case of the results obtained in this paper as far as only covariant realizations (i.e. the operators P_0, P_1 and P_3 are represented by the commuting shifts only) were considered and the sets of dependent and independent variables were fixed before the calculations.

Table 3. Realizations of $p(1,2)$ in two variables

Complementary vector space	Subalgebra	Realization
$\{J_{01}+J_{12}, J_{02}\}$	$\langle P_0, P_1, P_2, J_{01}-J_{12}\rangle$	$P_0=0,\ P_1=0,\ P_2=0,$ $J_{01}=\frac{1}{2}(1-x_1^2)\partial_1+x_1\partial_2,$ $J_{02}=-x_1\partial_1+\partial_2,\ J_{12}=\frac{1}{2}(1+x_1^2)\partial_1-x_1\partial_2$
$\{J_{01}, J_{12}\}$	$\langle P_0, P_1, P_2, J_{02}\rangle$	$P_0=0,\ P_1=0,\ P_2=0,$ $J_{01}=\partial_1,\ J_{02}=-\cosh x_1\tan x_2\partial_1-\sinh x_1\partial_2,$ $J_{12}=\sinh x_1\tan x_2\partial_1+\cosh x_1\partial_2$
$\{J_{01}, J_{02}\}$	$\langle P_0, P_1, P_2, J_{12}\rangle$	$P_0=0,\ P_1=0,\ P_2=0,\ J_{01}=\partial_1,$ $J_{02}=-\sinh x_1\tanh x_2\partial_1+\cosh x_1\partial_2,$ $J_{12}=\cosh x_1\tanh x_2\partial_1-\sinh x_1\partial_2$
$\{P_0-P_2, J_{01}+J_{12}\}$	$\langle P_0+P_2, P_1, J_{01}-J_{12}, J_{02}\rangle$	$P_0=\frac{1}{2}(1+x_2^2)\partial_1,\ P_1=-x_2\partial_1,$ $P_2=-\frac{1}{2}(1-x_2^2)\partial_1,\ J_{01}=-x_1x_2\partial_1+\frac{1}{2}(1-x_2^2)\partial_2,$ $J_{02}=-x_1\partial_1-x_2\partial_2,\ J_{12}=x_1x_2\partial_1+\frac{1}{2}(1+x_2^2)\partial_2$

Table 4. Realization of $p(1,2)$ in one variable

Complementary vector space	Subalgebra	Realization
$\{J_{01}+J_{12}\}$	$\langle P_0, P_1, P_2, J_{02}, J_{01}-J_{12}\rangle$	$P_0=0,\ P_1=0,\ P_2=0,\ J_{01}=\frac{1}{2}(1-x_1^2)\partial_1,$ $J_{02}=-x_1\partial_1,\ J_{12}=\frac{1}{2}(1+x_1^2)\partial_1$

4. Deformations of $p(1,1)$ and $p(1,2)$

Deformations of Lie algebras appear in mathematical theories when it is necessary to classify non-isomorphic structures or to implement an analytical structure on the variety of such objects. In physics the deformations are used for quantization or for integration of theories with different symmetries into one model, etc., see also [15] for review.

Here we use the notion of one-parametric deformation proposed by Gerstenhaber [16]. Let $\mathfrak{g}=(V,[\ ,\ ])$ be a Lie algebra over V, then its *deformation* is the one-parametric family of Lie algebras $\mathfrak{g}_t=(V,[\ ,\ ]_t)$, $t\in\mathbb{R}$, where $[x,y]_t=[x,y]+t\varphi_1(x,y)+t^2\varphi_2(x,y)+\cdots$, and $\varphi_i\colon V\times V\to V$ are bilinear, antisymmetric and satisfy the Jacobi identity.

Construction of all possible deformations of a given Lie algebra requires the study of respective cohomology algebra, in particular, for the existence of infinitesimal deformation it is necessary that $\mathcal{H}^2(\mathfrak{g},\mathfrak{g})\neq 0$. However for the Poincaré Lie algebra $p(1,1)$ we can construct one of it's infinitesimal deformations directly from the classification of three-dimensional Lie algebras using the isomorphism $p(1,1)\sim A_{3.4}^{-1}$:

$$[P_0,J_{01}]_{t'}=P_1+t'(P_0-P_1),\quad [P_1,J_{01}]_{t'}=P_0+t'(P_1-P_0),\quad t'\in\mathbb{R}.$$

Another deformation of the Poincaré algebra can be constructed using the fact that there exists a contraction from the de Sitter algebras to the Poincaré one

$$[P_0,J_{01}]_t=P_1,\quad [P_1,J_{01}]_t=P_0,\quad [P_0,P_1]_t=tJ_{01},\quad t\in\mathbb{R}.$$

Journal of Physics: Conference Series **621** (2015) 012009 doi:10.1088/1742-6596/621/1/012009

The same scheme works for all higher-dimensional Poincaré algebras, in particular for $p(1,2)_t$:

$$\begin{array}{lll}
[P_0, J_{01}]_t = P_1, & [P_1, J_{01}]_t = P_0, & [P_2, J_{02}]_t = P_0, \\
[P_0, J_{02}]_t = P_2, & [P_1, J_{12}]_t = -P_2, & [P_2, J_{12}]_t = P_1, \\
[P_1, P_2]_t = tJ_{12}, & [P_1, P_2]_t = tJ_{01}, & [P_0, P_2]_t = tJ_{02}, \\
[J_{01}, J_{02}]_t = -J_{12}, & [J_{02}, J_{12}]_t = J_{01}, & [J_{01}, J_{12}]_t = -J_{02}.
\end{array}$$

Using the same method as in previous sections we construct the generic realizations $R(p(1,1)_{t'})$ and $R(p(1,1)_t)$ of the above deformations, before each realization we present the complementary vector space that was used for the calculation of this realization:

$$\begin{aligned}
&\{P_0, P_1, J_{01}\}, \quad R(p(1,1)_{t'})\colon P_0 = \partial_1, \quad P_1 = \partial_2, \\
&\qquad J_{01} = (t'x_1 + (1-t')x_2)\partial_1 + ((1-t')x_1 + t'x_2)\partial_2 + \partial_3; \\
&\{P_0 + P_1, P_0 - P_1, J_{01}\}, \quad R(p(1,1)_t)\colon P_0 = \tfrac{1}{2}(1 - tx_1^2)\partial_1 + (\tfrac{1}{2} + tx_1x_2)\partial_2 - tx_1\partial_3, \\
&\qquad P_1 = \tfrac{1}{2}(1 + tx_1^2)\partial_1 - (\tfrac{1}{2} + tx_1x_2)\partial_2 + tx_1\partial_3, \\
&\qquad J_{01} = x_1\partial_1 - x_2\partial_2 + \partial_3.
\end{aligned}$$

To construct the generic realization of the deformed algebra $p(1,2)_t$ we use the complementary space to the empty subalgebra with the basis $\{P_0 + P_2, P_0 - P_2, P_1, J_{01} + J_{12}, J_{01} - J_{12}, J_{02}\}$ and obtain the following result:

$$\begin{aligned}
P_0 &= -\tfrac{1}{2}B\partial_1 + C\partial_2 + tx_1x_4\partial_4 - tx_1\partial_5, \\
P_1 &= -Tx_1F\partial_1 - (1 + 2tx_1x_2)Tx_2F\partial_2 + (1 + 2tx_1x_2)\partial_3 + \\
&\quad + \frac{1}{E}((2x_2C - 2tx_1x_2x_4TD - x_1{x_4}^2)\partial_4 + (2tx_1x_4 + tx_2TD)\partial_5 + (tx_1e^{-x_5})\partial_6), \\
P_2 &= \tfrac{1}{2}A\partial_1 - C\partial_2 - tx_1x_4\partial_4 + tx_1\partial_5, \\
J_{01} &= \frac{AF}{2T}\partial_1 + \frac{(1 + 2tx_2(x_1 + x_2A))F}{2T}\partial_2 + (x_1 + x_2A)\partial_3 - \frac{A({x_4}^2 - t{x_2}^2) - 2C}{2E}\partial_4 + \\
&\quad + x_2x_4TAF\partial_4 + \frac{A(x_2TD + x_4)}{E}\partial_5 + \frac{Ae^{-x_5}}{2E}\partial_6, \\
J_{02} &= x_1\partial_1 - x_2\partial_2 - x_4\partial_4 + \partial_5, \\
J_{12} &= \frac{BF}{2T}\partial_1 + \frac{(1 + 2tx_2(x_1 + x_2B))F}{2T}\partial_2 + (x_1 + x_2B)\partial_3 - \frac{B({x_4}^2 - t{x_2}^2) - 2C}{2E}\partial_4 + \\
&\quad + x_2x_4TBF\partial_4 + \frac{(x_2TD + x_4)B}{E}\partial_5 + \frac{Be^{-x_5}}{2E}\partial_6.
\end{aligned}$$

To improve the presentation of the above realization we used the shortcuts $A = tx_1^2 + 1$, $B = tx_1^2 - 1$, $C = tx_1x_2 + \frac{1}{2}$, $D = \sinh((-t)^{1/2}x_3)$, $E = \cosh((-t)^{1/2}x_3)$, $F = \tanh((-t)^{1/2}x_3)$, $T = (-t)^{1/2}$.

Note, that each of the last three realizations represent not only one Lie algebra, but the family of non-isomorphic Lie algebras at the same time. As soon as the set of dependent and independent variables will be fixed all the obtained realizations can be used for the construction of different types of equations and systems invariant with respect to the underlying symmetry groups.

Acknowledgments
The author gratefully acknowledges the support and invitation of the organizing committee of the 7th Workshop "Group Analysis of Differential Equations and Integrable Systems". The author also thanks the referees for their important remarks and helpful comments.

References

[1] Olver P 2000 *Applications of Lie Groups to Differential Equations* (New York: Springer)
[2] Fushchych W and Nikitin A 1994 *Symmetries of Equations of Quantum Mechanics* (New York: Allerton Press Inc.)
[3] Hernádez Heredero R and Olver P 1996 Classification of invariant wave equations *J. Math. Phys.* **37** 6419
[4] Fushchych W, Tsyfra I and Boyko V 1994 Nonlinear representations for Poincaré and Galilei algebras and nonlinear equations for electromagnetic field *J. Nonlinear Math. Phys.* **1** 210 (*Preprint* arXiv:math-ph/0207024)
[5] Fushchych V and Zhdanov R 1997 *Symmetries and Exact Solutions of Nonlinear Dirac Equations* (Kyiv: Mathematical Ukraina Publisher) (*Preprint* arXiv:math-ph/0609052)
[6] Popovych R, Boyko V, Nesterenko M and Lutfullin M 2003 Realizations of real low-dimensional Lie algebras *J. Phys.* A **36** 7337 (*Preprint* arXiv:math-ph/0301029)
[7] Magazev A, Mikheyev V, Shirokov I 2013 Evaluation of composition function along with invariant vector fields on Lie group using structural constants of corresponding Lie algebra (*Preprint* arXiv:1312.0362)
[8] Fushchych W, Barannyk L and Barannyk A 1991 *Subgroup Analysis of the Galilei and Poincaré Groups and Reduction of Nonlinear Equations* (Kiev: Naukova Dumka) [in Russian]
[9] Rideau G and Winternitz P 1990 Nonlinear equations invariant under the Poincaré, similitude and conformal groups in two-dimensional space-time *J. Math. Phys.* **31** 1095
[10] Rideau G and Winternitz P 1993 Evolution equations invariant under two-dimensional space-time Schrödinger group *J. Math. Phys.* **34** 558
[11] Fushchych W, Zhdanov R and Lahno V 1994 On linear and nonlinear representations of the generalized Poincaré groups in the class of Lie vector fields *J. Nonlinear Math. Phys.* **1** 295
[12] Yehorchenko I 1992 Nonlinear representation of the Poincaré algebra and invariant equations, in *Symmetry Analysis of Equations of Mathematical Physics*, ed W I Fushchych (Kyiv: Institute of Mathematics) p 62
[13] Zhdanov R, Lahno V and Fushchych W 2000 On new realizations of the Poincaré groups $P(1,2)$ and $P(2,2)$ *Comm. Math. Phys.* **212** 535
[14] Boyko V, Nesterenko M and Popovych R Realizations of Poincaré algebras (in preparation)
[15] Nesterenko M and Popovych R 2006 Contractions of Low-dimensional Lie Algebras *J. Math. Phys.* **47** 123515 (*Preprint* arXiv:math-ph/0608018)
[16] Gerstenhaber M 1964 On the deformation of rings and algebras *Ann. Math.* **79** 59

Laplace–Runge–Lentz vectors for arbitrary spin and arbitrary dimension

Anatoly G Nikitin
Institute of Mathematics of NAS of Ukraine, 3 Tereshchenkivska Str., Kyiv 01601, Ukraine
E-mail: nikitin@imath.kiev.ua

Abstract. Laplace–Runge–Lentz (LRL) vector is a cornerstone of celestial mechanics. It also plays an important role in quantum mechanics, being an integral of motion for the Hydrogen atom and some other systems. However, the majority of models of non-relativistic systems admitting LRL vector ignore the spin of orbital particles. In this survey a new collection of QM systems admitting LRL vector with spin is presented. It includes $2d$ and $3d$ systems with arbitrary spin, as well as systems of arbitrary dimension with spins 0, 1/2, and 1. All these systems are superintegrable and can be solved exactly. They emulate neutral particles with non-trivial multipole momenta (in particular, the neutron) interacting with a central external field.

1. Introduction

Some quantum mechanical systems are characterized by very attractive global properties: *exact solvability, super integrability,* and (or) *supersymmetry.*

Exact solvability means that all energy levels can be calculated algebraically; and the corresponding wave functions are polynomials multiplied by an overall gauge factor.

Integrability of a d-dimensional system presupposes the existence of $d-1$ integrals commuting with the Hamiltonian and amongst each other. If the system admits at least one more integral of motion (not necessary commuting with the other ones) it is called superintegrable. The system is called maximally superintegrable, if it admits $2d-1$ integrals of motion including the Hamiltonian. Moreover, d of them should commute with each other.

Supersymmetry in quantum mechanics has various faces, but we will discuss only one of them, namely, the shape invariance. That is a specific symmetry of the Hamiltonian with respect to the Darboux transform.

Some systems, like the Hydrogen atom or isotropic harmonic oscillator, are both superintegrable, supersymmetric and exactly solvable. Notice that there exists a tight coupling between superintegrability and supersymmetry [1]. A more contemporary discussion of such coupling can be found in [2–4]. However, we will see that this connection is missing for the case of more general systems with arbitrary spins.

Superintegrability and (or) supersymmetry usually cause the exact solvability of a QM system. However, the exact solvability even of maximally supereintegrable systems never was proved rigorously and can be considered only as a suggestion, see, e.g., [5].

The classification of superintegrable models of quantum mechanics which was started with papers [6, 7] has an interesting and inspiring history. We will not discuss it (see survey [8]),

Journal of Physics: Conference Series **621** (2015) 012010 doi:10.1088/1742-6596/621/1/012010

but restrict ourselves to contemporary results obtained for superintegrable system with spin. Moreover, just the systems which admit analogues of the Laplace–Runge–Lenz (LRL) vector will be considered.

The LRL vector is a corner stone of celestial mechanics. It has a great value also in quantum mechanics. In particular, using this vector, Pauli found the spectrum of the Hydrogen atom before the Schrödinger equation was discovered [9].

Till the year 2013 there were only few known examples of $3d$ quantum mechanical systems with spin admitting the LRL vector. They are the dyon with gyromagnetic ratio $g = 4$, interacting with a magnetic monopole field plus a Coulomb plus a fine-tuned inverse-square potential [10], and a neutral particle with a non-trivial dipole momentum [11]. In addition, a system with spin-orbit interaction and a special inverse-square potential with *fixed* coupling constant was presented in [12], but it is equivalent to the direct sum of two scalar Hydrogen like systems [11]. All the mentioned systems include particles with spin 1/2.

However in 2013 and 2014 a number of new systems with spin admitting the LRL vector have been presented in papers [13–15]. In fact, the complete description of an extended class of such systems has been made, which includes systems with arbitrary spin and arbitrary dimension. Let us mention that the two dimension systems with arbitrary spin had been classified a bit earlier in papers [16] and [17].

In this presentation the results of papers [11, 13–17] are summarized. In other words a survey of the contemporary results concerning the quantum mechanical systems admitting the LRL vector with arbitrary spin in arbitrary dimension space is presented. In addition, supersymmetric aspects of the these systems are discussed using the classification of shape invariant matrix potentials presented in papers [18] and [19].

Mathematically, the subject of our analysis are systems of coupled Schrödinger equations of the following form

$$H\psi = E\psi, \tag{1}$$

where

$$H = -\frac{\Delta_d}{2m} + V(\mathbf{x}). \tag{2}$$

Here ψ is a *multicomponent* wave function, $\mathbf{x} = (x_1, x_2, \dots, x_d)$, Δ_d is Laplace operator in d-dimensional Euclidean space, and $V(\mathbf{x})$ is a *matrix potential*. Physically, just equations of generic form (1) for $d = 3$ are requested to construct models of neutral particles which have nontrivial dipole moments. A perfect example of such particle is the neutron.

Symmetries of systems (1) including two or three equations have been investigated in [20] and [21]. However, it was done only for diagonal potentials $V(\mathbf{x})$ depending on one spatial variable. The systems classified in the present paper include hermitian matrices $V(\mathbf{x})$ of general form and arbitrary dimension.

2. Planar systems

2.1. Pron'ko–Stroganov model

The first (historically) model with Hamiltonian (2) including spin, which admits an analogue of the LRL vector, was proposed by Pron'ko and Stroganov [22]. This $2d$ model is based on the following Hamiltonian

$$\mathcal{H} = \frac{p_1^2 + p_2^2}{2m} + \frac{\mu(\mathbf{n})}{r}, \tag{3}$$

where

$$\mu(\mathbf{n}) = \lambda\frac{\sigma_1 x_2 - \sigma_2 x_1}{r}, \quad r = \sqrt{x_1^2 + x_2^2}, \quad \mathbf{n} = \frac{\mathbf{x}}{r}, \quad p_1 = -\mathrm{i}\frac{\partial}{\partial x_1}, \quad p_2 = -\mathrm{i}\frac{\partial}{\partial x_2}, \tag{4}$$

and describes a neutral spinor with a non-trivial dipole momentum, interacting with the magnetic field generated by a constant straight line currents. In (4) σ_1 and σ_2 are Pauli matrices and λ is the integrated coupling constant.

Let us note that making the unitary transformation $\mathcal{H} \to U\mathcal{H}U^\dagger$ with $U = \frac{1}{\sqrt{2}}(1 - \mathrm{i}\sigma_3)$ it is possible to reduce matrix $\mu(\mathbf{n})$ and potential $V = \frac{\mu(\mathbf{n})}{r}$ to the following equivalent forms:

$$\mu(\mathbf{n}) \to U\mu(\mathbf{n})U^\dagger = \sigma_1 n_1 + \sigma_2 n_2, \quad V \to \hat{V} = \frac{\sigma_1 x_1 + \sigma_2 x_2}{r^2}. \tag{5}$$

Hamiltonian (3) commutes with the total orbital momentum

$$J_3 = x_1 p_2 - x_2 p_1 + S_3, \tag{6}$$

where $S_3 = \frac{1}{2}\sigma_3$. There are two more integrals of motion for (3), namely:

$$\begin{aligned} K_1 &= \frac{1}{2}(J_3 p_1 + p_1 J_3) + \frac{m}{r}\mu(\mathbf{n})x_2, \\ K_2 &= \frac{1}{2}(J_3 p_2 + p_2 J_3) - \frac{m}{r}\mu(\mathbf{n})x_1. \end{aligned} \tag{7}$$

Operators (7) commute with Hamiltonian $\mathcal{H}$ and satisfy the following relations:

$$[J_3, K_1] = \mathrm{i}K_2, \quad [J_3, K_2] = -\mathrm{i}K_1, \quad [K_1, K_2] = -2\mathrm{i}mJ_3\mathcal{H}.$$

Changing Hamiltonian $\mathcal{H}$ by its eigenvalue E we come to the Lie algebra isomorphic to $\mathfrak{so}(3)$ provided $E < 0$, to $e(2)$ provided $E = 0$, and to $\mathfrak{so}(1,2)$ if E is positive.

Thus there are three integrals of motion for a system with two degrees of freedom, but only two of them are functionally independent. It means that the Pron'ko–Stroganov system is maximally superintegrable. Moreover, operators K_1 and K_2 can be treated as components of the two-dimensional analogue of the LRL vector.

Using these symmetries it is possible to find eigenvalues $\mathcal{E}$ for coupled states algebraically. First we rescale operators K_1 and K_2 and introduce the following terms J_1 and J_2:

$$J_1 = \frac{K_1}{\sqrt{-2mE}}, \quad J_2 = \frac{K_2}{\sqrt{-2mE}}, \tag{8}$$

which, together with J_3, realize a representation of algebra $\mathfrak{so}(3)$. This representation is irreducible if wave function ψ satisfies the following condition:

$$C\psi \equiv (J_1^2 + J_2^2 + J_3^2)\psi = n(n+1)\psi, \tag{9}$$

where n are half integers. In addition, ψ is supposed to be an eigenvector of operator J_3 commuting with $\mathcal{H}$ and C:

$$J_3\psi = k\psi, \quad k = \pm\frac{1}{2}, \pm\frac{3}{2}, \ldots, \pm n. \tag{10}$$

Substituting (6) and (7), (8) into (9) and using (10) we obtain the energy spectrum in the following form:

$$E = -\frac{m\lambda^2}{2N^2}, \tag{11}$$

where $N = 2n + 2k + 1 = 1, 2, \ldots$.

The spectrum (11) and the corresponding eigenvectors can be found in another way, using tools of supersymmetric quantum mechanics. Indeed, introducing the polar coordinates and expanding solutions via eigenfunctions of J we reduce the related equation (1) to the following eigenvalue problem for radial functions:

$$\mathcal{H}_\kappa\psi_\kappa \equiv \left(-\frac{\partial^2}{\partial r^2} + \kappa(\kappa-\sigma_3)\frac{1}{r^2} + \sigma_1\frac{\lambda}{r}\right)\psi_\kappa = \mathcal{E}\psi_\kappa,$$

where $\mathcal{E} = 2mE$.

The effective potential

$$V_\kappa(r) = \kappa(\kappa-\sigma_3)\frac{1}{r^2} + \sigma_1\frac{\lambda}{r}$$

is *shape invariant*. To prove this statement we represent V_κ as

$$V_k(x) = -\frac{\partial W_k}{\partial r} + W_k^2 - \frac{\lambda^2}{(2\kappa+1)^2} \tag{12}$$

and solve this Riccati equation for W_κ. We obtain

$$W_\kappa = \frac{1}{2r}\sigma_3 - \frac{\lambda}{2\kappa+1}\sigma_1 - \frac{2\kappa+1}{2r}. \tag{13}$$

Using (13) we can define the superpartner for potential V_κ

$$V_k^+ = \frac{\partial W_k}{\partial r} + W_k^2 = (\kappa+1)(\kappa+1-\sigma_3)\frac{1}{r^2} + \sigma_1\frac{\lambda}{r} + \frac{\lambda^2}{(2\kappa+1)^2}$$

which appears to be shape invariant, i.e.,

$$V_\kappa^+ = V_{\kappa+1} + C_\kappa, \quad C_\kappa = \frac{\lambda^2}{(2\kappa+1)^2} - \frac{\lambda^2}{(2\kappa+3)^2}.$$

Thus the Pron'ko–Stroganov system is shape invariant and can be solved using the standard technique of supersymmetric quantum mechanics, see Section 2.4.

2.2. 2d systems with arbitrary spin

The Pron'ko–Stroganov model describes a particle with spin 1/2. Let us show how it can be generalized to the case of arbitrary spin.

By definition a system with arbitrary spin should be invariant w.r.t. the rotations, i.e., commute with the total orbital momentum (6), where S_3 is a matrix of spin s, which can be chosen in the diagonal form:

$$S_3 = \text{diag}(s, s-1, s-2, \dots, -s). \tag{14}$$

We will search for Hamiltonians of generic form (3), which commute with J_3 and have two additional integrals of motion given by equation (7), where $\mu(\mathbf{n})$ are unknown matrices of dimension $(2s+1)\times(2s+1)$.

Operators (7) commute with Hamiltonian (3) provided matrix $\mu(\mathbf{n})$ satisfies the following equations [17]:

$$[\mu_s(\mathbf{n}), J_3] = 0, \tag{15}$$

$$\mu_s(\mathbf{n})S_z + S_z\mu_s(\mathbf{n}) = 0. \tag{16}$$

The condition (15) is satisfied iff $\mu_s(\mathbf{n})$ is a function of $\mathbf{S}\cdot\mathbf{n} = S_1n_1+S_2n_2$, $\mathbf{S}\times\mathbf{n} = S_1n_2-S_2n_1$ and S_3, where S_1 and S_2 are spin matrices realizing (together with S_3 (14)) irreducible representation $D(s)$ of algebra $\mathfrak{so}(3)$. First we consider matrices $\mu_s(\mathbf{n})$ which are polynomials in $\mathbf{S}\cdot\mathbf{n}$ and expand them via projectors:

$$\mu_s(\mathbf{n}) = \sum_{\nu=-s}^{s} c_\nu\Lambda_\nu, \quad \Lambda_\nu = \prod_{\nu'\neq\nu} \frac{\mathbf{S}\cdot\mathbf{n}-\nu'}{\nu-\nu'}. \tag{17}$$

Matrix (17) satisfies condition (15) by construction. Substituting (17) into (16) and using the identities

$$\begin{aligned} S_3\mu_s(\mathbf{n}) + \mu_s(\mathbf{n})S_3 &= 2S_3\mu_s(\mathbf{n}) + [\mu_s(\mathbf{n}), S_3], \\ [\Lambda_\nu, S_3] &= \frac{1}{2}S_3(2\Lambda_\nu - \Lambda_{\nu+1} - \Lambda_{\nu-1}) + \frac{\mathrm{i}}{2}(n_1S_2 - n_2S_1)(\Lambda_{\nu+1} - \Lambda_{\nu-1})] \end{aligned} \tag{18}$$

we obtain [17]:

$$\mu_s(\mathbf{n}) = \lambda\sum_\nu(-1)^{[\nu]}\Lambda_\nu, \tag{19}$$

where $[\nu]$ is the entire part of ν, $\nu = s, s-1, \dots, -s$.

Formula (19) is defined for arbitrary spin s and gives a particular solution of equations (15). It is possible to show [17] that the most general solution can be represented as:

$$\tilde{\mu}_s(\mathbf{n}) = \sum_{\nu\geq 0}\lambda_\nu(\Lambda_\nu + \Lambda_{-\nu})\mu_s(\mathbf{n}), \tag{20}$$

where $\mu_s(\mathbf{n})$ is matrix (19), λ_ν are arbitrary real parameters, and summation is imposed over all non negative values of $\nu = s, s-1, \dots$.

Thus the Hamiltonian for superintegrable model of arbitrary spin can be represented in the following form:

$$\mathcal{H}_s = \frac{p_x^2 + p_y^2}{2m} + \frac{1}{r}\tilde{\mu}_s(\mathbf{n}), \tag{21}$$

where $\hat{\mu}_s(\mathbf{n})$ is the matrix defined by equations (19) and (20). Hamiltonian (21) admits integrals of motion given by equations (6) and (7) with $\mu(\mathbf{n}) \to \tilde{\mu}_s(\mathbf{n})$. In the particular case $s = 1/2$ operator (21) is equivalent to the Pron'ko–Stroganov Hamiltonian (3), (4).

2.3. 2d system with spin 1

Consider one more particular case of the presented systems which corresponds to spin 1. The entries of the related spin matrices can be chosen in the form:

$$S_1 = \frac{1}{\sqrt{2}}\begin{pmatrix} 0 & 1 & 0 \\ 1 & 0 & 1 \\ 0 & 1 & 0 \end{pmatrix}, \quad S_2 = \frac{\mathrm{i}}{\sqrt{2}}\begin{pmatrix} 0 & -1 & 0 \\ 1 & 0 & -1 \\ 0 & 1 & 0 \end{pmatrix}, \quad S_3 = \begin{pmatrix} 1 & 0 & 0 \\ 0 & 0 & 0 \\ 0 & 0 & 1 \end{pmatrix}. \tag{22}$$

while matrix $\tilde{\mu}_s(\mathbf{n})$ is reduced to the following one:

$$\tilde{\mu}_s(\mathbf{n}) = \mu_1(\mathbf{n}) = \nu(2(\mathbf{S}\times\mathbf{n})^2 - 1) + \lambda(2(\mathbf{S}\cdot\mathbf{n})^2 - 1),$$

where ν and λ are arbitrary parameters.

Let us consider in more detail the case $\nu = 0$, $\lambda = \omega^2 > 0$. The corresponding Hamiltonian (21) takes the following form:

$$\mathcal{H}_1 = \frac{p_1^2 + p_2^2}{2m} + \omega^2 \left(\frac{2(\mathbf{S}\cdot\mathbf{x})^2}{r^3} - \frac{1}{r} \right) \tag{23}$$

and can be rewritten as

$$\mathcal{H}_1 = \frac{p_1^2 + p_2^2}{2m} + \omega^2 Q_{ab} \frac{\partial E_a}{\partial x_b} + \frac{\omega^2}{3} \operatorname{div} \mathbf{E}, \tag{24}$$

where

$$Q_{ab} = S_a S_b + S_b S_a - \frac{2}{3} s(s+1)\delta_{ab}$$

is the quadruple interaction tensor, and

$$\mathbf{E} = \left(\frac{x_1}{r}, \frac{x_2}{r}, 0 \right). \tag{25}$$

Operator (24) can be interpreted as a Hamiltonian of spin-one particle, which includes a quadruple and Darwin interaction with the external electric field (25).

2.4. *Shape invariance and exact solutions*

Thus we present a countable set of quantum mechanical systems with arbitrary spins, any of which admits three integrals of motion given by relations (6), (7) and (17). All these systems are superintagrable.

Let us show that the presented systems are also supersymmetric. Consider the eigenvalue problem (1) for Hamiltonian (1), (19). We introduce polar coordinates and expand solutions via eigenvectors of operator J_3 (6) where S_3 is matrix (14). These eigenvectors can be represented as:

$$\psi_\kappa = \frac{1}{\sqrt{r}} \exp(\mathrm{i}(\kappa - S_z)\theta)\Phi_\kappa(r), \quad \Phi_\kappa(r) = \mathrm{column}(\phi_s, \phi_{s-1}, \ldots, \phi_{-s}).$$

As a result we obtain the following equation for radial functions Φ_k:

$$\hat{\mathcal{H}}_\kappa \Phi_\kappa \equiv \left(-\frac{\partial^2}{\partial r^2} + V_\kappa \right) \Phi_\kappa = \mathcal{E}\Phi_\kappa, \tag{26}$$

where $\mathcal{E} = 2mE$ and

$$V_\kappa = \left((k - S_z)^2 - \frac{1}{4} \right) \frac{1}{r^2} + \tilde{\mu}_s \frac{1}{r}, \quad \tilde{\mu}_s = 2m\mu_s(\mathbf{n})|_{n_y=0}.$$

Hamiltonian $\hat{\mathcal{H}}_\kappa$ and matrix S_3^2 commute with each other and so have a mutual system of eigenfunctions. The eigenvalue problem (26) can be decoupled to the following equations:

$$\hat{\mathcal{H}}_{\kappa,\nu}\psi_{\kappa,\nu} \equiv \left(-\frac{\partial^2}{\partial r^2} + V_\kappa \right) \psi_{\kappa,\nu} = \mathcal{E}\psi_{\kappa,\nu}, \tag{27}$$

where $\psi_{\kappa,\nu} = \begin{pmatrix} \phi_\nu \\ \phi_{-\nu} \end{pmatrix}$ are two-component functions, and

$$V_{\kappa,\nu} = \frac{(\kappa - \nu\sigma_3)^2 - \frac{1}{4}}{r^2} + \frac{\tilde{\lambda}}{r}\sigma_1, \quad \tilde{\lambda} = 2m\lambda, \quad \nu = s, s-1, \ldots, \neq 0. \tag{28}$$

Journal of Physics: Conference Series **621** (2015) 012010 doi:10.1088/1742-6596/621/1/012010

For $\nu = 0$ the corresponding reduced potential is one-dimensional:

$$V_{\kappa,0} = \frac{\kappa^2 - \frac{1}{4}}{r^2} + \frac{\tilde{\lambda}}{r}. \tag{29}$$

Potentials (28) were discussed in paper [18], where parameter ν was denoted as $\mu + 1/2$. These potentials are shape invariant and can be expressed via superpotentials as follows:

$$V_{\kappa,\nu} = W_{\kappa,\nu}^2 - W'_{\kappa,\nu} + c_\kappa, \tag{30}$$

where

$$W_{\kappa,\nu} = \frac{\nu}{r}\sigma_3 - \frac{\tilde{\lambda}}{2\kappa+1}\sigma_1 - \frac{2\kappa+1}{2r}, \quad c_\kappa = -\frac{\tilde{\lambda}^2}{(2\kappa+1)^2}. \tag{31}$$

One-dimensional potential (29) is shape invariant too. It can be represented in the standard form (30) with

$$W_{\kappa,0} = -\frac{\tilde{\lambda}}{2\kappa+1} - \frac{2\kappa+1}{2r}. \tag{32}$$

Alternatively, potentials (30) can be expressed via another superpotential:

$$V_{\kappa,\nu} = W_{\nu,\kappa}^2 - W'_{\nu,\kappa} + c_\nu, \quad c_\nu = -\frac{\tilde{\lambda}^2}{(2\nu+1)^2}, \tag{33}$$

where $W_{\nu,\kappa}$ is the superpotential obtained from $W_{\kappa,\nu}$ by the change $\nu \to \kappa, \kappa \to \nu$. In other words, potential (28) appears to be shape invariant w.r.t. the shifts of two parameters, i.e., κ and ν. This is a particular case of the dual shape invariance phenomena discovered in [18].

Thus equations (27) are shape invariant and can be solved exactly using tools of SUSY quantum mechanics. The ground state vectors $\psi^0_{\kappa,\nu} = \text{column}(\phi^0_{\kappa,\nu}, \phi^0_{\kappa,-\nu})$, $\nu = s, s-1, s-2, \ldots, \nu > 0$ should solve the first order equation,

$$\left(\frac{\partial}{\partial x} + W_{\kappa,\nu}\right)\psi^0_{\kappa,\nu} = 0,$$

and have the following form [17]:

$$\begin{aligned} \phi^0_{\kappa,\nu} &= d_\nu r^{\kappa+1} K_{\nu+\frac{1}{2}}\left(\frac{\tilde{\lambda}r}{2\kappa+1}\right), \quad \kappa \geq \nu, \\ \phi^0_{\kappa,-\nu} &= d_\nu (-1)^{\nu-\frac{1}{2}} r^{\kappa+1} K_{\nu-\frac{1}{2}}\left(\frac{\tilde{\lambda}r}{2\kappa+1}\right), \end{aligned} \tag{34}$$

for superpotential (31), and

$$\begin{aligned} \phi^0_{\kappa,\nu} &= d_\nu r^{\nu+1} K_{\kappa+\frac{1}{2}}\left(\frac{\tilde{\lambda}r}{2\nu+1}\right), \quad 0 \leq \kappa < \nu, \\ \phi^0_{\kappa,-\nu} &= d_\nu (-1)^{\kappa-\frac{1}{2}} r^{\nu+1} K_{\kappa-\frac{1}{2}}\left(\frac{\tilde{\lambda}r}{2\nu+1}\right) \end{aligned} \tag{35}$$

for the alternative potential $W_{\nu,\kappa}$. Here $K_{\cdot}(\cdot)$ are modified Bessel functions, and d_ν are integration constants. Solution for $\nu = 0$, i.e., the component ϕ^0_0, is given by the following equation:

$$\phi^0_{\kappa,0} = du^{\kappa+\frac{1}{2}} \exp\left(-\tilde{\lambda}r\right), \quad \kappa = 0, 1, 2, \ldots.$$

Vectors for exited states and the corresponding energy levels are given by the following formulae

$$\psi^n_{\kappa,\nu}(r) = a^+_{\kappa,\nu} a^+_{\kappa+1,\nu} \cdots a^+_{\kappa+n-1,\nu} \psi^0_{\kappa+n,\nu}(r) \tag{36}$$

and

$$\mathcal{E}_n = -\frac{\tilde{\lambda}^2}{(2\kappa + 2n + 1)^2}, \tag{37}$$

where $a^+_{\kappa,\nu} = -\frac{\partial}{\partial r} + W_{\kappa,\nu}$. In accordance with the analysis presented in [18] all such vectors are square integrable and vanish at $r = 0$.

In complete analogy with the above we can solve the eigenvalue problem for the more general Hamiltonian (21) including $[s+1]$ arbitrary parameters $\tilde{\lambda}_\nu$. Actually, to this effect it is sufficient to change $\tilde{\lambda} \to \tilde{\lambda}\lambda_\nu$ in formulae (28), (31)–(35) and (37).

The presented solutions are valid for models with arbitrary spin s, including the particular cases $s = 1/2$ and $s = 1$ discussed in Sections 2.1, 2.2 and 2.3.

3. Superintegrable systems in 3*d* space

3.1. Runge–Lenz vector for Hydrogen atom

Let us start with the most prominent system admitting the LRL vector, i.e., with the Hydrogen atom. This system is specified by the following Hamiltonian:

$$H = \frac{p^2}{2m} + V(x), \tag{38}$$

where $p^2 = p_1^2 + p_2^2 + p_3^2$, $p_1 = -i\frac{\partial}{\partial x_1}$, $V = -\frac{\alpha}{x}$, $x = \sqrt{x_1^2 + x_2^2 + x_3^2}$, $\alpha = e^2 > 0$.

Hamiltonian (38) commutes with generators of group SO(3) which are components of the orbital momentum $\mathbf{L} = \mathbf{x} \times \mathbf{p}$. There is also the additional vector constant of motion:

$$\mathbf{K} = \frac{1}{2m}(\mathbf{p} \times \mathbf{L} - \mathbf{L} \times \mathbf{p}) + \mathbf{x}V \tag{39}$$

whose components satisfy the following commutation relations:

$$[L_a, L_b] = \mathrm{i}\varepsilon_{abc}L_c, \quad [K_a, L_b] = \mathrm{i}\varepsilon_{abc}K_c, \quad [K_a, K_b] = -\tfrac{2\mathrm{i}}{m}\varepsilon_{abc}L_c H. \tag{40}$$

Changing H in (40) by its eigenvalue E we obtain the Lie algebra isomorphic to $\mathfrak{so}(4)$ provided $E < 0$ or to $\mathfrak{so}(1,3)$ if E is positive.

This symmetry causes the maximal superintegrability of the Hydrogen atom. Moreover, using the presented integrals of motion it is possible to find the energy spectrum of the Hydrogen atom algebraically. It has been done by Pauli as long as in 1926 [9].

3.2. Runge–Lenz vector for 3d systems with arbitrary spin

The non-relativistic model of the Hydrogen atom ignores the spin of the orbital electron. To introduce a spin it is necessary to change the angular momentum $\mathbf{L}$ by the total angular momentum

$$\mathbf{L} \to \mathbf{J} = \mathbf{L} + \mathbf{S}, \tag{41}$$

where $\mathbf{S}$ is a spin vector whose components satisfy

$$[S_a, S_b] = \mathrm{i}\varepsilon_{abc}S_c, \quad S_1^2 + S_2^2 + S_3^2 = s(s+1)I.$$

By analogy with (39) we will search for the LRL vector with spin in the following form:

$$\hat{\mathbf{K}} = \frac{1}{2m}(\mathbf{p} \times \mathbf{J} - \mathbf{J} \times \mathbf{p}) + \mathbf{x}\hat{V}. \tag{42}$$

By definition vector (42) should commute with the Hamiltonian

$$H = \frac{p^2}{2m} + \hat{V}.$$

It is the case iff potential $\hat{V}$ satisfies the following conditions:

$$[\hat{V}, \mathbf{J}] = \mathbf{0}, \quad (43)$$

$$\mathbf{x} \cdot \nabla\hat{V} + \hat{V} = 0, \quad (44)$$

$$\mathbf{S} \times \nabla\hat{V} - \nabla\hat{V} \times \mathbf{S} = 0, \quad (45)$$

where $\cdot$ and $\times$ are the symbols of the scalar and vector products and ∇ is the gradient vector.
In accordance with (43) and (44) $\hat{V}$ should be a rotational scalar of dimension $\frac{1}{x}$,

$$\hat{V} = \lambda\frac{\mu_s(\mathbf{n})}{x}, \quad (46)$$

where $\mu_s(\mathbf{n})$ is a scalar and dimensionless matrix which, without loss of generality, can be represented in form form (17) with $\mathbf{S} \cdot \mathbf{n} = S_1 n_1 + S_2 n_2 + S_3 n_3$ and $n_a = \frac{x_a}{x}$. Substituting (17) into (45) and using the $3d$ analogue of relation (18) we come to an algebraic equation for coefficients c_ν which is easy solvable. As a result we obtain the following expressions for potentials $\hat{V}$ [13]:

$$\hat{V} = \frac{\lambda}{x}\Lambda_0 \quad \text{for integer spins}, \quad (47)$$

$$\hat{V} = \frac{\lambda}{x}\sum_{\nu=-s}^{s}\frac{1}{\nu}\Lambda_\nu \quad \text{for half integer spins}. \quad (48)$$

Thus we find $3d$ systems with arbitrary spin which admit the LRL vector. The most important systems which correspond to $s = 1/2$ and $s = 1$ are considered in more detail in the following sections.

3.3. 3d system with spin 1/2

Let $s = 1/2$ then, in accordance with (46) and (48), the potential is reduced to the following form

$$\hat{V} = \lambda\frac{\boldsymbol{\sigma} \cdot \mathbf{x}}{x^2} \Rightarrow H = \frac{p^2}{2m} + \lambda\frac{\boldsymbol{\sigma} \cdot \mathbf{x}}{x^2}. \quad (49)$$

This Hamiltonian admits two vector integrals of motion given by equations (41) and (42) which satisfy the same commutation relations (40) as in the case of the Hydrogen atom. Thus, like in the Hydrogen atom, the bound states of our system has a hidden symmetry w.r.t. algebra $\mathfrak{so}(4)$. This symmetry makes it possible to find the spectrum of Hamiltonian H (49) algebraically [11]. Moreover, the admissible eigenvalues of E for bound states are [11]

$$E = -\frac{m\lambda^2}{2N^2} \quad (50)$$

where $N = 2n + 2j + 1/2$, $n = 0, 1, 2, \ldots$. In contrast with the Hydrogen atom, the main quantum number N should be half integer.

The considered system with spin 1/2 inherits one more nice property of the Hydrogen atom. It is supersymmetric (shape invariant) and can be integrated with using tools of SUSY quantum mechanics. Indeed, the corresponding radial equations are

$$\mathcal{H}_j\Phi_j \equiv \left(-\frac{\partial^2}{\partial r^2} + V_j\right)\Phi_j = \mathcal{E}\Phi_j,$$

where V_j is the matrix potential

$$V_j = \left(j(j+1) + \frac{1}{4} - \sigma_3\left(j + \frac{1}{2}\right)\right)\frac{1}{r^2} - \sigma_1\frac{\omega}{r}, \quad \omega = 2m\lambda.$$

This potential belongs to the list of shape invariant matrix potentials found in [18]. It can be expressed via the superpotential (compare with (12)):

$$V_j(x) = -\frac{\partial W_j}{\partial x} + W_j^2 - \frac{\lambda^2}{(2j+1)^2}, \tag{51}$$

where

$$W_j = \frac{1}{2x}\sigma_3 - \frac{\lambda}{2j+1}\sigma_1 - \frac{2j+1}{2x}.$$

Thus to find state vectors corresponding to eigenvalues (50) we again can use tools of SUSY quantum mechanics. The ground state vector Φ_j^0 looks as follows

$$\Phi_j^0 = C_j x^{j+\frac{3}{2}} \begin{pmatrix} K_1\left(\frac{\lambda x}{2(j+1)}\right) \\ K_0\left(\frac{\lambda x}{2(j+1)}\right) \end{pmatrix} \tag{52}$$

with normalization constants C_j given by the following formulae

$$C_j = 2^{-2(j+\frac{1}{2})}\left(G_{00}^{22}\left(1\middle|^{-1,0}_{j,j+1}\right) + G_{00}^{22}\left(1\middle|^{0,0}_{j+1,j+1}\right)\right)^{-1},$$

where $G_{00}^{22}(\cdot|^{\cdot,\cdot}_{\cdot,\cdot})$ are the Meijer functions. The nth excited state can be calculated in complete analogy with (36).

Comparing (49) with (5), and (50) with (11) we see that the $3d$ system with spin 1/2 is a rather straightforward generalization of the Pron'ko–Stroganov system. However, in contrast with the $2d$ case, hamiltonian (49) does not describe interaction of a fermion with the magnetic field. On the contrary, potential $V(\mathbf{x})$ represents a Pauli type interaction with the external vector field

$$\mathbf{E} \sim \frac{\mathbf{x}}{x^2} \tag{53}$$

which can be interpreted as the electric field which can be realized experimentally at least on finite interval $a < x < b, a > 0$ [11]. Notice that functions (53) also solve equations of axion electrodynamics [23, 24].

3.4. 3d system with spin 1

Consider now a model for vector particle. The corresponding potential is given by equation (47) for $s = 1$, i.e.,

$$\hat{V} = \frac{\alpha}{x}\left(1 - (\mathbf{S}\cdot\mathbf{n})^2\right),$$

where $\mathbf{S}$ is the spin vector whose components are given by equation (22). The related eigenvalue problem takes the form

$$H\psi \equiv \left(\frac{p^2}{2m} + \frac{\alpha}{x}(1 - (\mathbf{S}\cdot\mathbf{n})^2\right)\psi = E\psi. \tag{54}$$

The eigenvectors $\psi = \psi(\mathbf{x})$ are supposed to be normalizable and vanishing at $\mathbf{x} = 0$.

This rather complicated system of three coupled Schrödinger equations can be integrated in closed form thanks to the sufficient number of its integrals of motion. The admissible energy levels for bound states are given by equation (50) where $N = n + j + 1$, $n = 0, 1, \dots$ [13]. The corresponding eigenvectors was found in the following form

$$\psi = \frac{1}{r}\sum_{j,\kappa,\lambda}\psi_{j,\kappa,\lambda}(r)\Omega^s_{j,\kappa,\lambda}(\varphi,\theta),$$

where $r = \sqrt{-2mE}x$, $s = 1$, $j = 0, 1, \dots$, $\kappa = -j, -j+1, \dots, j$, $\lambda = 1, 0, -1$, and $\Omega^s_{j,\kappa,\lambda}$ are spherical harmonics for spin 1. In addition, the radial wave functions take the form [13]:

$$\psi_{j,\kappa,-1} = C_{jn}\sqrt{j}r^{j+2}\exp(-r)\left((j+n)(j+1)\mathcal{F}(-n, 2j+2, 2r) - n\mathcal{F}(1-n, 2j+3, 2r)\right),$$
$$\psi_{j,\kappa,+1} = C_{jn}\sqrt{j+1}r^{j+1}\exp(-r)\left(nr\mathcal{F}(1-n, 2j+3, 2r)\right.$$
$$\left. - (j+1)(2j+1+(n+j)r)\mathcal{F}(-n, 2j+2, 2r)\right), \quad \psi_{j,\kappa,0} = 0,$$

where $\mathcal{F}(\cdot,\cdot,\cdot,2r)$ are confluent hypergeometric functions ${}_1F_1$, and C_{jn} are normalization constants.

In contrast to $s = 1/2$ case, $3d$ Hamiltonian (54) is by no means a straightforward generalization of the $2d$ Hamiltonian for $s = 1$ discussed in Section 2.3. In particular, as distinct from operator (23) Hamiltonian (54) is not shape invariant.

4. Superintegrable systems in any dimension

4.1. General analysis

Consider a generic d-dimensional stationary Schrödinger equation with a matrix potential given by equations (1) and (2).

We suppose that Hamiltonian H is invariant with respect to the rotation group SO(d) whose generators can be chosen in the standard form

$$J_{\mu\nu} = x_\mu p_\nu - x_\nu p_\mu + S_{\mu\nu}, \tag{55}$$

where indices μ and ν run over the values $1, 2, \dots, d$, and $S_{\mu\nu}$ are matrices satisfying the familiar $\mathfrak{so}(d)$ commutation relations:

$$[S_{\mu\nu}, S_{\lambda\sigma}] = \mathrm{i}(\delta_{\mu\lambda}S_{\nu\sigma} + \delta_{\nu\sigma}S_{\mu\lambda} - \delta_{\mu\sigma}S_{\nu\lambda} - \delta_{\nu\lambda}S_{\mu\sigma}),$$

where $\delta_{\mu\lambda}$ is the Kronecker symbol. By definition the matrix potential $V(\mathbf{x})$ should commute with generators (55):

$$[V, J_{\mu\nu}] = 0. \tag{56}$$

We search for equations (1) admitting integrals of motion:

$$K_\mu = \frac{1}{2m}(p_\nu J_{\mu\nu} + J_{\mu\nu}p_\nu) + x_\mu V. \tag{57}$$

Operators K_μ commute with H iff

$$x_\nu \nabla_\nu V + V = 0, \tag{58}$$
$$S_{\mu\nu} \nabla_\nu V + \nabla_\nu V S_{\mu\nu} = 0, \tag{59}$$

where $\nabla_\nu = \frac{\partial}{\partial x_\nu}$ and summation from 1 to d is imposed over the repeating index ν.

If conditions (56), (58) and (59) are fulfilled then operators $J_{\mu\nu}$ and K_μ form a basis of algebra $\mathfrak{so}(d+1)$, $\mathfrak{so}(1,d)$ or $\mathfrak{e}(d)$ for $E<0$, $E>0$ and $E=0$ respectively. In other words, all systems whose potentials satisfy conditions (56), (58) and (59) admit a hidden symmetry of Fock type. The corresponding integral of motion (57) is an analogue of the LRL vector for d-dimensional space.

4.2. Scalar systems

Let matrices $S_{\mu\nu}$ be trivial. The corresponding potential should satisfy

$$[\hat{V}, L_{\mu\nu}] = 0, \quad x_\nu \nabla_\nu \hat{V} = -\hat{V},$$

where $L_{\mu\nu} = x_\mu \partial_\nu - x_\nu \partial_\mu$. The general solution of these equations is the d-dimensional Coulomb potential [15]

$$V = -\frac{\alpha}{r}, \tag{60}$$

where α is a constant.

The corresponding equation (1) is superintegrable and admits a d-dimensional analogue of the LRL vector, given by formula (57) with $J_{\mu\nu} = L_{\mu\nu}$.

Thus we recover the known result [25] concerning the generalization of the LRL vector in d dimensions. Moreover, in [15] a formal proof is presented that the only scalar potential which is compatible with the d-dimensional LRL vector is the Coulomb potential.

To solve equations (1) we use the hyper-spherical variables connected with the Cartesian variables via the following relations:

$$\begin{array}{l} x_d = r\cos\theta_{d-1}, \\ x_{d-1} = r\sin\theta_{d-1}\cos\theta_{d-2}, \\ x_{d-2} = r\sin\theta_{d-1}\sin\theta_{d-2}\cos\theta_{d-3}, \\ \cdots\cdots\cdots\cdots\cdots\cdots\cdots\cdots\cdots \\ x_2 = r\sin\theta_{d-1}\sin\theta_{d-2}\cdots\sin\theta_2\cos\theta_1, \\ x_1 = r\sin\theta_{d-1}\sin\theta_{d-2}\cdots\sin\theta_2\sin\theta_1. \end{array} \tag{61}$$

Expanding wave function ψ via hyper-spherical harmonics Y^l_λ

$$\Psi = r^{\frac{1-d}{2}} \psi_{l\lambda} Y^l_\lambda \tag{62}$$

and substituting (60), (61) and (62) into (1), we come to the following equation for radial functions $\psi_{l\lambda}$:

$$H_\mu \psi_{l\lambda}(r) \equiv \left(-\frac{\partial^2}{\partial r^2} + \frac{\mu(\mu+1)}{r^2} + \hat{V} \right) \psi_{l\lambda}(r) = \epsilon \psi_{l\lambda}(r), \tag{63}$$

where $\hat{V} = 2mV = -\frac{2m\alpha}{r}$, $\epsilon = 2mE$ and

$$\mu = l + \frac{d-3}{2}, \quad l = 0, 1, 2, \ldots .$$

Up to the admissible values of μ equation (63) coincides with the radial equation for the 3d Hydrogen atom. Negative eigenvalues ϵ and the corresponding values of E are given by the following formulae

$$\epsilon = -\frac{m^2\alpha^2}{N^2}, \quad E = -\frac{m\alpha^2}{2N^2}, \tag{64}$$

where

$$N = n + l + \frac{d-1}{2}, \quad n = 0, 1, 2, \dots .$$

For $d = 3$ equation (64) is reduced to the familiar Balmer formula.

4.3. d-dimensional systems with $s = 1/2$

Consider the case when matrices $S_{\mu\nu}$ realize irreducible representation $D(\frac{1}{2}, \frac{1}{2}, \dots, \frac{1}{2})$ of algebra $\mathfrak{so}(d)$ for even d or representation $D(\frac{1}{2}, \frac{1}{2}, \dots, -\frac{1}{2})$ for odd d. Making reduction of these representations on subalgebra $\mathfrak{so}(3)$ we obtain a direct sum of representations $D(\frac{1}{2})$. Thus the corresponding equations (1) present a model of a particle with spin $\frac{1}{2}$.

The considered matrices $S_{\mu\nu}$ admit the following uniform representation

$$S_{\mu\nu} = \frac{\mathrm{i}}{4}\left(\gamma_\mu\gamma_\nu - \gamma_\nu\gamma_\mu\right), \tag{65}$$

where γ_μ are basis elements of the Clifford algebra satisfying the following relations

$$\gamma_\mu\gamma_\nu + \gamma_\nu\gamma_\mu = 2\delta_{\mu\nu}.$$

The dimension of irreducible matrices (65) is equal to $2^{\left[\frac{d}{2}\right]}$, where $\left[\frac{d}{2}\right]$ is the integer part of $\frac{d}{2}$.

The general solution of relations (56), (58) and (59) where $S_{\mu\nu}$ are matrices (65) are given by the following formula [15]

$$\hat{V} = \frac{\alpha}{r}\gamma_\nu n_\nu, \tag{66}$$

where $n_\nu = \frac{x_\nu}{r}$.

The system with potential (66) is invariant with respect to group SO(d) and admits the generalized LRL vector (57). In addition, it is shape invariant and so can be solved algebraically [15].

To separate variables in equation (1) with potential (66) we use hyper-spherical variables (61) and expand solutions as

$$\Psi(\mathbf{r}) = r^{\frac{d-1}{2}}\psi_{j\varrho\lambda}(r)\Omega^j_{\varrho\lambda}, \tag{67}$$

where $\Omega^j_{\varrho\lambda}$ are hyper-spherical spinors satisfying the conditions

$$\frac{1}{2}J_{\mu\nu}J_{\mu\nu}\Omega^j_{\varrho\lambda} = \left(j(j+d-2) + \frac{1}{8}(d-2)(d-3)\right)\Omega^j_{\varrho\lambda},$$

$$D\Omega^j_{\varrho\lambda} \equiv \left(\frac{1}{2}\gamma_\mu\gamma_\nu L_{\mu\nu} + \frac{d-1}{2}\right)\Omega^j_{\varrho\lambda} = \varrho\Omega^j_{\varrho\lambda}$$

with $j = \frac{1}{2}, \frac{3}{2}, \dots,$ and $\varrho = \pm\left(j + \frac{d-2}{2}\right)$.

Substituting (66) and (67) into (1) we come to the following equation for radial wave functions

$$H_\varrho\phi \equiv \left(-\frac{\partial^2}{\partial r^2} + \frac{\varrho^2 + \sigma_3\varrho}{r^2} + \frac{\omega}{r}\sigma_1\right)\phi = \varepsilon\phi, \quad \phi = \begin{pmatrix} \psi_{j|\varrho|\lambda}(r) \\ \psi_{j-|\varrho|\lambda}(r) \end{pmatrix}, \tag{68}$$

where $\omega = 2m\alpha$.

Hamiltonian $\mathcal{H}_j$ includes matrix potential which again belongs to the list of shape invariant potentials found in [18]. Thus equation (68) can be solved in the way used in Section 3.3. To have this effect it is sufficient simply to change $j \to \rho - \frac{1}{2}$ in equations (50), (51) and (52).

4.4. Systems with spin 1

Let us present a bosonic d-dimensional system admitting generalized LRL vector (57). We suppose the corresponding matrices $S_{\mu\nu}$ are irreducible and realize representation $D(1,0,0,\ldots,0)$ of algebra $\mathfrak{so}(d)$, where the symbols in brackets are the Gelfand–Tsetlin numbers. Up to equivalence, their entries $(S_{\mu\nu})_{ab}$ can be represented in the following form:

$$(S_{\mu\nu})_{ab} = \mathrm{i}(\delta_{\mu a}\delta_{\nu b} - \delta_{\nu a}\delta_{\mu b}), \quad \mu, \nu, \ a, b = 1, 2, \ldots, d, \quad d > 2. \tag{69}$$

The general solution of the corresponding equations (56), (58) and (59) has the following form [15]

$$V = \frac{\alpha}{(d-2)r}\left((d-1)(d-4) + 2S_{\mu\nu}n_\nu S_{\mu\lambda}n_\lambda\right), \quad d \neq 2 \tag{70}$$

and so the potential includes the following entries

$$V_{\mu\nu} = \frac{\alpha}{2r}\left((d-3)\delta_{\mu\nu} + 2n_\mu n_\nu\right).$$

Hamiltonian (2) with potentials (70) commutes with the total orbital momentum (55) and the LRL vector (57), where $S_{\mu\nu}$ are matrices (69). It means that the corresponding Schrödinger equations (1) possess symmetries of Fock type. They are exactly solvable for arbitrary d. We will not reproduce the cumbersome procedure of construction of the exact solutions which is presented in [15], but give the explicit formulae for the energy spectrum and the corresponding eigenvalues:

$$E = -\frac{m\alpha^2(d-1)^2}{2(2n+2l+d-1)^2}, \quad n = 0, 1, 2, \ldots, \quad l = 0, 1, 2, \ldots,$$

and $\Psi = \mathrm{column}(\Psi_1, \Psi_2, \ldots, \Psi_d)$ with

$$\Psi_\mu = \varphi^1_{l\lambda}(r)\frac{x_\mu}{r}Y^l_\lambda + \varphi^2_{l\lambda}(r) r \nabla_\mu Y^l_\lambda,$$

where Y^l_λ are hyper-spherical harmonics, and the radial functions are:

$$\varphi^2_{l\lambda} = C_{ln} z^{l+d-1}\exp(-z)\mathcal{F}(-n, l+d-1, 2z),$$

$$\varphi^1_{l\lambda} = C_{ln}\left(z^{l+d-1}\exp(-z)\left((l+d+(k-1)z)\mathcal{F}(-n, l+d-1, 2z) - \frac{2kz}{l+d-1}\mathcal{F}(-n+1, l+d, 2z)\right)\right),$$

where C_{ln} are integration constants, $z = \frac{m\alpha r}{k}$, $k = \frac{2n+2l+d-1}{d-1}$, and $\mathcal{F}(-n, l+d-1, 2z)$ are confluent hypergeometric functions ${}_1F_1$. The presented solutions are normalizable and vanish at $z = 0$.

5. Discussion

The main goal of this presentation is to extend the field of known superintegrable quantum mechanical systems admitting LRL vector. Originally this vector was specified for three-dimensional space. Then it was generalized to the case of d dimensions [25], but it was done only for scalar QM systems.

The first example of LRL vector with spin was proposed apparently in paper [22], where a $2d$ superintegrable system was discovered. LRL vectors with spin in three-dimensional space were discussed in [11]. Then these results were extended to the case of arbitrary spin [13] and arbitrary dimension [15].

The d-dimensional systems with spin admitting LRL keep all basic properties of the systems with $d = 3$. They are superintegrable, and their extended symmetries present effective tools for finding exact solutions.

The discussed systems with spin 1/2 and arbitrary dimension are supersymmetric, i.e., the corresponding radial equations are shape invariant. Moreover, they belong to the list of shape invariant systems classified in [18] and [19].

The systems with spin $s > 1/2$ and $d = 2$ are also supersymmetric. However, for $d > 2$ they are neither supersymmetric nor shape invariant. Nevertheless, these systems are superintegrable and exactly solvable, see Section 4.4 and paper [15] for their exact solutions for arbitrary d. Exact solutions for $d = 2$ and $d = 3$ are given in the present paper in more details.

Thus the tight connection between superintegrability and shape invariance discussed in [1–4] is missing in cases when $s > 1/2$ and simultaneously $d > 2$.

To find superintegrable systems with arbitrary spin and dimension we restrict ourselves to a priori specified forms (42) and (57) of the generalized LRL vectors. However, at least for $3d$ systems with spin 1/2 this restriction is not essential, and potential (49) can be obtained requiring only for the rotation invariance of equation (1) [26].

The discussed superintegrable systems with $d = 2$ and $d = 3$ have a clear physical meaning and describe neutral particles with non-trivial multipole momenta interacting with the external electromagnetic fields. A perfect example of such particle is the neutron. For more detailed physical interpretation see papers [11], [13] and [17].

Let us mention that there are known examples of *relativistic* (super)integrable systems with spin, see, e.g., [11, 26, 27]. The complete classification of such systems at least for spin 1/2 is a rather inspiring challenge.

One more challenge is an extension of the results of papers [11, 13, 14] to the case of non-stationary Schrödinger equations, like it was done in [28] for scalar systems.

Acknowledgments

The author is grateful for the hospitality and financial support provided by the University of Cyprus.

References

[1] Lyman J M and Aravind P K 1993 Deducing the Lenz vector of the hydrogen atom from supersymmetry *J. Phys.* A **26** 3307
[2] Marquette I 2009 Supersymmetry as a method of obtaining new superintegrable systems with higher order integrals of motion *J. Math. Phys.* **50** 122102
[3] Mota R D, Garcia J and Granados V D 2001 Laplace–Runge–Lenz vector, ladder operators and supersymmetry *J. Phys.* A **34** 2041
[4] Ragnisco O and Riglioni D 2010 A family of exactly solvable radial quantum systems on space of non-constant curvature with accidental degeneracy in the spectrum *SIGMA* **6** 097
[5] Tempesta P, Turbiner A V and Winternitz P 2001 *J. Math. Phys.* **42** 4248
[6] Winternitz P, Smorodinsky Ya A, Uhlír M and Friš I 1966 Symmetry groups in classical and quantum mechanics *Yad. Fiz.* **4** 625 [in Russian]; English transl. in 1967 *Sov. J. Nucl. Phys.* **4** 444
[7] Makarov A, Smorodinsky J, Valiev Kh and Winternitz P 1967 A systematic search for non-relativistic systems with dynamical symmetries *Nuovo Cimento A* **52** 1061
[8] Miller W Jr, Post S and Winternitz P 2013 Classical and quantum superintegrability with applications *J. Phys.* A **46** 423001
[9] Pauli W Jr 1926 Über das Wasserstoffspektrum vom Standpunkt der neuen Quantenmechanik *Z. Physik* **36** 336
[10] D'Hoker E and Vinet L 1986 Constants of Motion for a Spin-1/2 Particle in the Field of a Dyon *Phys. Rev. Lett.* **55** 1043
[11] Nikitin A G 2012 New exactly solvable systems with Fock symmetry *J. Phys.* A **45** 485204
[12] Désilets J-F, Winternitz P and Yurduşen I 2012 Superintegrable systems with spin and second-order integrals of motion *J. Phys.* A **45** 475201

[13] Nikitin A G 2013 Laplace–Runge–Lenz vector for arbitrary spin *J. Math. Phys.* **54** 123506
[14] Nikitin A G 2013 Superintegrable systems with arbitrary spin *Ukr. J. Phys.* **58** 1046
[15] Nikitin A G 2014 Laplace–Runge–Lenz vector with spin in any dimension *J. Phys.* A **47** 375201
[16] Pronko G P 2007 Quantum superintegrable systems for arbitrary spin *J. Phys.* A **40** 13331
[17] Nikitin A G 2012 Matrix superpotentials and superintegrable systems for arbitrary spin *J. Phys.* A **45** 225205
[18] Nikitin A G and Karadzhov Yu 2011 Matrix superpotentials *J. Phys.* A **44** 305204
[19] Nikitin A G and Karadzhov Yu 2011 Enhanced classification of matrix superpotentials *J. Phys.* A **44** 445202
[20] Beckers J, Debergh N and Nikitin A G 1992 More on supersymmetries of the Schrödinger equation *Modern Phys. Lett.* A **7** 1609
[21] Beckers J, Debergh N and Nikitin A G 1993 More on parasupersymmetries of the Schrödinger equation *Modern Phys. Lett.* A **8** 435
[22] Pron'ko G P and Stroganov Y G 1977 New example of quantum mechanical problem with hidden symmetry *Sov. Phys. JETP* **45** 1075
[23] Nikitin A G and Kuriksha O 2012 Invariant solutions for equations of axion electrodynamics *Commun. Nonlinear Sci. Numer. Simulat.* **17** 4585
[24] Nikitin A G and Kuriksha O 2012 Symmetries of field equations of axion electrodynamics *Phys. Rev.* D **86** 025010
[25] Sudarshan E C G, Mukunda N and ORaifeartaigh L 1965 Group theory of the Kepler problem *Phys. Lett.* **19** 322
[26] Ferraro E, Messina N and Nikitin A G 2010 Exactly solvable relativistic model with the anomalous interaction *Phys. Rev.* A **81** 042108
[27] Katsura H and Aoki H 2006 Exact supersymmetry in the relativistic hydrogen atom in general dimensions, supercharge and the generalized Johnson-Lippmann operator *J. Math. Phys.* **47** 032301
[28] Fushchich W I and Nikitin A G 1997 Higher symmetries and exact solutions of linear and nonlinear Schrödinger equation *J. Math. Phys.* **38** 5944

Journal of Physics: Conference Series **621** (2015) 012011 doi:10.1088/1742-6596/621/1/012011

Equivalence groupoid of generalized potential Burgers equations

Oleksandr A Pocheketa

Institute of Mathematics of NAS of Ukraine, 3 Tereshchenkivska Str., 01601 Kyiv, Ukraine

E-mail: pocheketa@yandex.ua

Abstract. We find the equivalence groupoid of a class of $(1+1)$-dimensional second-order evolution equations, which are called generalized potential Burgers equations. This class is related via potentialization with two classes of variable-coefficient generalized Burgers equations. Its equivalence groupoid is of complicated structure and is described via partitioning the entire class into three normalized subclasses such that there are no point transformations between equations from different subclasses. For each of these subclasses we construct its equivalence group of an appropriate kind.

1. Introduction

The classical Burgers equation $u_t + uu_x - u_{xx} = 0$ was generalized in various ways including variable-coefficient generalizations. Some of them were studied in the framework of symmetry analysis [1–9].

In this paper we study admissible transformations of the class of $(1+1)$-dimensional second-order evolution equations of the general form

$$v_t + v_x^2 + f(t,x)v_{xx} = 0, \quad f \neq 0. \tag{1}$$

We denote an equation of the form (1) with a fixed value of the arbitrary element f by $\mathcal{P}_f$, and call it a *generalized potential Burgers equation* for two reasons. The first one is that $\mathcal{P}_f$ is a potential equation for the equation

$$u_t + 2uu_x + \big(f(t,x)u_x\big)_x = 0 \tag{2}$$

with the same $f \neq 0$. For any value of the arbitrary element f, the space of conservation laws of the corresponding equation $\mathcal{C}_f$ of the form (2) is one-dimensional and is spanned by the obvious conservation law with the characteristic 1. Hence the associated potential system has the form $v_x = u$, $v_t = -u^2 - fu_x$, which leads to the equation $\mathcal{P}_f$ for the potential v.

The second reason is that an equation $\mathcal{P}_{\hat f}$ in hatted variables, $\hat v_{\hat t} + \hat v_{\hat x}^2 + \hat f(\hat t,\hat x)\hat v_{\hat x\hat x} = 0$, with $\hat f_{\hat x\hat x\hat x} = 0$ is similar with respect to a point transformation to the potential equation for an equation $\mathcal{L}_f$ with $f_{xxx} = 0$ from the class

$$u_t + uu_x + f(t,x)u_{xx} = 0, \quad f \neq 0, \tag{3}$$

which is called the class of *generalized Burgers equations* [7]. The equation $\mathcal{L}_f$ has nonzero conservation laws if and only if $f_{xxx} = 0$, and in this case the space of conservation laws is one-dimensional. It is spanned by the conservation law with the characteristic $\lambda = e^{\int f_{xx}\,\mathrm{d}t}$, which depends only on t, see [10]. Here and in what follows an integral with respect to t means a fixed antiderivative. The potential system constructed for $\mathcal{L}_f$ with respect to the canonical conserved current of the above conservation law is

$$v_x = \lambda u, \quad v_t = -\lambda\left(\frac{1}{2}u^2 + fu_x - f_x u\right),$$

and the corresponding potential equation is

$$v_t + \frac{1}{2\lambda}v_x^2 + fv_{xx} - f_x v_x = 0.$$

Under the assumption $f_{xxx} = 0$, the last equation is similar, with respect to the transformation

$$\hat{t} = \frac{1}{2}\int \lambda\,\mathrm{d}t, \quad \hat{x} = \lambda x + \int f^1\lambda\,\mathrm{d}t, \quad \hat{v} = v,$$

to the equation $\mathcal{P}_{\hat{f}}$, where $\hat{f}(\hat{t},\hat{x}) = 2\lambda f(t,x)$ and $f^1 = f_x - xf_{xx}$ is a function of t only.

Note that the intersection of the classes (2) and (3) is their subclass consisting of the equations with $f = f(t)$. Traditionally, such equations have received more attention due to their appearance in physical models (see [11–16] and also [17, Chapter 4]). We point out that it is the first class of differential equations that was considered in order to determine its equivalence groupoid [3]. The symmetry analysis of this class was carried out in [2,5] and was enhanced and completed in [18].

In general, the further potentialization of equations from the class (1) is not possible since any equation $\mathcal{P}_f$ with $f \neq \mathrm{const}$ has no nonzero conservation laws. At the same time, equations with constant values of the arbitrary element f are singular in the class (1) from the point of view of conservation laws and transformational properties, and they constitute the orbit of the potential Burgers equation $\mathcal{P}_{-1}$. It is well known that for each equation $\mathcal{P}_f$ with $f = \mathrm{const}$, its maximal invariance algebra is infinite-dimensional [19, Example 2.42]. The space of conservation laws of $\mathcal{P}_f$ is also infinite-dimensional and is spanned by conservation laws with the characteristics $\mu = \psi\exp(\frac{v}{2f})$, where the parameter-function $\psi = \psi(t,x)$ runs through the solution set of the linear heat equation $\psi_t = f\psi_{xx}$ [20]. This is explained by the fact that the equation $\mathcal{P}_f$ with $f = \mathrm{const}$ can be linearized to the "backward" heat equation $\tilde{v}_t + f\tilde{v}_{xx} = 0$ by means of the variable change $\tilde{v} = e^{v/f}$. Moreover, this essentially complicates the structure of the equivalence groupoid of the entire class (1), cf. Proposition 2 below, and makes the study challenging.

Admissible transformations considered with the natural operation of transformation composition constitute the so-called equivalence groupoid of the class (1). The notion of equivalence groupoid was suggested in [21,22]. In order to describe the equivalence groupoid of the class (1) we apply the technique based on partitioning the class (1) into normalized (in the usual, or in the generalized, or in the generalized extended sense) subclasses, cf. [23,24]. Equations of different subclasses are not similar with respect to point transformations. For each of the subclasses, we construct the corresponding equivalence group of the same kind (usual, generalized, or generalized extended) as the kind of normalization of the subclass. We also comprehensively compare the equivalence groupoids of the classes (1), (2) and (3).

Definitions of admissible transformations, normalized and semi-normalized classes of differential equations, various kinds of equivalence groups and other concepts and techniques related to group classification problems can be found in [24–26] and references therein.

Journal of Physics: Conference Series **621** (2015) 012011 doi:10.1088/1742-6596/621/1/012011

2. Equivalence groupoid

Transformational properties in the class (1) are exhaustively described by the following chain of assertions. These assertions result from a long calculation procedure, and thus it is convenient to place their proofs in a separate section.

Proposition 1. *The usual equivalence group $G^{\sim}_{\rm pot}$ of the class (1) consists of the transformations*

$$\tilde t=\alpha t+\beta,\quad \tilde x=\kappa(x+\mu_1 t+\mu_0),\quad \tilde v=\frac{\kappa^2}{\alpha}\left(v+\frac{\mu_1}{2}x+\frac{\mu_1^2}{4}t+\nu\right),\quad \tilde f=\frac{\kappa^2}{\alpha}f \tag{4}$$

where α, β, κ, μ_1, μ_0, ν are arbitrary constants with $\alpha\neq 0$ and $\kappa\neq 0$.

We point out that the generalized extended equivalence group of this class coincides with its usual equivalence group.

Proposition 2. *The generalized equivalence group of the subclass $\mathcal P^3=\{\mathcal P_f\mid f=\mathrm{const}\}$ of the class (1) is formed by the transformations*

$$\tilde t=\frac{\alpha t+\beta}{\gamma t+\delta},\quad \tilde x=\kappa\frac{x+\mu_1 t+\mu_0}{\gamma t+\delta},\quad \tilde v=\frac{\kappa^2 f}{\alpha\delta-\beta\gamma}\ln\left|F^1\left(e^{v/f}+F^2\right)\right|,\quad \tilde f=\frac{\kappa^2}{\alpha\delta-\beta\gamma}f, \tag{5}$$

where α, β, γ, δ, κ, μ_1, μ_0 are arbitrary constants with $\alpha\delta-\beta\gamma\neq 0$ and $\kappa\neq 0$; the tuple $(\alpha,\beta,\gamma,\delta,\kappa)$ is defined up to a nonzero multiplier;

$$F^1=\begin{cases} k\sqrt{|\gamma t+\delta|}\exp\left(-\dfrac{(\gamma x-\mu_1\delta+\mu_0\gamma)^2}{4f\gamma(\gamma t+\delta)}\right), & \gamma\neq 0,\\[2ex] k\exp\dfrac{2\mu_1 x+\mu_1^2 t}{4f}, & \gamma=0,\end{cases} \tag{6}$$

k is a nonzero constant, and F^2 is a solution of the linear equation

$$F^2_t+fF^2_{xx}=0. \tag{7}$$

The subclass $\mathcal P^3$ is normalized in the generalized sense.

The usual equivalence group of the subclass $\mathcal P^3$ coincides with $G^{\sim}_{\rm pot}$, but its generalized equivalence group is wider. Therefore, the subclass $\mathcal P^3$ is not normalized in the usual sense. Moreover, this also implies that in the subclass $\mathcal P^3$ there exist admissible transformations that are not induced by $G^{\sim}_{\rm pot}$, which is the generalized extended equivalence group of the class (1). This means that the class (1) is not normalized in any sense.

Proposition 2 shows that for any constant $f\neq 0$ the equation $\mathcal P_f$ has rather complicated point symmetry transformations that are not related to equivalence transformations of the class (1); one may substitute $\kappa^2=\alpha\delta-\beta\gamma$ into (5) to derive the form of symmetry transformations of $\mathcal P_f$ and observe the above fact. It is easy to check that any admissible transformation in the subclass $\mathcal P^3$ is induced by the composition of a symmetry transformation of the corresponding initial equation and a transformation from $G^{\sim}_{\rm pot}$. In other words, the subclass $\mathcal P^3$ is semi-normalized in the usual sense. This is a consequence of that the subclass $\mathcal P^3$ is the single orbit of any equation from $\mathcal P^3$ with respect to its usual equivalence group $G^{\sim}_{\rm pot}$, cf. the proof of [27, Proposition 2].

Proposition 3. *The usual equivalence group of the subclass $\mathcal P^1=\{\mathcal P_f\mid f_{xxx}\neq 0\}$ of the class (1) coincides with the usual equivalence group $G^{\sim}_{\rm pot}$ of the entire class (1) and, thus, it is governed by (4). The subclass $\mathcal P^1$ is normalized in the usual sense.*

Although the class (1) is not normalized, it can be partitioned into three subclasses that are normalized (in certain senses), namely $\mathcal{P}^1$, $\mathcal{P}^3$, and the complement of their union, $\mathcal{P}^2 = \overline{\mathcal{P}^1 \cup \mathcal{P}^3}$. The subclass $\overline{\mathcal{P}^1}$, which is singled out from the class (1), by the constraint $f_{xxx} = 0$ has complicated transformation properties and is not normalized. At the same time, subtracting the small subclass $\mathcal{P}^3$ from $\overline{\mathcal{P}^1}$ we obtain the normalized subclass $\mathcal{P}^2$.

Proposition 4. *The generalized extended equivalence group of the subclass* $\mathcal{P}^2 = \{\mathcal{P}_f \mid f_{xxx} = 0, f \neq \mathrm{const}\}$ *of the class* (1) *is constituted by the transformations of the form*

$$\tilde{t} = \frac{1}{c_0}\int (X^1)^2\,\mathrm{d}t + c_5, \quad \tilde{x} = X^1 x + c_1\int (X^1)^2\left(\int\frac{f^1}{\lambda}\,\mathrm{d}t + c_3\right)\mathrm{d}t + c_4,$$
$$\tilde{v} = c_0\left(v - \frac{c_1X^1}{4\lambda}x^2 + X^2x + \int (X^2)^2\,\mathrm{d}t + \frac{c_1}{2}\int\frac{f^0}{\lambda}X^1\,\mathrm{d}t\right) + c_5, \quad \tilde{f} = c_0 f, \tag{8}$$

where $f = f^2(t)x^2 + f^1(t)x + f^0(t)$,

$$X^1(t) := \left(c_1\int\frac{\mathrm{d}t}{\lambda} + c_2\right)^{-1}, \quad X^2(t) := \frac{c_1X^1}{2}\left(\int\frac{f^1}{\lambda}\,\mathrm{d}t + c_3\right), \quad \lambda(t) := e^{2\int f^2\,\mathrm{d}t},$$

and $c_0, \dots, c_5$ *are arbitrary constants with* $c_0 \neq 0$ *and* $(c_1, c_2) \neq (0, 0)$.

The subclass $\mathcal{P}^2$ *is normalized in the generalized extended sense.*

The usual equivalence group of the subclass $\mathcal{P}^2$ also coincides with $G^{\sim}_{\mathrm{pot}}$, but it does not generate all admissible transformations in this subclass.

Remark 1. Two approaches can be used here when we represent the generalized extended equivalence group of $\mathcal{P}^2$. On the one hand, if we consider the function f as the single arbitrary element of the subclass $\mathcal{P}^1$, then we should explicitly express the coefficients f^2, f^1 and f^0 of the polynomial $f = f^2(t)x^2 + f^1(t)x + f^0(t)$ in terms of f and its derivatives,

$$f^2 = \frac{1}{2}f_{xx}, \quad f^1 = f_x - xf_{xx}, \quad f^0 = f + \frac{x^2}{2}f_{xx} - xf_x,$$

and substitute these expressions into (8). On the other hand, we may assume that the three functions f^2, f^1 and f^0 (or, e.g., λ, f^1 and f^0) are the arbitrary elements of the class (1) and perform a reparametrization of this class. In particular, the equation $\tilde{f} = c_0 f$ should be considered as $\tilde{f}^2(\tilde{t})\tilde{x}^2 + \tilde{f}^1(\tilde{t})\tilde{x} + \tilde{f}^0(\tilde{t}) = c_0\big(f^2(t)x^2 + f^1(t)x + f^0(t)\big)$, and then it should be split with respect to x in order to derive explicit expressions for $\tilde{f}^2$, $\tilde{f}^1$ and $\tilde{f}^0$ in terms of f^2, f^1 and f^0. In this paper, we prefer to avoid any reparametrization, and thus the notation of λ serves only for minimizing the size of expressions.

Remark 2. Note that in the case when $f^2 = 0$ (i.e. $f = f^1(t)x + f^0(t)$), we have $\lambda = 1$. Hence the expressions (8) after redenoting the constants take the form

$$\tilde{t} = \frac{\alpha t + \beta}{\gamma t + \delta}, \quad \tilde{x} = \frac{\kappa(x + \nu\Delta^{-1}(\alpha t + \beta))}{\gamma t + \delta} + \gamma\kappa\int\frac{\int f^1\,\mathrm{d}t}{(\gamma t + \delta)^2}\,\mathrm{d}t + c_4,$$
$$\tilde{v} = \frac{\kappa^2}{\Delta}\left(v - \frac{\gamma x^2 + 2\big[\gamma\int f^1\,\mathrm{d}t + \nu\big]x}{4(\gamma t + \delta)} + \frac{1}{4}\int\left[\frac{\gamma\int f^1\,\mathrm{d}t + \nu}{(\gamma t + \delta)}\right]^2\mathrm{d}t + \frac{\gamma}{2}\int\frac{f^0\,\mathrm{d}t}{\gamma t + \delta}\right) + c_5, \quad \tilde{f} = \frac{\kappa^2}{\Delta}f,$$

where $\Delta = \alpha\delta - \beta\gamma \neq 0$ and the tuple $(\alpha, \beta, \gamma, \delta, \kappa)$ is defined up to a nonzero multiplier. Further simplification is possible if $f^1 = 0$.

Proposition 5. *There are no point transformations between any two equations from different subclasses $\mathcal{P}^1$, $\mathcal{P}^2$ or $\mathcal{P}^3$ of the class* (1).

Proposition 5 completes the description of the equivalence groupoid of the class (1), cf. [24, Section 3.4].

Recall that the equivalence group of a subclass of a class of differential equations is called a *conditional equivalence group* of the entire class [23,24]. Only maximal conditional equivalence groups are essential [24, Definition 7]. Therefore, Propositions 2 and 4 mean that the class (1) admits nontrivial conditional equivalence groups (in the generalized or the generalized extended sense). Taking into account the results of [24, Section 3.4] and Proposition 5 we can also claim that the maximal conditional equivalence groups of the class (1) are exhausted by the above conditional equivalence groups and the equivalence group of the entire class (1).

3. Proofs

Given an equation $\mathcal{P}_f$ of the form (1) and another ("tilded") equation from the same class, $\tilde v_{\tilde t}+\tilde v_{\tilde x}^2+\tilde f\tilde v_{\tilde x\tilde x}=0$, we consider point transformations between these equations. In view of [25, Lemma 1] these transformations are of the form $\tilde t=T(t)$, $\tilde x=X(t,x)$, $\tilde v=V(t,x,v)$, where $T_tX_xV_v\neq0$. Expressing in the "tilded" equation all the tilded entities in terms of the untilded ones,

$$\tilde v_{\tilde t}=\frac{1}{T_t}\left(V_t+V_vv_t-\frac{V_x+V_vv_xX_t}{X_x}\right),$$
$$\tilde v_{\tilde x}=\frac{V_x+V_vv_x}{X_x},\quad \tilde v_{\tilde x\tilde x}=\frac{V_{xx}+2V_{vx}v_x+V_{vv}v_x^2+V_vv_{xx}}{X_x^2}-\frac{X_{xx}V_x+X_{xx}V_vv_x}{X_x^3},$$

and confining it to the manifold defined by the equation $\mathcal{P}_f$ via substituting $v_t=-v_x^2-fv_{xx}$, we split the result with respect to derivatives of v. This gives $\tilde f=X_x^2f/T_t$ and the classifying equations

$$V_v^2-\frac{X_x^2}{T_t}V_v+f\frac{X_x^2}{T_t}V_{vv}=0, \tag{9}$$
$$2V_xV_v-\frac{X_tX_x}{T_t}V_v+2f\frac{X_x^2}{T_t}V_{xv}-f\frac{X_xX_{xx}}{T_t}V_v=0, \tag{10}$$
$$V_t\frac{X_x^2}{T_t}-\frac{X_tX_x}{T_t}V_x+V_x^2+f\frac{X_x^2}{T_t}V_{xx}-f\frac{X_xX_{xx}}{T_t}V_x=0. \tag{11}$$

In order to find the usual equivalence group for the class (1), we vary f and split (9)–(11) with respect to it. Solving the obtained system of determining equations proves Proposition 1.

However, we are interested in further investigation of transformational properties of the class (1), so we continue to analyze the equations (9)–(11). Integrating (9) we get $V=\tilde f\ln\left|F^1(t,x)(e^{v/f}+F^2(t,x))\right|$, where F^1 and F^2 are smooth functions of their arguments, and $F^1(t,x)\neq0$ since $V_v\neq0$. There are two essentially different cases for admissible transformations of the class (1) depending on whether $F^2(t,x)=0$.

Case 1. Under the assumption $F^2 \neq 0$, the functions 1, $e^{v/f}$, $ve^{v/f}$, $\ln|F^1(e^{v/f}+F^2)|$ and $\ln|F^1(e^{v/f}+F^2)|e^{v/f}$ are linearly independent as functions of v over the ring of smooth functions of (t,x). Substituting the obtained expression for V into the equation (10) and splitting it with respect to the above functions we obtain $f_x=0$ and $X_{xx}=0$, so we can represent X as $X=X^1(t)x+X^0(t)$, $X^1\neq 0$. In the same way, from the determining equation (11) we have

$$f_t=0,\quad 2X^1_t=\frac{T_{tt}}{T_t}X^1,\quad 2fX^1F^1_x=(X^1_tx+X^0_t)F^1,$$
$$X^1(F^1_t+fF^1_{xx})=(X^1_tx+X^0_t)F^1_x,\quad X^1((F^1F^2)_t+f(F^1F^2)_{xx})=(X^1_tx+X^0_t)(F^1F^2)_x,$$

where the last three equations are already simplified in view of the condition $f=\text{const}$.

Integrating the second and the third equations we obtain

$$T_t=c_0(X^1)^2,\quad F^1=K(t)\exp\left(\frac{X^1_tx^2+2X^0_tx}{4fX^1}\right),$$

where c_0 is a nonzero constant and K is a nonzero smooth function of t to be defined. Thus, the fourth equation takes the form

$$\frac{K_t}{K}=-\frac{X^1}{4f}\left(\frac{X^1_t}{(X^1)^2}\right)_t x^2-\frac{X^1}{2f}\left(\frac{X^0_t}{(X^1)^2}\right)_t x+\frac{1}{4f}\left(\frac{X^0_t}{X^1}\right)^2-\frac{X^1_t}{2X^1}. \tag{12}$$

Since the ratio K_t/K in the left-hand side is a function of t, we conclude that the coefficients of x and of x^2 in (12) necessarily vanish. As a result, we derive the forms of X^1, X^0 and T,

$$X^1(t)=\frac{\kappa}{\gamma t+\delta},\quad X^0(t)=\kappa\frac{\mu_1t+\mu_0}{\gamma t+\delta},\quad T(t)=\frac{\alpha t+\beta}{\gamma t+\delta},$$

where α, β, γ, δ, μ_0, μ_1 and κ are arbitrary constants satisfying conditions of Proposition 2. Then equation (12) is rewritten as

$$\frac{K_t}{K}=\frac{(\mu_1\delta-\mu_0\gamma)^2}{4f(\gamma t+\delta)^2}+\frac{\gamma}{2(\gamma t+\delta)},$$

which can be easily integrated. The general solution is

$$K=\begin{cases}k\sqrt{|\gamma t+\delta|}\exp\left(-\dfrac{(\mu_1\delta-\mu_0\gamma)^2}{4f\gamma(\gamma t+\delta)}\right), & \gamma\neq 0,\\ k\exp\left(\dfrac{\mu_1^2}{4f}t\right), & \gamma=0,\end{cases}$$

where the constant k is nonzero. Hence F^1 is of the form (6) and F^2 is a solution of (7). Thus, Proposition 2 holds.

Case 2. If $F^2=0$, then V can be represented in the form $V=\dfrac{X_x^2}{T_t}v+V^0$, where V^0 is a smooth function of (t,x), which can be expressed in terms of F^1 (the precise expression is not essential). Equation (9) becomes an identity. Splitting equations (10)–(11) with respect to v we obtain

$$X_{xx}=0,\quad \text{hence}\quad X=X^1(t)x+X^0(t),$$
$$2X^1_tT_t-X^1T_{tt}=0,\quad\text{hence}\quad (X^1)^2=c_0T_t,\quad c_0\neq 0,$$
$$V^0_x=\frac{X^1}{2T_t}(X^1_tx+X^0_t),\quad\text{hence}\quad V^0=\frac{c_0X^1_t}{4X^1}x^2+\frac{c_0X^0_t}{2X^1}x+V^{00}(t),$$
$$(X^1X^1_{tt}-2(X^1_t)^2)x^2+2(X^1X^0_{tt}-2X^1_tX^0_t)x+2fX^1X^1_t+4T_tV^{00}_t-(X^0_t)^2=0. \tag{13}$$

If $f_{xxx}\neq 0$, then the functions 1, x, x^2 and f are linearly independent as functions of x over the ring of smooth functions of t. Thereby the last equation of system (13) admits splitting with respect to these functions. Integrating the complete system of its consequences we get $X^1=\kappa$, $X^0=\kappa(\mu_1 t+\mu_0)$, $T(t)=\alpha t+\beta$, $V=\frac{\kappa^2}{\alpha}\big(v+\frac{\mu_1}{2}x+\frac{\mu_1^2}{4}t+\nu\big)$ and $\tilde f=\frac{\kappa^2}{\alpha}f$, where $\alpha:=\frac{\kappa^2}{c_0}$, β, μ_1, μ_0, ν and κ are arbitrary constants with $\alpha\kappa\neq 0$. As a result, Proposition 3 is derived.

In the case $f_{xxx}=0$ we represent f as $f=f^2(t)x^2+f^1(t)x+f^0(t)$, where f^2, f^1 and f^0 are smooth functions of t, and the last equation of (13) after splitting with respect to x gives

$$2f^2+\frac{X^1_{tt}}{X^1_t}-2\frac{X^1_t}{X^1}=0,\quad \left(\frac{X^0_{tt}}{X^1_t}\right)_t=-\frac{f^1X^1}{(X^1)^2},\quad 4(X^1)^2=c_0(X^0)^2-2c_0f^0X^1X^1_t.$$

These equations can be easily integrated by quadratures, and the form of T can be found from the second line of (13). Consequently, we obtain Proposition 4.

The above consideration implies that for each admissible transformation of the class (1) the transformation component for f is of the form $\tilde f=c_0f$, where c_0 is a nonzero constant, the transformation component for t depends only of t and the transformation component for x is affine in x. Therefore, the constraints that single out the subclasses $\mathcal P^1$, $\mathcal P^2$ and $\mathcal P^3$ (i.e. $f_{xxx}\neq 0$; $f_{xxx}=0$ and $f\neq\mathrm{const}$; or $f=\mathrm{const}$) are preserved by any admissible transformation of the class (1). This proves Proposition 5.

4. Comparison of equivalence groupoids

The equivalence groupoid for the class (2) was presented in [8]. Presenting its description in a way similar to Section 2 we can state the following assertions.

Proposition 6. *The usual equivalence group of the class* (2) *consists of the transformations*

$$\tilde t=\alpha t+\beta,\quad \tilde x=\kappa(x+\mu_1t+\mu_0),\quad \tilde u=\frac{\kappa}{\alpha}(u+\mu_1),\quad \tilde f=\frac{\kappa^2}{\alpha}f, \tag{14}$$

where α, β, κ, μ_1, μ_0 are arbitrary constants with $\alpha\neq 0$ and $\kappa\neq 0$.

Proposition 7. *The usual equivalence group of the subclass $\mathcal C^1=\{\mathcal C_f\mid f_{xxx}\neq 0\}$ of the class* (2) *coincides with the usual equivalence group of the entire class* (2) *and is governed by* (14). *The subclass $\mathcal C^1$ is normalized in the usual sense.*

Proposition 8. *The generalized extended equivalence group of the subclass $\mathcal C^2=\{\mathcal C_f\mid f_{xxx}=0\}$ of the class* (2) *consists of the transformations*

$$\tilde t=\frac{1}{c_0}\int(X^1)^2\,\mathrm dt+c_5,\quad \tilde x=X^1x+\int(X^1)^2\left(c_1\int\frac{f^1}{\lambda}\,\mathrm dt+c_3\right)\mathrm dt+c_4,$$
$$\tilde u=c_0\left(\frac{1}{X^1}u-\frac{c_1}{\lambda}x+c_1\int\frac{f^1}{\lambda}\,\mathrm dt+c_3\right),\quad \tilde f=c_0f,$$

where $c_0,\dots,c_4$ are arbitrary constants with $c_0\neq 0$ and $(c_1,c_2)\neq(0,0)$,

$$f=f^2(t)x^2+f^1(t)x+f^0(t),\quad X^1(t):=\left(c_1\int\frac{\mathrm dt}{\lambda}+c_2\right)^{-1},\quad \lambda(t):=\mathrm e^{2\int f^2\,\mathrm dt}.$$

Proposition 9. *There are no point transformations between equations from the subclasses $\mathcal C^1$ and $\mathcal C^2$.*

Propositions 6–9 give the exhaustive description of the equivalence groupoid of the class (2). In view of [24, Section 3.4] they also imply that the class (2) has exactly two maximal conditional equivalence groups, namely the equivalence group of the whole class (2) and the generalized extended equivalence group of the subclass $\mathcal{C}^2$.

The equivalence groupoids of the classes (1) and (2) have similar structure. Propositions 6, 7, 8 and 9 are the counterparts of Propositions 1, 3, 4 and 5, respectively. This in particular establishes a correspondence between maximal conditional equivalence groups of these classes, or, equivalently, equivalence groups (of appropriate kind) of its subclasses, except the subclass $\mathcal{P}^3$ of the class (1), which has no counterpart in the class (2).

The class (3) is normalized in the usual sense. Its usual equivalence group consists of the transformations

$$\tilde t=\frac{\alpha t+\beta}{\gamma t+\delta},\quad \tilde x=\frac{\kappa x+\mu_1 t+\mu_0}{\gamma t+\delta},\quad \tilde u=\frac{\kappa(\gamma t+\delta)u-\kappa\gamma x+\mu_1\delta-\mu_0\gamma}{\alpha\delta-\beta\gamma},\quad \tilde f=\frac{\kappa^2}{\alpha\delta-\beta\gamma}f,$$

where the constant tuple $(\alpha,\beta,\gamma,\delta,\kappa,\mu_1,\mu_0)$ is defined up to a nonzero multiplier, $\alpha\delta-\beta\gamma\neq 0$ and $\kappa\neq 0$ [7]. Therefore, the class (3) admits no nontrivial generalizations of the equivalence group and no nontrivial conditional equivalence groups.

The equivalence group of the class (3) is more complicated than the equivalence groups of the classes (1) and (2) since it additionally contains conformal transformations and hence it is not solvable. The situation is not the same from the point of view of admissible transformations. The equivalence groupoid of the class (3) is of the simplest structure (among all possible ones), since this class is normalized with respect to its finite-dimensional usual equivalence group, which is obviously not the case for the classes (1) and (2). An explanation for why the class (3) so differs from the classes (1) and (2) in transformational properties is that an equation of the form (3) can be potentialized only if $f_{xxx}=0$.

It is interesting to emphasize that the intersection of the classes (2) and (3), which is their subclass consisting of equations of the form $u_t+uu_x+f(t)u_{xx}=0$ with $f\neq 0$, inherits transformational properties of the class (3) but not of the class (2). This subclass is also normalized in the usual sense and has the same equivalence group as the whole class (3). The restriction of the generalized extended equivalence group given in Proposition 8 to the subset of arbitrary elements f not depending on x coincides with the equivalence group for the class (3).

5. Conclusion

In this paper we have considered the class of generalized potential Burgers equations of the form (1). Its equivalence groupoid appeared to be of a complicated structure. In order to describe this structure in the optimal way, the class (1) has been represented as the disjoint union of three subclasses that are not related by point transformations, namely

- $\mathcal{P}^1=\{\mathcal{P}_f\mid f_{xxx}\neq 0\}$, which is normalized in the usual sense;
- $\mathcal{P}^2=\{\mathcal{P}_f\mid f_{xxx}=0,\ f\neq\mathrm{const}\}$, which is normalized in the generalized extended sense;
- $\mathcal{P}^3=\{\mathcal{P}_f\mid f=\mathrm{const}\}$, which is normalized in the generalized sense.

The partition of the class (1) into the three subclasses $\mathcal{P}^1$, $\mathcal{P}^2$ and $\mathcal{P}^3$ results in the partition of its equivalence groupoid into the equivalence groupoids of these subclasses.

The subclass $\mathcal{P}^3$ is in fact the orbit of the single equation of this subclass with $f=-1$, which is the potential Burgers equation. The complexity of the equivalence group of the subclass $\mathcal{P}^3$ reflects the complexity of the point symmetry group of a potential Burgers equation of the form (1) with any constant f. At the same time, the equivalence group of the class (1), which coincides with the usual equivalence group of the subclass $\mathcal{P}^1$, is rather simple.

Comprehensive analysis has shown that the equivalence groupoid of the class (1) has several similar features with the equivalence groupoid of the class (2) and completely differs from the equivalence groupoid of the class (3).

Any contact admissible transformation between two equations from the class (1) is the first prolongation of a point transformation between these equations [28, Proposition 2]. This is why the problem of describing contact transformations in the class (1) is reduced to that for point transformations. Thus, the knowledge of the (point) equivalence groupoid provides us with the exhaustive description of the contact equivalence groupoid.

Knowing transformational properties of equations from the class (1), which have been collected in Propositions 1–4, is useful for solving the group classification problem for the class (1). This problem will be a subject of a forthcoming paper [10].

Acknowledgements
The author thanks Prof. R.O. Popovych for formulation of the present problem and collaborating in the course of study a great number of problems related to generalized Burgers equations.

References

[1] Katkov V L 1965 Group classification of solutions of the Hopf equation *Zh. Prikl. Mekh. i Tekhn. Fiz.* **6** 105
[2] Doyle J and Englefield M J 1990 Similarity solutions of a generalized Burgers equation *IMA J. Appl. Math.* **44** 145
[3] Kingston J G and Sophocleous C 1991 On point transformations of a generalised Burgers equation *Phys. Lett.* A **155** 15
[4] Boyko V M 1997 On new generalizations of the Burgers and Korteweg–de Vries equations *Proc. 2nd Int. Conf. Symmetry in Nonlinear Mathematical Physics (Kyiv, 1997)* (Kyiv: Institute of Mathematics) **1** p 122
[5] Wafo Soh C 2004 Symmetry reductions and new exact invariant solutions of the generalized Burgers equation arising in nonlinear acoustics *Internat. J. Engng. Sci.* **42** 1169
[6] Sophocleous C 2004 Transformation properties of a variable-coefficient Burgers equation *Chaos Solitons Fractals* **20** 1047
[7] Pocheketa O A and Popovych R O 2012 Reduction operators and exact solutions of generalized Burgers equations *Phys. Lett.* A **376** 2847 (*Preprint* arXiv:1112.6394)
[8] Pocheketa O A 2013 Normalized classes of generalized Burgers equations *Proc. 6th Int. Workshop on Group Analysis of Differential Equations and Integrable Systems (Protaras, 2012)* (Nicosia: University of Cyprus) p 170
[9] Vaneeva O O, Sophocleous C and Leach P G L 2015 Lie symmetries of generalized Burgers equations: application to boundary-value problems *J. Eng. Math.* **91** 165 (*Preprint* arXiv:1303.3548)
[10] Pocheketa O A and Popovych R O 2015 Extended symmetry analysis of generalized Burgers equations *In preparation*
[11] Crighton D G and Scott J F 1979 Asymptotic solutions of model equations in nonlinear acoustics *Philos. Trans. Roy. Soc. London Ser.* A **292** 101
[12] Scott J F 1981 The long time asymptotics of solutions to the generalized Burgers equation *Proc. Roy. Soc. London Ser.* A **373** 443
[13] Hammerton P W and Crighton D G 1989 Approximate solution methods for nonlinear acoustic propagatioin over long ranges *Proc. R. Soc. Lond.* A **426** 125
[14] Hammerton P W and Crighton D G 1989 Old-age behaviour of cylindrical and spherical nonlinear waves: numerical and asymptotic results *Proc. R. Soc. Lond.* A **422** 387
[15] Sioñoid P N and Cates A T 1994 The generalized Burgers and Zabolotskaya–Khokhlov equations: transformations, exact solutions and qualitative properties *Proc. Roy. Soc.* A **447** 253
[16] Sachdev P L, Joseph K T and Nair K R C 1994 Exact N-wave solutions for the non-planar Burgers equation *Proc. R. Soc. London Ser.* A **445** 501
[17] Leibovich S and Seebass A R 1974 *Nonlinear Waves, Based on a series of seminars held at Cornell University (Ithaca, N.Y., 1969)* (Ithaca, N.Y. – London: Cornell University Press)
[18] Pocheketa O A, Popovych R O and Vaneeva O O 2014 Group classification and exact solutions of variable-coefficient generalized Burgers equations with linear damping *Appl. Math. Comput* **243** 232 (*Preprint* arXiv:1308.4265)

[19] Olver P J 1993 *Applications of Lie Groups to Differential Equations* 2nd ed. Graduate Texts in Mathematics, vol 107 (New York: Springer–Verlag)

[20] Popovych R O and Ivanova N M 2005 Hierarchy of conservation laws of diffusion-convection equations *J. Math. Phys.* **46** 043502 (*Preprint* arXiv:math-ph/0407008)

[21] Popovych R O and Bihlo A 2012 Symmetry preserving parameterization schemes *J. Math. Phys.* **53** 073102 (*Preprint* arXiv:1010.3010)

[22] Bihlo A, Cardoso-Bihlo E D S and Popovych R O 2012 Complete group classification of a class of nonlinear wave equations *J. Math. Phys.* **53** 123515 (*Preprint* arXiv:1106.4801)

[23] Popovych R O 2006 Classification of admissible transformations of differential equations *Collection of Works of Institute of Mathematics* vol 3 (Kyiv: Institute of Mathematics) p 239

[24] Popovych R O, Kunzinger M and Eshraghi H 2010 Admissible transformations and normalized classes of nonlinear Schrödinger equations *Acta Appl. Math.* **109** 315 (*Preprint* arXiv:math-ph/0611061)

[25] Ivanova N M, Popovych R O and Sophocleous C 2010 Group analysis of variable coefficient diffusion–convection equations. I. Enhanced group classification *Lobachevskii J. Math.* **31** 100 (*Preprint* arXiv:0710.2731)

[26] Vaneeva O O, Popovych R O and Sophocleous C 2009 Enhanced group analysis and exact solutions of variable coefficient semilinear diffusion equations with a power source *Acta Appl. Math.* **106** 1 (*Preprint* arXiv:0708.3457)

[27] Boyko V M, Popovych R O and Shapoval N M 2014 Equivalence groupoids of classes of linear ordinary differential equations and their group classification *Preprint* arXiv:1403.6062

[28] Popovych R O and Samoilenko A M 2008 Local conservation laws of second-order evolution equations *J. Phys.* A **41** 362002 (*Preprint* arXiv:0806.2765)

Journal of Physics: Conference Series **621** (2015) 012012 doi:10.1088/1742-6596/621/1/012012

Canonical forms for matrices of Saletan contractions

Dmytro R Popovych

Faculty of Mechanics and Mathematics, National Taras Shevchenko University of Kyiv, building 7, 2, Academician Glushkov Av., 03127 Kyiv, Ukraine

E-mail: `deviuss@gmail.com`

Abstract. We show that each Saletan (linear) contraction can be realized, up to change of bases of the initial and the target Lie algebras, by a matrix-function that is completely defined by a partition of the dimension of Fitting component of its value at the limit value of the contraction parameter. The codimension of the Fitting component and this partition constitute the signature of the Saletan contraction. We study Saletan contractions with Fitting component of maximal dimension and trivial one-part partition. All contractions of such kind in dimension three are completely classified.

1. Introduction

Historically, the first extensively studied kind of contractions of Lie algebras, after Segal introduced the general notion of contractions [1], was the class of Saletan (linear) contractions. Contractions of Lie algebras became known as a tool of theoretical physics after the famous papers by Inönü and Wigner [2,3] on an important specific subclass of linear contractions. Note that Inönü and Wigner planned to consider the whole class of linear contractions but they erroneously claimed in [2] that any linear contraction is diagonalizable. Even though Inönü and Wigner corrected their considerations in the next paper [3], they proceeded to exclusively study diagonalizable linear contractions, which due to their contribution are now called *Inönü–Wigner contractions*. The effectiveness of such contractions in applications is ensured by their close connection to subalgebras of initial algebras. More precisely, in modern terms the main result of [2], which is Theorem 1 at [2, p. 513], can be reformulated in the following way: Any Inönü–Wigner contraction of a Lie algebra $\mathfrak{g}$ to a Lie algebra $\mathfrak{g}_0$ is associated with a subalgebra of $\mathfrak{g}$, say $\mathfrak{s}$, and starting with an arbitrary subalgebra of the algebra $\mathfrak{g}$ one can construct an Inönü–Wigner contraction of this algebra. In the contracted algebra $\mathfrak{g}_0$ there exists an Abelian ideal $\mathfrak{i}$ such that the factor-algebra $\mathfrak{g}_0/\mathfrak{i}$ is isomorphic to $\mathfrak{s}$.

A thorough study of linear contractions was conducted by Saletan in the course of preparation of his doctoral thesis and was published in [4]. In particular, Saletan obtained a simplified form for matrices of linear contractions up to reparametrization and basis change, derived a criterion for a linear matrix-function to be a contraction matrix, and gave the expression for the Lie bracket of the contracted Lie algebra. He also studied iterated linear contractions, related characteristics of the contraction matrix with the subalgebraic structure of the initial algebra, and discussed linear contractions of representations of Lie algebras.

Further studies by other authors extended rather than deepened Saletan's results. Thus, Inönü–Wigner contractions of three- and four-dimensional Lie algebras were classified [5,6] due to the subalgebraic structure of these algebras being known [7]. Following Saletan, contractions realized by matrix-functions of the generalized form $A\varepsilon + B\varepsilon^p$, where A and B are constant matrices and ε is the contraction parameter, were considered in a similar fashion [8–10]. Linear contractions of general algebraic structures were studied in [11].

In contrast to the above studies, this paper is aimed to enhance the original results by Saletan. We find the canonical form of Saletan contraction matrices, which creates the basis for introducing the notion of Saletan contraction's signature, for developing an algorithm for computation of Saletan contractions, and for posing new problems in this field.

The structure of the paper is the following: Basic notions and results on contractions and, specifically, on Saletan contractions are presented in Section 2. The main result of the paper, Theorem 1, which deals with the canonical form of Saletan contraction matrices, is proved in Section 3. After defining the notion of Saletan signature, we relate the signature of a Saletan contraction with the nested chain of subalgebras of the initial algebra that corresponds to this contraction. In Section 4 we carry out a preliminary study of Saletan contractions associated with chains of nested subalgebras of the maximal possible length, which coincides with the algebra dimension. Then we exhaustively describe such contractions between three-dimensional Lie algebras over the complex (resp. real) field. In the final section, we discuss obtained results and propose new problems for the further investigation.

2. Basic notions and auxiliary results

Given a finite-dimensional vector space V over the field $\mathbb{F} = \mathbb{R}$ or $\mathbb{F} = \mathbb{C}$, by $\mathcal{L}_n = \mathcal{L}_n(\mathbb{F})$ we denote the set of all possible Lie brackets on V, where $n = \dim V < \infty$. Each element μ of $\mathcal{L}_n$ corresponds to a Lie algebra with the underlying space V, $\mathfrak{g} = (V, \mu)$. Fixing a basis $\{e_1, \dots, e_n\}$ of V leads to a bijection between $\mathcal{L}_n$ and the set of structure constant tensors

$$\mathcal{C}_n = \big\{ (c_{ij}^k) \in \mathbb{F}^{n^3} \mid c_{ij}^k + c_{ji}^k = 0,\ c_{ij}^{i'} c_{i'k}^{k'} + c_{ki}^{i'} c_{i'j}^{k'} + c_{jk}^{i'} c_{i'i}^{k'} = 0 \big\}.$$

The structure constant tensor $(c_{ij}^k) \in \mathcal{C}_n$ associated with a Lie bracket $\mu \in \mathcal{L}_n$ is given by the formula $\mu(e_i, e_j) = c_{ij}^k e_k$. Here and in what follows, the indices i, j, k, i', j', k', p and q run from 1 to n and the summation convention over repeated indices is assumed. The right action of the group $\mathrm{GL}(V)$ on $\mathcal{L}_n$, which is conventional for the physical literature, is defined as

$$(U \cdot \mu)(x, y) = U^{-1}\big(\mu(Ux, Uy)\big) \quad \forall\, U \in \mathrm{GL}(V),\ \forall\, \mu \in \mathcal{L}_n,\ \forall\, x, y \in V.$$

Definition 1. Given a Lie bracket $\mu \in \mathcal{L}_n$ and a continuous matrix function $U\colon (0,1] \to \mathrm{GL}(V)$, we construct the parameterized family of Lie brackets $\mu_\varepsilon = \mu \circ U_\varepsilon$, $\varepsilon \in (0,1]$. Each Lie algebra $\mathfrak{g}_\varepsilon = (V, \mu_\varepsilon)$ is isomorphic to $\mathfrak{g} = (V, \mu)$. If the limit

$$\lim_{\varepsilon \to +0} \mu_\varepsilon(x, y) = \lim_{\varepsilon \to +0} U_\varepsilon{}^{-1} \mu(U_\varepsilon x, U_\varepsilon y) =: \mu_0(x, y)$$

exists for any $x, y \in V$, then μ_0 is a well-defined Lie bracket. The Lie algebra $\mathfrak{g}_0 = (V, \mu_0)$ is called a *one-parametric continuous contraction* (or simply a *contraction*) of the Lie algebra $\mathfrak{g}$. We call a limiting process that provides $\mathfrak{g}_0$ from $\mathfrak{g}$ with a matrix function a *realization* of the contraction $\mathfrak{g} \to \mathfrak{g}_0$.

The notion of contraction is extended to the case of an arbitrary algebraically closed field in terms of orbit closures in the variety of Lie brackets, see, e.g., [12–16].

If a basis $\{e_1, \dots, e_n\}$ of V is fixed, then the operator U_ε can be identified with its matrix $U_\varepsilon \in \mathrm{GL}_n(\mathbb{F})$, which is denoted by the same symbol, and Definition 1 can be reformulated in terms of structure constants. Let $C = (c_{ij}^k)$ be the tensor of structure constants of the algebra $\mathfrak{g}$ in the basis chosen. Then the tensor $C_\varepsilon = (c_{\varepsilon,ij}^k)$ of structure constants of the algebra $\mathfrak{g}_\varepsilon$ in this basis is the result of the action by the matrix U_ε on the tensor C, $C_\varepsilon = C \circ U_\varepsilon$. In term of components this means that

$$c^k_{\varepsilon,ij} = (U_\varepsilon)^{i'}_i (U_\varepsilon)^{j'}_j (U_\varepsilon{}^{-1})^k_{k'} c^{k'}_{i'j'}.$$

Then Definition 1 is equivalent to that the limit

$$\lim_{\varepsilon\to+0} c^k_{\varepsilon,ij} =: c^k_{0,ij}$$

exists for all values of i, j and k and, therefore, $c^k_{0,ij}$ are components of the well-defined structure constant tensor C_0 of the Lie algebra $\mathfrak{g}_0$. The parameter ε and the matrix-function U_ε are called a *contraction parameter* and a *contraction matrix*, respectively.

The following useful assertion is obvious.

Lemma 1. *If the matrix U_ε of a contraction $\mathfrak{g} \to \mathfrak{g}_0$ is represented in the form $U_\varepsilon = \hat U_\varepsilon \check U_\varepsilon$, where $\hat U$ and $\check U$ are continuous functions from $(0, 1]$ to $\mathrm{GL}_n(\mathbb{F})$ and the function $\check U$ has a limit $\check U_0 \in \mathrm{GL}_n(\mathbb{F})$ at $\varepsilon \to +0$, then $\hat U_\varepsilon \check U_0$ also is a matrix of the contraction $\mathfrak{g} \to \mathfrak{g}_0$.*

Remark 1. Lemma 1 implies that $\hat U_\varepsilon$ is a matrix of the equivalent contraction $\mathfrak{g} \to \tilde{\mathfrak{g}}_0$, where $\tilde{\mathfrak{g}}_0 = (V, \mu_0 \circ \check U_0^{-1})$ is the algebra isomorphic to $\mathfrak{g}_0$ with respect to the matrix $\check U_0^{-1}$.

Historically, the first contractions studied were the ones realized by linear matrix-functions.

Definition 2. A realization of a contraction with a matrix-function that is linear in the contraction parameter is called a *Saletan (linear) contraction* [4].

This class of contractions includes the Inönü–Wigner contractions [2–4].

The matrix of any linear contraction has a well-defined limit at $\varepsilon = 0$. This is why in contrast to the general definition of contractions, in the case of a linear contraction its matrix-function U_ε can be assumed to be defined on the closed interval $[0, 1]$. Then it is convenient to represent the matrix U_ε in the form $U_\varepsilon = (1 - \varepsilon)U_0 + \varepsilon U_1$, where U_0 and U_1 are the values of U_ε at $\varepsilon = 0$ and $\varepsilon = 1$, respectively [4]. By the definition of contraction matrix, the matrix U_1 is nonsingular, and, for proper contractions, the matrix U_0 is necessarily singular.

There exist specific reparametrizations that preserve the class of Saletan contractions [4]. Let $U_\varepsilon = B + \varepsilon A$ be the matrix of a Saletan contraction. We fix $\lambda > -1$ and consider the matrix-function U_ε on the interval $[0, (1 + \lambda)^{-1}]$ instead of $[0, 1]$. Then

$$B + \varepsilon A = (1 - \lambda\varepsilon)B + \varepsilon(A + \lambda B) = (1 - \lambda\varepsilon)\left(B + \frac{\varepsilon}{1 - \lambda\varepsilon}(A + \lambda B)\right).$$

The multiplier $(1-\lambda\varepsilon)$ is not essential since its limit at $\varepsilon = 0$ equals 1. Removing this multiplier and denoting $\varepsilon/(1 - \lambda\varepsilon)$ by $\tilde\varepsilon$, we obtain the well-defined linear matrix-function

$$\tilde U_{\tilde\varepsilon} = B + \tilde\varepsilon(A + \lambda B), \quad \tilde\varepsilon \in [0, 1],$$

which realizes the same Saletan contraction as U_ε.

3. Canonical forms of Saletan contraction matrices

We denote the $m \times m$ unit matrix by E^m, and $m \times m$ Jordan block with an eigenvalue λ by J^m_λ.

Theorem 1. *Up replacing the algebras $\mathfrak{g}$ and $\mathfrak{g}_0$ with isomorphic ones, every Saletan contraction $\mathfrak{g} \to \mathfrak{g}_0$ is realized by a matrix of the canonical form*

$$E^{n_0} \oplus J^{n_1}_\varepsilon \oplus \cdots \oplus J^{n_s}_\varepsilon, \quad \textit{or, equivalently,} \quad E^{n_0} \oplus J^{n_1}_0 \oplus \cdots \oplus J^{n_s}_0 + \varepsilon E^n, \tag{1}$$

where $n_0 + \cdots + n_s = n$.

Proof. The initial steps of the proof follow [4]. As the contraction matrix U_ε is linear in ε, it admits the representation $U_\varepsilon = (1-\varepsilon)U_0 + \varepsilon U_1$, where U_0 and U_1 are the values of U_ε at $\varepsilon = 0$ and $\varepsilon = 1$, respectively. Taking the algebra $\mathfrak{g}_1 = (V, \mu_1)$ instead of $\mathfrak{g}$ as the initial algebra of the contraction leads to the new matrix $U_1 = E^n$. We restrict the range of the parameter ε to $[0, 1/2]$ and factor out the multiplier $1-\varepsilon$, which can be canceled due to Lemma 1 as it has limit 1 at $\varepsilon = 0$. We reparameterize the contraction matrix by introducing the new parameter $\tilde{\varepsilon} = \varepsilon/(1-\varepsilon)$, whose range is $[0,1]$, and we omit the tilde over ε. As a result, the contraction matrix takes the form $U_0 + \varepsilon E^n$, which was derived in [4].

We carry out the Fitting decomposition of the space V relative to the operator U_0. More specifically, we partition the space V into the direct sum of subspaces V_0 and V_1, $V = V_1 \oplus V_0$ so that the restriction W_0 of the operator U_0 on the Fitting null component V_0 is nilpotent and the restriction W_1 of this operator on the Fitting one component V_1 is nonsingular. The partition of the space V induces the partition of the new operator U_ε,

$$U_\varepsilon = U_0 + \varepsilon E^n = (W_1 + \varepsilon E^{n_0}) \oplus (W_0 + \varepsilon E^{n-n_0}),$$

where $n_0 = \dim V_1$ is the dimension of the Fitting one component relative to U_0.

Consider the matrices $\hat{U}_\varepsilon = E^{n_0} \oplus (W_0 + \varepsilon E^{n-n_0})$ and $\check{U}_\varepsilon = (W_1 + \varepsilon E^{n_0}) \oplus E^{n-n_0}$. The representation $U_\varepsilon = \hat{U}_\varepsilon \check{U}_\varepsilon$ satisfies the conditions of Lemma 1. Hence in view of Remark 1 $\hat{U}_\varepsilon$ is a matrix of the equivalent contraction $\mathfrak{g}_1 \to \tilde{\mathfrak{g}}_0$, where $\tilde{\mathfrak{g}}_0 = (V, \mu_0 \circ \check{U}_0^{-1})$ is the algebra isomorphic to $\mathfrak{g}_0$ with respect to the matrix $\check{U}_0^{-1}$.

Replacing the basis elements spanning the subspace V_0, we can reduce the nilpotent matrix W_0 to its Jordan form, which is $J_0^{n_1} \oplus \cdots \oplus J_0^{n_s}$ for some $n_1, \ldots, n_s$ with $n_1 + \cdots + n_s = \dim V_0 = n - n_0$. Then the matrix $W_0 + \varepsilon E^{n-n_0}$ takes its Jordan form $J_\varepsilon^{n_1} \oplus \cdots \oplus J_\varepsilon^{n_s}$, which is equivalent to representing the contraction matrix $\hat{U}_\varepsilon$ in the first canonical form.

In view of Lemma 1, instead of the matrix $\hat{U}_\varepsilon$ we can consider the matrix $E^{n_0} \oplus W_0 + \varepsilon E^n$, which differs from $\hat{U}_\varepsilon$ by the right multiplier $(1+\varepsilon)E^{n_0} \oplus E^{n-n_0}$ with the unit matrix E^n as its limit at $\varepsilon = 0$. Then the above basis change in V_0 results in the second canonical form of the linear contraction matrix U_ε. □

Definition 3. Theorem 1 means that any Saletan contraction can be realized by a matrix of the form $AS_\varepsilon B$, where A and B are constant nonsingular matrices and the linear matrix-function S_ε is in the canonical form (1). Then the tuple $(n_0; n_1, \ldots, n_s)$, where $n_1, \ldots, n_s$ constitute a partition of the dimension $n - n_0$ of the Fitting null component relative to U_0 and $n_0 \in \{0, \ldots, n\}$, is called the *signature* of this Saletan contraction.

Due to containing a partition of $n - n_0$, the signature of a Saletan contraction is defined up to permutation of its parts excluding the first one. Saletan contraction with signature (n) is improper, i.e., the contracted algebra is isomorphic to the initial one. So, for a proper Saletan contraction we necessarily have $n_0 < n$. Inönü–Wigner contractions are associated with Saletan signatures of the form $(n_0; 1, \ldots, 1)$. The Saletan signature $(0; 1, \ldots, 1)$ corresponds to the trivial contraction to the Abelian algebra.

The necessary and sufficient condition for the algebra $\mathfrak{g}$ to be contracted by the linear matrix-function U_ε [4] is

$$U_0^2[x,y]^0 - U_0[U_0x, y]^0 - U_0[x, U_0y]^0 + [U_0x, U_0y]^0 = 0 \quad \forall x, y \in V. \tag{2}$$

Here and in what follows $[\cdot,\cdot]^0$ and $[\cdot,\cdot]^1$ denote the projections of the Lie brackets $[\cdot,\cdot]$ on the subspaces V_0 and V_1, respectively, which are not, in general, Lie brackets. Then the contracted Lie bracket is defined by

$$[x,y]_0 = W_1^{-1}[U_0x, U_0y]^1 - W_0[x,y]^0 + [U_0x, y]^0 + [x, U_0y]^0 \quad \forall x, y \in V.$$

The use of the canonical form of U_ε simplifies analysis of both the necessary and sufficient conditions and properties of the contracted Lie bracket. In particular, then $W_1^{-1} = E^{n-n_0}$. We would like to emphasize that changing the basis of the underlying space without applying Lemma 1 can simplify the matrix W_1 only to its Jordan form.

Remark 2. If U_0 is the value of the matrix of a well-defined Saletan contraction of the Lie algebra $\mathfrak{g}$ at $\varepsilon = 0$, then each power of U_0 is the value of the matrix of another well-defined Saletan contraction $\mathfrak{g}$ at $\varepsilon = 0$. The image $\operatorname{im} U_0$ of U_0 is a subalgebra of $\mathfrak{g}$. Combining the above two claims, we have that for each $m = 0, 1, 2, \ldots$ the image $\mathfrak{s}_m := \operatorname{im} U_0^m$ of U_0^m is also a subalgebra of $\mathfrak{g}$, and $\mathfrak{s}_m = V_1$ if $m \geqslant m_0 := \max(n_1, \ldots, n_s)$ [4]. In other words, the matrix of any Saletan contraction is associated with the chain of nested subalgebras

$$\mathfrak{s}_0 := \mathfrak{g} \supset \mathfrak{s}_1 \supset \mathfrak{s}_2 \supset \cdots \supset \mathfrak{s}_{m_0} = V_1.$$

The dimensions of these subalgebras are completely defined by the contraction signature,

$$\dim \mathfrak{s}_m = n - l_1 - \cdots - l_m, \quad m = 0, \ldots, m_0, \quad \text{where} \quad l_m := |\{n_i \mid n_i \geqslant m,\ i = 1, \ldots, s\}|.$$

In particular, $\dim \mathfrak{s}_{m_0} = n_0$. The above relation establishes necessary conditions of consistency between the structure of a Lie algebra and signatures of its Saletan contractions.

Example 1. Consider the real three-dimensional orthogonal algebra so(3) with the canonical commutation relations $[e_1, e_2] = e_3$, $[e_2, e_3] = e_1$, $[e_3, e_1] = e_2$. The algebra so(3) has no two-dimensional subalgebras. Therefore, the only possible signature of a proper Saletan contraction of so(3) is $(1, 1, 1)$. The first canonical form of the contraction matrix with this signature is $E^1 \oplus J_\varepsilon^1 \oplus J_\varepsilon^1$, which realizes the single Inönü–Wigner contraction of so(3), which is to the Euclidean algebra e(2) defined by the nonzero commutation relations $[e_1, e_3] = -e_2$, $[e_2, e_3] = e_1$. This implies that the other proper contraction of so(3), which is to the Heisenberg algebra h(1) with the nonzero commutation relations $[e_2, e_3] = e_1$, cannot be realized as a Saletan contraction, cf. [4].

4. Saletan contractions with signature $(0; n)$

There are two different ways of studying Saletan contractions. Given a fixed pair of Lie algebras, one can check whether there exists a Saletan contraction between these algebras and then try to describe all possible Saletan contractions for this pair. The other way is to describe all possible contractions which are realized by contraction matrices with certain signature. A disadvantage of this way is the necessity of classifying Lie algebras that satisfy specific constraints.

In this section we consider contractions with the signature $(0; n)$. Choosing this signature poses the most restrictive constraints on the structure of the initial Lie algebra $\mathfrak{g}$ compared to other Saletan signatures, cf. equation (2). In particular, the algebra $\mathfrak{g}$ should contain a nested chain of n nonzero Lie subalgebras, and, in general, this condition on $\mathfrak{g}$ is not sufficient.[1]

For the signature $(0; n)$ we have $V_0 = V$ and we can set

$$U_0 = J_0^n.$$

Then $[\cdot,\cdot]^0 = [\cdot,\cdot]$ and the Saletan condition (2) takes the form

$$[U_0x, U_0y] - U_0[U_0x, y] = U_0([x, U_0y] - U_0[x, y]) \quad \forall\, x, y \in V.$$

or, equivalently, $[\![\operatorname{ad}_{U_0x}, U_0]\!] = U_0[\![\operatorname{ad}_x, U_0]\!]$ for any $x \in V$. Here and in what follows, $[\![A, B]\!]$ denotes the commutator of operators A and B, $[\![A, B]\!] := AB - BA$. Specifying this condition for

[1] This strongly differs from the case of Inönü–Wigner contractions, for which there exists a one-to-one correspondence with proper subalgebras of $\mathfrak{g}$.

basis elements, for which $U_0 = J_0^n$, we derive $[\![\mathrm{ad}_{e_i}, U_0]\!] = U_0[\![\mathrm{ad}_{e_{i+1}}, U_0]\!] = \cdots = U_0^{n-i}[\![\mathrm{ad}_{e_n}, U_0]\!]$, i.e.,

$$[\![\mathrm{ad}_{e_i}, U_0]\!] = U_0^{n-i}[\![A, U_0]\!], \tag{3}$$

where we use the notation $A = (a_{ij}) := \mathrm{ad}_{e_n}$. For each fixed i, the equation (3) can be considered as inhomogeneous linear system of algebraic equations with respect to entries of the matrix ad_{e_i}. A particular solution of this system is given by $U_0^{n-i}A$, since $[\![U_0^kA, U_0]\!] = U_0^kAU_0 - U_0^{k+1}A = U_0^k[\![A, U_0]\!]$. The solution space of the corresponding homogeneous system $[\![\mathrm{ad}_{e_i}, U_0]\!] = 0$ coincides with the space of matrices commuting with U_0, which is spanned by the powers of U_0 due to U_0 being a single Jordan block. Therefore, the general solution of the system (3) is

$$\mathrm{ad}_{e_i} = U_0^{n-i}A + b_{ij}U_0^{n-j}$$

with parameters b_{ij}, where $b_{nj} = 0$ as $\mathrm{ad}_{e_n} = A$ by definition and $a_{in} = 0$ as $Ae_n = [e_n, e_n] = 0$. Recall that we assume the summation convention over repeated indices. The Lie bracket is skew-symmetric, which implies

$$[e_i, e_n] + [e_n, e_i] = \mathrm{ad}_{e_i}e_n + \mathrm{ad}_{e_n}e_i = b_{ij}e_j + a_{ji}e_j = 0,$$

i.e., $b_{ij} + a_{ji} = 0$. In other words, the commutation relations of the algebra $\mathfrak{g}$ are

$$\begin{aligned}[e_i, e_j] = \mathrm{ad}_{e_i}e_j = U_0^{n-i}Ae_j - a_{ki}U_0^{n-k}e_j &= a_{kj}e_{k+i-n} - a_{ki}e_{k+j-n} \\ &= (a_{p+n-i,j} - a_{p+n-j,i})e_p. \end{aligned}\tag{4}$$

Here and in what follows, if an index goes beyond the index interval $\{1, \dots, n\}$, then the corresponding object is assumed zero. Thus, in view of (4) the skew-symmetric property of the Lie bracket obviously holds for any pair of elements of $\mathfrak{g}$. Note that the number of essential parameters in the above commutation relations does not exceed $n(n-1)$. The Jacobi identity imposes more constraints in the form of a system of quadratic equations with respect to entries of the matrix A,

$$\begin{aligned}&(a_{p+n-i,j} - a_{p+n-j,i})(a_{q+n-k,p} - a_{q+n-p,k}) + \\ &(a_{p+n-j,k} - a_{p+n-k,j})(a_{q+n-i,p} - a_{q+n-p,i}) + \\ &(a_{p+n-k,i} - a_{p+n-i,k})(a_{q+n-j,p} - a_{q+n-p,j}) = 0.\end{aligned}$$

Unfortunately, we were not able to solve this system for an arbitrary dimension of the underlying space.

The contracted Lie bracket is defined by $[x, y]_0 = [U_0x, y] + [x, U_0y] - U_0[x, y]$ for all $x, y \in V$. Hence, the commutation relations of the contracted algebra $\mathfrak{g}_0$ are

$$\begin{aligned}[e_i, e_j]_0 &= [e_{i-1}, e_j] + [e_i, e_{j-1}] - U_0[e_i, e_j] \\ &= (a_{p+n-i+1,j} - a_{p+n-j,i-1})e_p + (a_{p+n-i,j-1} - a_{p+n-j+1,i})e_p \\ &\quad - (a_{p+n-i,j} - a_{p+n-j,i})e_{p-1} \\ &= (a_{p+n-i,j-1} - a_{p+n-j,i-1})e_p.\end{aligned}$$

In particular, $[e_n, e_j]_0 = (a_{p,j-1} - a_{p+n-j,n-1})e_p$. Consider the matrix $A_0 = (a_{0,ij})$, where $a_{0,ij} = a_{i,j-1} - a_{i+n-j,n-1}$. In terms of A and J_0^n we have the representation

$$A_0 = AJ_0^n - \sum_{i=0}^{n-1} a_{n-i,n-1}(J_0^n)^i.$$

Journal of Physics: Conference Series **621** (2015) 012012 doi:10.1088/1742-6596/621/1/012012

Roughly speaking, the matrix A_0 is obtained from the matrix A by shifting the columns of A to the right, filling of the first column by zeros and subtracting a specific linear combination of powers of J_0^n that gives zeros in the last column of A_0. The structure of the algebra $\mathfrak{g}_0$ is defined in terms of the matrix A_0 in the same way as the structure of the algebra $\mathfrak{g}$ is defined in terms of the matrix A since $a_{p+n-i,j-1} - a_{p+n-j,i-1} = a_{0,p+n-i,j} - a_{0,p+n-j,i}$. This is consistent with Lemma 3 of [4]. Indeed, as the algebra $\mathfrak{g}_0$ can be contracted by the same matrix $U = J_\varepsilon^n$, its structure constants satisfy the same constraints imposed by the Saletan conditions (2). Lemma 3 of [4] also implies that n iterations of this contraction leads to the Abelian algebra.

We exhaustively study the case $n = 3$. There are three essential relations among the commutation relations (4) with $n = 3$,

$$[e_3, e_1] = a_{p1}e_p,$$
$$[e_3, e_2] = a_{p2}e_p,$$
$$[e_1, e_2] = (a_{32} - a_{21})e_1 - a_{31}e_2,$$

and the single Jacobi identity $[e_1, [e_2, e_3]] + [e_2, [e_3, e_1]] + [e_3, [e_1, e_2]] = 0$. Collecting coefficients of basis elements in the Jacobi identity and making additional arrangements, we obtain the following system of equations on entries of the matrix A:

$$\begin{gathered} a_{31}a_{21} = 0, \quad a_{31}a_{12} = 0, \quad a_{31}(a_{11} - a_{22}) = 0, \quad a_{21}(2a_{32} - a_{21}) = 0, \\ a_{32}(a_{11} - a_{22}) + a_{12}a_{22} = 0. \end{gathered} \tag{5}$$

A consequence of the system is $a_{21}(a_{11} + a_{22}) = 0$.

In order to simplify the form of the matrix A, we can use the transition to a Lie algebra isomorphic to $\mathfrak{g}$ or, equivalently, changing the basis of the underlying space. In view of problem's statement, admitted basis changes are those whose matrices commute with the matrix $U_0 = J_0^n$. Therefore, each of such matrices is a linear combination of powers of U_0,

$$S = \gamma(E + \alpha U_0 + \beta U_0^2),$$

where α, β and γ are arbitrary constants with $\gamma \neq 0$ and E is the 3×3 identity matrix. The inverse of S is

$$S^{-1} = \gamma^{-1}(E - \alpha U_0 + (\alpha^2 - \beta)U_0^2).$$

The expressions for entries of the transformed matrix $\tilde{A}$ follow from those for the transformed Lie brackets $[e_3, e_1]^\sim$ and $[e_3, e_2]^\sim$. We have

$$\begin{aligned} [e_3, e_1]^\sim &= S^{-1}[Se_3, Se_1] = \gamma(a_{11} - \alpha a_{32} - \beta a_{31})e_1 + \gamma a_{21}e_2 + \gamma a_{31}e_3, \\ [e_3, e_2]^\sim &= S^{-1}[Se_3, Se_2] = \gamma(a_{12} + \alpha(a_{11} - a_{22}) - \beta a_{21})e_1 \\ &\quad + \gamma(a_{22} + \alpha(a_{21} - a_{32}) - \beta a_{31})e_2 + \gamma(a_{32} + \alpha a_{31})e_3, \end{aligned}$$

i.e.,

$$\begin{aligned} \tilde{a}_{11} &= \gamma(a_{11} - \alpha a_{32} - \beta a_{31}), & \tilde{a}_{12} &= \gamma(a_{12} + \alpha(a_{11} - a_{22}) - \beta a_{21}), \\ \tilde{a}_{21} &= \gamma a_{21}, & \tilde{a}_{22} &= \gamma(a_{22} + \alpha(a_{21} - a_{32}) - \beta a_{31}), \\ \tilde{a}_{31} &= \gamma a_{31}, & \tilde{a}_{32} &= \gamma(a_{32} + \alpha a_{31}). \end{aligned}$$

The contracted algebra $\mathfrak{g}_0$ is defined by the commutation relations

$$[e_3, e_1]_0 = -a_{32}e_1,$$
$$[e_3, e_2]_0 = (a_{11} - a_{22})e_1 + (a_{21} - a_{32})e_2 + a_{31}e_3,$$
$$[e_1, e_2]_0 = a_{31}e_1.$$

We study possible cases of the solutions of the system (5) up to allowed basis changes.

If $a_{31} \neq 0$, then the system (5) implies that $a_{21} = a_{12} = 0$ and $a_{11} = a_{22}$. Selecting certain values of the parameters α, β, and γ of the basis transformation, we can set $a_{32} = 0$, $a_{11} = a_{22} = 0$ and $a_{31} = -1$. In other words, the commutation relations of the algebra $\mathfrak{g}$ take the form $[e_1, e_2] = e_2$, $[e_1, e_3] = e_3$ and $[e_2, e_3] = 0$. Hence the basis elements e_2 and e_3 span the maximal Abelian ideal of the algebra $\mathfrak{g}$, and the element e_1 acts on this ideal as the identity operator, i.e. the algebra $\mathfrak{g}$ is the almost Abelian algebra associated with the identity operator, which is denoted by $\mathfrak{g}_{3.3}$ in Mubarakzyanov's classification of three-dimensional Lie algebras [17]. In contrast to Example 1, in what follows we mostly use Mubarakzyanov's notations.[2] For the contracted algebra $\mathfrak{g}_0$, the commutation relations are: $[e_3, e_1]_0 = 0$, $[e_2, e_3]_0 = e_3$, $[e_2, e_1]_0 = e_1$. Therefore, this algebra is isomorphic to the initial algebra $\mathfrak{g}$. An isomorphism is established by a permutation of the basis elements. This means that the contraction is improper.

Suppose that $a_{31} = 0$ and $a_{21} \neq 0$. The solution of the system (5) gives $a_{32} = \frac{1}{2}a_{21}$, $a_{21} = -a_{11}$, $a_{22} = -a_{11}$ and $(a_{21} - a_{12})a_{11} = 0$. The constants a_{11}, a_{22}, a_{12} and a_{21} can be set to 0, 0, 0, and -2, respectively, by changing the basis with an appropriate matrix S. As a result, we obtain the canonical commutation relations of the algebra $\mathrm{sl}(2, \mathbb{R})$, $[e_1, e_2] = e_1$, $[e_2, e_3] = e_3$, $[e_1, e_3] = 2e_2$. The contracted algebra $\mathfrak{g}_0$ is isomorphic to the algebra $\mathfrak{g}_{3.3}$, which can be seen from its commutation relations, $[e_1, e_2]_0 = 0$, $[e_2, e_3]_0 = e_2$, $[e_1, e_3]_0 = e_1$.

In the case $a_{31} = a_{21} = 0$ and $a_{32} \neq 0$ the system (5) is reduced to the single equation $a_{32}(a_{11} - a_{22}) + a_{12}a_{22} = 0$. Carrying out an admitted basis transformation, we select certain values of the parameters α and γ of the transformation matrix S in order to set $a_{22} = 0$ and $a_{32} = -1$. Then the above equation implies that $a_{11} = 0$. Finally, the commutation relations of $\mathfrak{g}$ take the form $[e_1, e_3] = 0$, $[e_2, e_3] = e_3 - a_{12}e_1$, $[e_2, e_1] = e_1$, i.e., $\mathfrak{g}$ is an almost Abelian algebra associated with the matrix

$$\begin{pmatrix} 1 & -a_{12} \\ 0 & 1 \end{pmatrix}.$$

The contracted algebra has the same commutation relations as in the previous case, $\mathfrak{g}_0 \sim \mathfrak{g}_{3.3}$. If $a_{12} = 0$, then the contraction $\mathfrak{g} \to \mathfrak{g}_0$ is improper since $\mathfrak{g} \sim \mathfrak{g}_{3.3}$. For $a_{12} \neq 0$, the contraction is equivalent to the unit fall[3] the matrix associated with the algebra $\mathfrak{g} \sim \mathfrak{g}_{3.2}$, and the resulting matrix defines the algebra $\mathfrak{g}_0 \sim \mathfrak{g}_{3.3}$.

The last case is given by $a_{31} = a_{21} = a_{32} = 0$. The single equation remaining in the system (5) is $a_{12}a_{22} = 0$. The commutation relations of the initial and the contracted algebras are respectively

$$\begin{aligned} &[e_3, e_1] = a_{11}e_1, && [e_3, e_1]_0 = 0, \\ &[e_3, e_2] = a_{12}e_1 + a_{22}e_2, && [e_3, e_2]_0 = (a_{11} - a_{22})e_1, \\ &[e_1, e_2] = 0, && [e_1, e_2]_0 = 0. \end{aligned}$$

Consider subcases depending on values of the remaining parameters. If $a_{11} \neq a_{22}$, then by selecting a proper value of α in the transformation matrix S we can set $a_{12} = 0$. The parameter β is not essential here, and we can choose the zero value for it. The parameter γ can be used for scaling a nonzero linear combination of a_{11} and a_{22} (e.g., $a_{11} - a_{22}$) to the unity. As a result, we have the contraction of the almost Abelian algebra $\mathfrak{g} = \mathfrak{g}_{3.4}$ associated with the diagonal (but not proportional to the identity matrix) matrix to the three-dimensional Heisenberg algebra $\mathrm{h}(1) = \mathfrak{g}_{3.1}$. If $a_{11} = a_{22}$, the contracted algebra is Abelian, i.e., we have the

[2] The classical Lie algebras $\mathrm{h}(1)$, $\mathrm{e}(2)$, $\mathrm{sl}(2, \mathbb{R})$ and $\mathrm{so}(3)$ are denoted by Mubarakzyanov as $\mathfrak{g}_{3.1}$, $\mathfrak{g}^0_{3.5}$, $\mathfrak{g}_{3.6}$ and $\mathfrak{g}_{3.7}$, respectively.

[3] In the case of 2×2 Jordan blocks, the only possible unit fall is the replacement of the value 1 in the $(1, 2)$th entry by 0.

trivial contraction of an almost Abelian Lie algebra (one of $\mathfrak{g}_{3.1}$, $\mathfrak{g}_{3.2}$, $\mathfrak{g}_{3.3}$ and $3\mathfrak{g}_1$, depending on values of $a_{11} = a_{22}$ and a_{12}).

Proposition 1. *Saletan contractions with the signature* $(0; 3)$ *realize only the following contractions between three-dimensional Lie algebras: the proper contractions* $\mathrm{sl}(2, \mathbb{R}) \to \mathfrak{g}_{3.3}$, $\mathfrak{g}_{3.2} \to \mathfrak{g}_{3.3}$, $\mathfrak{g}_{3.4} \to \mathfrak{g}_{3.1}$, *and* $\mathfrak{g}_{2.1} \oplus \mathfrak{g}_1 \to \mathfrak{g}_{3.3}$, *the trivial contractions of* $\mathfrak{g}_{3.1}$, $\mathfrak{g}_{3.2}$ *and* $\mathfrak{g}_{3.3}$ *to* $3\mathfrak{g}_1$, *as well as the improper contractions* $\mathfrak{g}_{3.3} \to \mathfrak{g}_{3.3}$ *and* $3\mathfrak{g}_1 \to 3\mathfrak{g}_1$.

5. Conclusion

The main result of the paper is Theorem 1, which describes the canonical form of Saletan contractions. The proved existence of the canonical form for each Saletan contraction gives a specific finite tuple of non-negative integers which corresponds to this contraction and is called its signature. The signature of a Saletan contraction completely defines its canonical form. Introducing the notion of signature leads to posing several interesting problems, which are related to Saletan contractions.

Thus, for each Lie algebra the set of its possible Saletan contractions is partitioned into the subsets corresponding to different Saletan signatures. This allows us to pose the problem on describing Saletan contractions with a fixed signature. The well-known Inönü–Wigner contractions constitute a subclass of Saletan contractions, which is singled out by Saletan signatures of the form $(n_0; 1, \dots, 1)$. Therefore, the study of Saletan contractions includes, as its simplest part, the study of Inönü–Wigner contractions. Inönü–Wigner contractions of three- and four-dimensional Lie algebras were exhaustively classified in [5] and [6], respectively. The Saletan contractions with other signatures do not have a connection with algebraic structure of initial and contracted algebras as direct as Inönü–Wigner contractions do. This is why the description of general Saletan contractions is a much more difficult problem.

Given a Lie algebra, another problem is finding the tuples of non-negative integers that can be signatures of Saletan contractions of this algebra. As shown in Remark 2, powers of the value of Saletan matrix at limit value of the contraction parameter form a nested chain of subalgebras of the initial algebra and signature components are expressed in the terms of the dimensions of these subalgebras. This claim relates the signatures to the subalgebraic structure of the initial algebra. At the same time, the presence of a nested chain of subalgebras does not imply the existence of the Saletan contraction associated with this chain. Additional constraints that admit no clear algebraic interpretation should be taken into account. Furthermore, even provided that a corresponding contraction exists, there is no known procedure to construct this contraction from the chain of subalgebras. This significantly differs from Inönü–Wigner contractions since there exists an algorithm to construct a well-defined Inönü–Wigner contraction starting from any subalgebra of the initial algebra. The study of Saletan signatures resembles the study of signatures of generalized Inönü–Wigner contractions [18–22]. Recall that the signature components of a generalized Inönü–Wigner contraction are diagonal entries of a diagonal differentiation of the algebra to be contracted, but the converse is not true.

The notion of Saletan signature may serve as a basis for an algorithm of exhaustive classification of Saletan contractions, at least in the case of lowest dimensions. It is known [22–25] that all contractions between three-dimensional complex (resp. real) Lie algebras (except the only contraction $\mathrm{so}(3) \to \mathrm{h}(1)$ in the real case) are realized by usual Inönü–Wigner contractions. The contraction $\mathrm{so}(3) \to \mathrm{h}(1)$ is realized as a generalized Inönü–Wigner contraction, but not as a Saletan one. In dimension four, the number of contractions that cannot be realized as usual Inönü–Wigner contractions increases crucially. Moreover, there is one (resp. two) contraction between four-dimensional complex (resp. real) Lie algebras that cannot be realized as generalized Inönü–Wigner contractions. Thus, the question whether these contractions can be realized as Saletan contractions is the most interesting problem on Saletan contractions of four-dimensional Lie algebras.

Acknowledgments
The author is grateful for the hospitality and financial support provided by the University of Cyprus. The author thanks Roman Popovych for productive and helpful discussions. The research was supported by the Austrian Science Fund (FWF), project P25064.

References
[1] Segal I E 1951 A class of operator algebras which are determined by groups *Duke Math. J.* **18** 221
[2] Inönü E and Wigner E P 1953 On the contraction of groups and their representations *Proc. Nat. Acad. Sci. U. S. A.* **39** 510
[3] Inönü E and Wigner E P 1954 On a particular type of convergence to a singular matrix *Proc. Nat. Acad. Sci. U. S. A.* **40** 119
[4] Saletan E J 1961 Contraction of Lie groups *J. Math. Phys.* **2** 1
[5] Conatser C W 1972 Contractions of the low-dimensional Lie algebras *J. Math. Phys.* **13** 196
[6] Huddleston P L 1978 Inönü–Wigner contractions of the real four-dimensional Lie algebras *J. Math. Phys.* **19** 1645
[7] Patera J and Winternitz P 1977 Subalgebras of real three and four-dimensional Lie algebras *J. Math. Phys.* **18** 1449
[8] Kupczyński M 1969 On the generalized Saletan contractions *Comm. Math. Phys.* **13** 154
[9] Lévy-Nahas M 1967 Deformation and contraction of Lie algebras *J. Math. Phys.* **8** 1211
[10] Mimura F and Ikushima A 1978 Structure of contracted Lie algebras *Bull. Kyushu Inst. Tech. Math. Natur. Sci.* no 25, 1
[11] Cariñena J F, Grabowski J and Marmo G 2001 Contractions: Nijenhuis and Saletan tensors for general algebraic structures *J. Phys. A* **34** 3769
[12] Burde D 1999 Degenerations of nilpotent Lie algebras *J. Lie Theory* **9** 193
[13] Burde D 2005 Degenerations of 7-dimensional nilpotent Lie algebras *Comm. Algebra* **33** 1259 (*Preprint* arXiv:math.RA/0409275)
[14] Burde D and Steinhoff C 1999 Classification of orbit closures of 4-dimensional complex Lie algebras *J. Algebra* **214** 729
[15] Grunewald F and O'Halloran J 1988 Varieties of nilpotent Lie algebras of dimension less than six *J. Algebra* **112** 315
[16] Lauret J 2003 Degenerations of Lie algebras and geometry of Lie groups *Differential Geom. Appl.* **18** 177
[17] Mubarakzjanov G M 1963 On solvable Lie algebras *Izv. Vysš. Učehn. Zaved. Matematika* no 1(32) 114
[18] Doebner H D and Melsheimer O 1967 On a class of generalized group contractions *Nuovo Cimento A (10)* **49** 306
[19] Hegerfeldt G C 1967 Some properties of a class of generalized Inönü–Wigner contractions *Nuovo Cimento A (10)* **51** 439
[20] Lyhmus J H 1969 Limit Lie groups (contracted Lie groups) *Second Summer School on the Problems of the Theory of Elementary Particles (Otepää, 1967), Part IV* (Inst. Fiz. i Astronom. Akad. Nauk Èston. SSR, Tartu) p 3 [in Russian]
[21] Popovych D R and Popovych R O 2009 Equivalence of diagonal contractions to generalized IW-contractions with integer exponents *Linear Algebra Appl.* **431** 1096 (*Preprint* arXiv:0812.4667)
[22] Popovych D R and Popovych R O 2010 Lowest dimensional example on non-universality of generalized Inonu-Wigner contractions *J. Algebra* **324** 2742 (*Preprint* arXiv:0812.1705)
[23] Campoamor-Stursberg R 2008 Some comments on contractions of Lie algebras *Adv. Stud. Theor. Phys.* **2** 865
[24] Nesterenko M and Popovych R 2006 Contractions of low-dimensional Lie algebras *J. Math. Phys.* **47** 123515 (*Preprint* arXiv:math-ph/0608018)
[25] Weimar-Woods E 1991 The three-dimensional real Lie algebras and their contractions *J. Math. Phys.* **32** 2028

GADEISVII IOP Publishing
Journal of Physics: Conference Series **621** (2015) 012013 doi:10.1088/1742-6596/621/1/012013

Generalized Bäcklund–Darboux transformation (GBDT): conservation laws, rational extensions and bispectrality

Alexander Sakhnovich

Fakultät für Mathematik, Universität Wien,
Oskar-Morgenstern-Platz 1, A-1090 Vienna, Austria

E-mail: oleksandr.sakhnovych@univie.ac.at

Abstract. Bäcklund–Darboux transformations are closely related to the integrability and symmetry problems. For the generalized Bäcklund–Darboux transformation (GBDT), we consider conservation laws, rational extensions and bispectrality. We use the case of the nonlinear optics equation (and its auxiliary linear system) as an example.

1. Introduction

Bäcklund–Darboux transformations and related commutation methods constitute one of the most fruitful approaches to the construction of explicit solutions of linear and nonlinear equations and to the explicit spectral theoretic results (see, e.g., [1–6] and numerous references therein). Similar to various symmetries studied in group analysis, Bäcklund–Darboux transformations (BDTs) transform solutions of linear differential equations of some fixed class into solutions of differential equations of the same (or another fixed) class and potentials into potentials, whereas potentials are often solutions of the corresponding integrable nonlinear equations. Some connections between Darboux transformations and potential symmetries are discussed in an interesting paper [7].

Though the idea of BDTs and commutation methods goes back to Bäcklund, Darboux and Jacobi, the notion of the so called *Darboux matrix* has a much later origin. According to J. Cieśliński [1], "all approaches to the construction of Darboux matrices originate in the dressing method". In order to explain the notion of the Darboux matrix, we consider some initial linear system

$$u_x = G(x,z)u, \quad u \in \mathbb{C}^m, \tag{1.1}$$

where $u_x := \frac{d}{dx}u$ ($u_x = \frac{\partial}{\partial x}u$ in the case of several variables), $G(x,z)$ is some $m \times m$ matrix function (i.e., $G(x,z) \in \mathbb{C}^{m\times m}$) and $\mathbb{C}$ stands for the complex plain. The *Darboux matrix* (matrix function) $\widehat{w}$ satisfies the equation

$$\widehat{w}_x = \widetilde{G}\widehat{w} - \widehat{w}G, \quad \widetilde{G}(x,z) \in \mathbb{C}^{m\times m}. \tag{1.2}$$

It is easy to see that if (1.1) and (1.2) hold, the product

$$\widetilde{u} = \widehat{w}u$$

satisfies the *transformed system*

$$\widetilde{u}_x = \widetilde{G}(x,z)\widetilde{u}.$$

The cases, where the Darboux matrix $\widehat{w}$ can be constructed explicitly, if u and G are given explicitly, are of special interest. Clearly, in those cases we construct explicitly $\widetilde{u}$ and also $\widetilde{G}$.

We are interested in a generalized version of the Bäcklund–Darboux transformation for which we use the acronym GBDT. In GBDT (for the case of one space variable) Darboux matrix is presented as the transfer matrix function in Lev Sakhnovich form:

$$\widehat{w}(z) = w_A(z) = I_m - \Pi_2^* S^{-1}(A_1 - zI_n)^{-1}\Pi_1; \quad \Pi_k \in \mathbb{C}^{n\times m}, \quad S \in \mathbb{C}^{n\times n}; \tag{1.3}$$

$$\text{where} \qquad A_1 S - S A_2 = \Pi_1 \Pi_2^*, \tag{1.4}$$

I_m is the $m \times m$ identity matrix. We note that an additional variable x appears in w_A, Π_1, Π_2 and S, so that we deal with matrix functions $\Pi_k(x)$, $S(x)$ and Darboux matrix $w_A(x,z)$. GBDT is applicable to all systems $u_x = G(x,z)u$, where G rationally depends on z, that is,

$$G(x,z) = \Big(\sum_{k=0}^{r} z^k q_k(x) + \sum_{s=1}^{j}\sum_{k=1}^{r_s}(z-c_s)^{-k}q_{sk}(x)\Big)$$

(see the paper [8], the reviews [9, 10] and Chapter 7 in [11] as well as numerous references therein). The coefficients $\widetilde{q}_k$ and $\widetilde{q}_{sk}$ of the transformed (rational) matrix function $\widetilde{G}(x,z)$ are expressed via the coefficients q_k and q_{sk} of $G(x,z)$ and matrix functions $S(x)$, $\Pi_1(x)$ and $\Pi_2(x)$. Matrix functions $\Pi_k(x)$ are constructed as solutions of systems related to the initial system.

Here, we shall consider in detail GBDT for the system

$$u_x = G(x,z)u, \quad G = \mathrm{i}zD - [D,\varrho], \quad D = \mathrm{diag}\,\{d_1, d_2, \dots, d_m\} = D^*; \tag{1.5}$$

$$\varrho^* = B\varrho B, \quad B = \mathrm{diag}\,\{b_1, b_2, \dots, b_m\} \quad (b_k = \pm 1); \quad [D,\varrho] := D\varrho - \varrho D, \tag{1.6}$$

where diag denotes a diagonal matrix. System (1.5) is an auxiliary system for the well-known N-wave (nonlinear optics) equation:

$$[D,\varrho_t] - [\widehat{D},\varrho_x] = \big[[D,\varrho],[\widehat{D},\varrho]\big], \tag{1.7}$$

$$\widehat{D} = \mathrm{diag}\,\{\widehat{d}_1, \dots, \widehat{d}_m\} = \widehat{D}^*. \tag{1.8}$$

More precisely, equation (1.7) is equivalent to the compatibility condition

$$G_t - F_x + [G,F] = 0 \tag{1.9}$$

of the auxiliary systems

$$u_x = G(x,t,z)u, \quad G = \mathrm{i}zD - [D,\varrho], \tag{1.10}$$

$$u_t = F(x,t,z)u, \quad F = \mathrm{i}z\widehat{D} - [\widehat{D},\varrho]. \tag{1.11}$$

We shall study conservation laws, rational extensions and bispectrality for the case of GBDT for system (1.5) and also for the N-wave equation (1.7).

The integrable model describing interaction of three wave packages [12] is the most well-known subcase of the N-wave equation (1.7). Putting, for instance, $m=3$, $B = I_3$ and $d_1 > d_2 > d_3$, and using transformations [13, Chapter 3]

$$\psi_1 = (\widehat{d}_2 - \widehat{d}_1)/(d_1 - d_2), \quad \psi_2 = (\widehat{d}_3 - \widehat{d}_2)/(d_2 - d_3), \quad \psi_3 = (\widehat{d}_3 - \widehat{d}_1)/(d_1 - d_3); \tag{1.12}$$

$$\phi_1 = -\mathrm{i}\sqrt{d_1 - d_2}\,\varrho_{12}, \quad \phi_2 = -\mathrm{i}\sqrt{d_2 - d_3}\,\varrho_{23}, \quad \phi_3 = -\mathrm{i}\sqrt{d_1 - d_3}\,\varrho_{13}, \tag{1.13}$$

Journal of Physics: Conference Series **621** (2015) 012013 doi:10.1088/1742-6596/621/1/012013

we rewrite (1.7) in the standard form of the corresponding 3-wave interaction:

$$(\phi_1)_t + \psi_1(\phi_1)_x = \mathrm{i}\varepsilon\overline{\phi}_2\phi_3, \quad (\phi_2)_t + \psi_2(\phi_2)_x = \mathrm{i}\varepsilon\overline{\phi}_1\phi_3, \quad (\phi_3)_t + \psi_3(\phi_3)_x = \mathrm{i}\varepsilon\phi_1\phi_2, \tag{1.14}$$

where $\overline{\phi}(x,t)$ stands for the function, which takes values complex conjugate to $\phi(x,t)$, and

$$\varepsilon = (d_1\widehat{d}_2 - d_2\widehat{d}_1 + d_2\widehat{d}_3 - d_3\widehat{d}_2 + d_3\widehat{d}_1 - d_1\widehat{d}_3)\big((d_1-d_2)(d_1-d_3)(d_2-d_3)\big)^{-1/2}. \tag{1.15}$$

If α is a scalar value or matrix, the notation $\overline{\alpha}$ stands for the scalar, which is complex conjugate to α, or the matrix with entries complex conjugate to the entries of α, respectively.

2. Preliminaries

GBDT for the system (1.5) and for the N-wave equation was described in [8] (see also [11, Subsection 1.1.3 and Section 7.1] and references therein). In this section, we give some necessary definitions and results from [8] and [11, Subsection 1.1.3 and Section 7.1]. We consider the case $x \geq 0$ for the system (1.5) and the case $x \geq 0$, $t \geq 0$ for the N-wave equation. We fix $n \times n$ matrices A and $S(0) = S(0)^*$ and an $n \times m$ matrix $\Pi(0)$ such that

$$AS(0) - S(0)A^* = \mathrm{i}\Pi(0)B\Pi(0)^*. \tag{2.1}$$

We introduce the $n \times m$ matrix function $\Pi(x)$ and the $n \times n$ matrix function $S(x)$ via initial values $\Pi(0)$ and $S(0)$ and differential equations

$$\Pi_x = -\mathrm{i}A\Pi D + \Pi[D, \varrho], \quad S_x = \Pi D B \Pi^*. \tag{2.2}$$

Relations (2.1) and (2.2) yield $AS(x) - S(x)A^* = \mathrm{i}\Pi(x)B\Pi(x)^*$. Then the following proposition describes GBDT determined by A, $S(0)$ and $\Pi(0)$.

Proposition 2.1 *Let system (1.5) (such that (1.6) holds) be given. Then (in the points of invertibility of $S(x)$) the matrix function*

$$w_A(x,z) = I_m - \mathrm{i}B\Pi(x)^*S(x)^{-1}(A - zI_n)^{-1}\Pi(x) \tag{2.3}$$

is a Darboux matrix of (1.5) and satisfies the equation

$$\frac{d}{dx}w_A(x,z) = \big(\mathrm{i}zD - [D, \widetilde{\varrho}(x)]\big)w_A(x,z) - w_A(x,z)\big(\mathrm{i}zD - [D, \varrho(x)]\big), \tag{2.4}$$

where

$$\widetilde{\varrho} = \varrho - B\Pi^*S^{-1}\Pi, \quad \widetilde{\varrho}^* = B\widetilde{\varrho}B. \tag{2.5}$$

Moreover, if $\det S(x) \neq 0$ for $x \geq 0$, a normalized fundamental solution $\widetilde{w}(x,z)$ of the transformed system

$$\widetilde{w}_x = (\mathrm{i}zD - [D, \widetilde{\varrho}])\widetilde{w}, \quad \widetilde{w}(0,z) = I_m \tag{2.6}$$

is given by the equality

$$\widetilde{w}(x,z) = w_A(x,z)w(x,z)w_A(0,z)^{-1}, \tag{2.7}$$

where w is the fundamental solution of the initial system (1.5) normalized by $w(0,z) = I_m$.

Next, we add the variable t and consider $\Pi(x,t)$, $S(x,t)$ and $w_A(x,t,z)$ which are determined by A, $\Pi(0,0)$ and $S(0,0)=S(0,0)^*$ via equations (2.2), (2.3) and

$$\Pi_t = -\mathrm{i}A\Pi\widehat{D} + \Pi[\widehat{D},\varrho], \quad S_t = \Pi\widehat{D}B\Pi^*. \tag{2.8}$$

Instead of (2.1) we assume that

$$AS(0,0) - S(0,0)A^* = \mathrm{i}\Pi(0,0)B\Pi(0,0)^*, \tag{2.9}$$

which implies that $AS(x,t) - S(x,t)A^* = \mathrm{i}\Pi(x,t)B\Pi(x,t)^*$. Proposition 2.1 yields:

Proposition 2.2 *Let an $m \times m$ matrix function ϱ ($\varrho^* = B\varrho B$) be continuously differentiable and satisfy the N-wave equation* (1.7). *Then $\widetilde{\varrho}$ of the form*

$$\widetilde{\varrho}(x,t) := \varrho(x,t) - B\Pi(x,t)^*S(x,t)^{-1}\Pi(x,t) \tag{2.10}$$

satisfies (in the points of invertibility of S) the equality $\widetilde{\varrho}^* = B\widetilde{\varrho}B$ and the N-wave equation.

Using [11, Therem 6.1] on wave functions, we easily obtain the next statement.

Remark 2.3 *If the conditions of Proposition* 2.2 *hold,* $\det S(x,t) \neq 0$ *on the semi-band* $0 \leq x < \infty$, $0 \leq t \leq \varepsilon$ *and $w(x,t,z)$ is the initial wave function (i.e., $w_x = Gw$, $w_t = Fw$ and $w(0,0,z) = I_m$), then the transformed wave function $\widetilde{w}(x,t,z)$ is given by the equality*

$$\widetilde{w}(x,t,z) = w_A(x,t,z)w(x,t,z)w_A(0,0,z)^{-1}. \tag{2.11}$$

Example 2.4 *The real-valued case is of special interest* [13, Section 3.4]. *It is immediate from Proposition* 2.2 *(and formulas defining $\widetilde{\varrho}$ considered in this proposition) that equalities*

$$\varrho = \overline{\varrho}, \quad A = -\overline{A}, \quad \Pi(0,0) = \overline{\Pi}(0,0), \quad S(0,0) = \overline{S}(0,0), \tag{2.12}$$

yield the equality $\widetilde{\varrho} = \overline{\widetilde{\varrho}}$. Assume that (2.12) *holds, and so $\widetilde{\varrho} = \overline{\widetilde{\varrho}}$. Then, in the subcase $m = 3$, $B = I_3$, $d_1 > d_2 > d_3$ of the 3-wave equation, we can rewrite* (1.14) *in the real-valued form and obtain solutions of the corresponding exact resonance equations. Namely, we set*

$$\varphi_1 = -\mathrm{i}\phi_1 = -\sqrt{d_1 - d_2}\,\widetilde{\varrho}_{12}, \quad \varphi_2 = -\mathrm{i}\phi_2 = -\sqrt{d_2 - d_3}\,\widetilde{\varrho}_{23}, \quad \varphi_3 = -\mathrm{i}\phi_3 = -\sqrt{d_1 - d_3}\,\widetilde{\varrho}_{13},$$

and, taking into account (1.14), *derive*

$$(\varphi_1)_t + \psi_1(\varphi_1)_x = \varepsilon\varphi_2\varphi_3, \quad (\varphi_2)_t + \psi_2(\varphi_2)_x = \varepsilon\varphi_1\varphi_3, \quad (\varphi_3)_t + \psi_3(\varphi_3)_x = -\varepsilon\varphi_1\varphi_2. \tag{2.13}$$

3. Conservation laws, rational extensions and bispectrality

3.1. Conservation laws

When ϱ is differentiable with respect to t (to x), we can differentiate both sides of the first relation in (2.2) (in (2.8)) with respect to t (to x). If ϱ is differentiable with respect to x and t and ϱ_t is continuous, then, according to the so called Clairaut's (or Schwarz's) theorem, we have $\Pi_{xt} = \Pi_{tx}$. Thus $\Pi_{xt} = \Pi_{tx}$ is the necessary condition of the compatibility of the first relations in (2.2) and (2.8). In a similar way, $S_{xt} = S_{tx}$ is a necessary condition of the compatibility of the second relations in (2.2) and (2.8). In other words, the necessary compatibility conditions are

$$(-\mathrm{i}A\Pi D + \Pi[D,\varrho])_t = (-\mathrm{i}A\Pi\widehat{D} + \Pi[\widehat{D},\varrho])_x, \tag{3.1}$$

$$(\Pi DB\Pi^*)_t = (\Pi\widehat{D}B\Pi^*)_x, \tag{3.2}$$

where Π_x should be substituted by $-\mathrm{i}A\Pi D + \Pi[D,\varrho]$ and Π_t should be substituted by $-\mathrm{i}A\Pi\widehat{D} + \Pi[\widehat{D},\varrho]$.

Proposition 3.1 *Let ϱ be differentiable and satisfy the N-wave equation* (1.7). *Then the equalities* (3.1) *and* (3.2) *hold.*

Proof. Substituting expressions for Π_t and Π_x we obtain

$$(-\mathrm{i}A\Pi D+\Pi[D,\varrho])_t=-\mathrm{i}A(-\mathrm{i}A\Pi\widehat{D}+\Pi[\widehat{D},\varrho])D+(-\mathrm{i}A\Pi\widehat{D}+\Pi[\widehat{D},\varrho])[D,\varrho]+\Pi[D,\varrho_t],$$
$$(-\mathrm{i}A\Pi\widehat{D}+\Pi[\widehat{D},\varrho])_x=-\mathrm{i}A(-\mathrm{i}A\Pi D+\Pi[D,\varrho])\widehat{D}+(-\mathrm{i}A\Pi D+\Pi[D,\varrho])[\widehat{D},\varrho]+\Pi[\widehat{D},\varrho_x].$$

Therefore, taking into account (1.7) we derive

$$(-\mathrm{i}A\Pi D+\Pi[D,\varrho])_t-(-\mathrm{i}A\Pi\widehat{D}+\Pi[\widehat{D},\varrho])_x=\Pi\big([D,\varrho_t]-[\widehat{D},\varrho_x]-\big[[D,\varrho],[\widehat{D},\varrho]\big]\big)=0,$$

and so (3.1) is proved. Using the same substitutions as before, we have also

$$(\Pi DB\Pi^*)_t=\big(-\mathrm{i}A\Pi\widehat{D}+\Pi[\widehat{D},\varrho]\big)DB\Pi^*+\Pi DB\big(\mathrm{i}\widehat{D}\Pi^*A^*+[\varrho^*,\widehat{D}]\Pi^*\big), \tag{3.3}$$
$$(\Pi\widehat{D}B\Pi^*)_x=\big(-\mathrm{i}A\Pi D+\Pi[D,\varrho]\big)\widehat{D}B\Pi^*+\Pi\widehat{D}B\big(\mathrm{i}D\Pi^*A^*+[\varrho^*,D]\Pi^*\big). \tag{3.4}$$

Equality (3.2) follows from (3.3), (3.4) and $B\varrho^*=\varrho B$. □

The sufficiency of the proved above equalities $\Pi_{xt}=\Pi_{tx}$ and $S_{xt}=S_{tx}$ for the compatibility of (2.2) and (2.8) is not self-evident but may be proved in a way similar to the proof of [11, Theorem 6.1].

Remark 3.2 *We note that relations* (3.1) *and* (3.2) *may be considered as conservation laws for* $\Pi(x,t)$.

3.2. Rational extensions

According to Propositions 2.1 and 2.2, in the case

$$\varrho\equiv 0,\quad \sigma(A)=\{0\}\quad (\sigma\,\text{stands for spectrum}), \tag{3.5}$$

the transformed potentials $\widetilde{\varrho}$ are rational matrix functions (*rational extensions* in the terminology of [14]). More precisely, if $\varrho\equiv 0$ and A is nilpotent, formulas (2.2), (2.3), (2.5), (2.7), (2.8), (2.10) and (2.11) imply the following proposition.

Proposition 3.3 *Let* (3.5) *hold. Then $\Pi(x)$ and $S(x)$ are matrix polynomials with respect to x, $\Pi(x,t)$ and $S(x,t)$ are matrix polynomials with respect to x and t; $\widetilde{\varrho}(x)=p(x)/\det S(x)$ and $\widetilde{\varrho}(x,t)=p(x,t)/\det S(x,t)$, where $p(x)$ is a matrix polynomial with respect to x, $p(x,t)$ is a matrix polynomial with respect to x and t, $\det S(x)$ is a polynomial with respect to x and $\det S(x,t)$ is a polynomial with respect to x and t. Moreover, for Darboux matrices w_A we have $w_A(x,z)=P(x,z)/\det S(x)$ and $w_A(x,t,z)=P(x,t,z)/\det S(x,t)$, where $P(x,z)$ ($P(x,t,z)$) is a matrix polynomial with respect to x and $1/z$ (x, t and $1/z$). Thus, the transformed fundamental solutions $\widetilde{w}(x,z)$ and wave functions $\widetilde{w}(x,t,z)$ are expressed via matrix polynomials and exponents of diagonal matrices:*

$$\widetilde{w}(x,z)=w_A(x,z)\mathrm{e}^{\mathrm{i}zxD}w_A(0,z)^{-1},\quad \widetilde{w}(x,t,z)=w_A(x,t,z)\mathrm{e}^{\mathrm{i}z(xD+t\widehat{D})}w_A(0,0,z)^{-1}. \tag{3.6}$$

Proof. Since $\varrho\equiv 0$, the first equation in (2.2) yields that the ith column f_i of Π is given by $f_i(x)=\mathrm{e}^{-\mathrm{i}d_ixA}f_i(0)=\sum_{k=0}^{n-1}\frac{1}{k!}(-\mathrm{i}d_ixA)^kf_i(0)$, and so $\Pi(x)$ is a matrix polynomial. (Here we used the equality $A^n=0$, which holds for all $n\times n$ nilpotent matrices.) Hence, the second equation in (2.2) yields that $S(x)$ is a matrix polynomial. In the same way we prove the polynomial form of $\Pi(x,t)$ and $S(x,t)$, and the required properties of $\widetilde{\varrho}$ follow. According to formula (1.84) from [11, p. 24] we have $w_A(z)Bw_A(\overline{z})^* = B$ and, in particular, the equalities

$$w_A(0,z)^{-1} = Bw_A(0,\overline{z})^*B, \quad w_A(0,0,z)^{-1} = Bw_A(0,0,\overline{z})^*B$$

are valid. Finally, the representations of $w_A(x,z)$ and $w_A(x,t,z)$ follow from the definition (2.3), properties of Π and S and the expansion $(A - zI_n)^{-1} = -z^{-1}\sum_{k=0}^{n-1}(z^{-1}A)^k$. □

Remark 3.4 *System* (1.5) *includes important subclasses: self-adjoint and skew-self-adjoint Dirac-type systems (which are also called Zakharov–Shabat or AKNS systems). Namely, putting*

$$D = j = \begin{bmatrix} I_{m_1} & 0 \\ 0 & -I_{m_2} \end{bmatrix} \quad (m_1 + m_2 = m), \quad V = \begin{bmatrix} 0 & v \\ v^* & 0 \end{bmatrix} \tag{3.7}$$

and $B = j$, *we rewrite* (1.5), (1.6) *in the form of the self-adjoint Dirac-type system*

$$u_x = \mathrm{i}(zj + jV)u, \quad v = 2\mathrm{i}\varrho_{12}, \tag{3.8}$$

where ϱ_{12} *is the upper right* $m_1 \times m_2$ *block of* ϱ. *Assuming* (3.7) *and* $B = I_m$, *we rewrite* (1.5), (1.6) *in the form of the skew-self-adjoint Dirac-type system*

$$u_x = (\mathrm{i}zj + jV)u, \quad v = -2\varrho_{12}. \tag{3.9}$$

3.3. Bispectrality

The notion of bispectrality was introduced in [15] (see also [16] and references therein). Matrix bispectrality for system (1.5) was studied in [17]. Namely, it was shown that for $\widetilde{\varrho}$ constructed in Proposition 3.3 and, correspondingly, for the solution $W(x,z) = w_A(x,z)\mathrm{e}^{\mathrm{i}zxD}$ of system

$$W_x = (\mathrm{i}zD - [D, \widetilde{\varrho}])W, \tag{3.10}$$

there is a non-degenerate matrix linear differential operator $\mathcal{B}$ with respect to the variable z such that $\mathcal{B}(z)W = 0$. It was suggested in [16] (and in some earlier works by F. Grünbaum and coauthors) that the bispectrality requirement $W\mathcal{B}(z) = \Theta(x)W$ is more meaningful than the requirement $\mathcal{B}(z)W = \Theta(x)W$. Unfortunately, the approach from [17] does not work properly for that case. Therefore, we should consider the system

$$W_x = (\mathrm{i}zD - [D, \widetilde{\varrho}])W - \mathrm{i}zWD \tag{3.11}$$

instead of system (1.5). In view of (2.4), the matrix function w_A satisfies this system in the case $\varrho \equiv 0$. Thus, we can deal in the same way as in [18], where Dirac systems were treated. Namely, if $\sigma(A)$ is concentrated at one point $\lambda \in \mathbb{C}$ we write down the resolvent $(A - zI_n)^{-1}$ in the form

$$(A - zI_n)^{-1} = -\sum_{k=1}^{n}(z-\lambda)^{-k}(A - \lambda I_n)^{k-1}. \tag{3.12}$$

By virtue of (2.3) and (3.12) the next proposition is immediate.

Proposition 3.5 *Assume that* $\varrho \equiv 0$, *the spectrum of* A *is concentrated at some point* $\lambda \in \mathbb{C}$ *and* $\widetilde{\varrho}$ *is given by* (2.5). *Then system* (3.11) *is bispectral in the sense of* [16].

Indeed, it is easy to choose some coefficients c_s so that the operator $B(z) = \sum_{s=1}^{n+1} c_s (z-\lambda)^s \frac{\partial^s}{\partial z^s}$ is non-degenerate and satisfies the equalities

$$B(z)(z-\lambda)^{-k} = 0 \qquad (1 \leq k \leq n). \tag{3.13}$$

According to formula (2.4) and identity $\varrho \equiv 0$, the matrix function $W = w_A(x,z)$ satisfies (3.11). Therefore, definition (2.3) of w_A and equalities (3.12) and (3.13) imply $B(z)W = 0$.

If we want to consider system (3.10), where $\varrho \equiv 0$ and $\sigma(A) = \{\lambda\}$, $\lambda \in \mathbb{R}$, we rewrite the first relation in (2.2) in the form

$$\left(\Pi(x)e^{i\lambda x D}\right)_x = -i(A - \lambda I_m)\left(\Pi(x)e^{i\lambda x D}\right) D. \tag{3.14}$$

Hence, taking into account the fact that $\sigma(A)$ is concentrated at λ, we see that $\Pi(x)e^{i\lambda x D} = P(x,\lambda) = P(x)$, where $P(x)$ is a matrix polynomial with respect to x. Substituting $\Pi(x) = P(x)e^{-i\lambda x D}$ into (2.2) and using the equalities $D = D^*$, $\lambda = \overline{\lambda}$, we obtain

$$S_x = P(x)e^{-i\lambda x D} D B e^{i\lambda x D} P(x)^* = P(x) D B P(x)^*, \quad \text{i.e.,} \quad S(x) = \widetilde{P}(x), \tag{3.15}$$

where $\widetilde{P}(x)$ a matrix polynomial with respect to x. Introduce an operator $\mathcal{B} = \mathrm{diag}\{\mathcal{B}_1, \ldots, \mathcal{B}_m\}$ such that $\mathcal{B}_k$ has the form

$$f\,\mathcal{B}_k = \sum_{\ell=1}^{N} c_{\ell k}(z) \frac{\partial^\ell}{\partial z^\ell}\left((z-\lambda)^m f\right), \quad N \in \mathbb{N}, \tag{3.16}$$

and notice that the equality

$$\sum_{\ell=1}^{N} c_{\ell k}(z) \frac{\partial^\ell}{\partial z^\ell}\left((z-\lambda)^m w_A(x,z) e^{izxD} e_k\right) = 0, \quad e_k := \{\delta_{ik}\}_{i=1}^m \quad (1 \leq k \leq m). \tag{3.17}$$

is equivalent to the relation

$$\sum_{\ell=1}^{N} c_{\ell k}(z) \left(\frac{\partial}{\partial z} + i d_k x\right)^\ell \left((z-\lambda)^m e^{i\lambda x D}\left(I_m - iBP(x)^* \widetilde{P}(x)^{-1}(A - zI_n)^{-1}P(x)\right) e^{-i\lambda x D} e_k\right) = 0$$

which, in turn, is equivalent to

$$\sum_{\ell=1}^{N} c_{\ell k}(z) \left(\frac{\partial}{\partial z} + i d_k x\right)^\ell$$
$$\times \left((\det \widetilde{P}(x))(z-\lambda)^m \left(I_m - iBP(x)^* \widetilde{P}(x)^{-1}(A - zI_n)^{-1}P(x)\right) e_k\right) = 0. \tag{3.18}$$

Open Problem. Are there some cases, where matrices D, A, $S(0)$ and $\Pi(0)$ generate nontrivial $\widetilde{\varrho}$ such that $W\mathcal{B}(z) = 0$ for the solution $W(x,z) = w_A(x,z)e^{izxD}$ of (3.10) and for $\mathcal{B} \neq 0$ of the form given above? In other words, are there D, A, $S(0)$, $\Pi(0)$ and $\{c_{lk}\}$ satisfying (3.18)?

Acknowledgments

The research was supported by the Austrian Science Fund (FWF) under Grant No. P24301.

References

[1] Cieśliński J L 2009 Algebraic construction of the Darboux matrix revisited *J. Phys.* A **42**:40 404003
[2] Gesztesy F 1993 A complete spectral characterization of the double commutation method *J. Funct. Anal.* **117**:2 401
[3] Gesztesy F and Teschl G 1996 On the double commutation method *Proc. Amer. Math. Soc.* **124**:6 1831
[4] Gu C H, Hu H and Zhou Z 2005 *Darboux Transformations in Integrable Systems. Theory and their Applications to Geometry* (Mathematical Physics Studies vol 26) (Dordrecht: Springer)
[5] Kostenko A, Sakhnovich A and Teschl G 2012 Commutation methods for Schrödinger operators with strongly singular potentials *Math. Nachr.* **285**:4 392
[6] Matveev V B and Salle M A 1991 *Darboux Transformations and Solitons* (Berlin: Springer)
[7] Popovych R O, Kunzinger M and Ivanova N M 2008 Conservation laws and potential symmetries of linear parabolic equations *Acta Appl. Math.* **100**:2 113
[8] Sakhnovich A L 1994 Dressing procedure for solutions of nonlinear equations and the method of operator identities *Inverse Problems* **10**:3 699
[9] Sakhnovich A L 2001 Generalized Bäcklund–Darboux transformation: spectral properties and nonlinear equations *J. Math. Anal. Appl.* **262**:1 274
[10] Sakhnovich A L 2010 On the GBDT version of the Bäcklund–Darboux transformation and its applications to linear and nonlinear equations and Weyl theory *Math. Model. Nat. Phenom.* **5**:4 340
[11] Sakhnovich A L, Sakhnovich L A and Roitberg I Ya 2013 *Inverse Problems and Nonlinear Evolution Equations. Solutions, Darboux Matrices and Weyl–Titchmarsh Functions* (de Gruyter Studies in Mathematics vol 47) (Berlin: De Gruyter)
[12] Zakharov V E and Manakov S V 1975 The theory of resonance interaction of wave packets in nonlinear media *Soviet Phys. JETP* **42**:5 842
[13] Novikov S, Manakov S V, Pitaevskii L P and Zakharov V E 1984 *Theory of Solitons. The Inverse Scattering Method*, Contemporary Soviet Mathematics (New York: Consultants Bureau [Plenum])
[14] Grandati Y 2011 Solvable rational extensions of the isotonic oscillator *Ann. Physics* **326**:8 2074
[15] Duistermaat J J and Grünbaum F A 1986 Differential equations in the spectral parameter *Commun. Math. Phys.* **103** 177
[16] Grünbaum F A 2014 Some noncommutative matrix algebras arising in the bispectral problem *SIGMA Symmetry Integrability Geom. Methods Appl.* **10** 078
[17] Sakhnovich A L and Zubelli J P 2001 Bundle bispectrality for matrix differential equations *Integral Eguations Operator Theory* **41** 472
[18] Gohberg I, Kaashoek M A and Sakhnovich A L 1998 Pseudo-canonical systems with rational Weyl functions: explicit formulas and applications *J. Differential Equations* **146**:2 375

GADEISVII IOP Publishing
Journal of Physics: Conference Series **621** (2015) 012014 doi:10.1088/1742-6596/621/1/012014

Symmetries of second-order PDEs and conformal Killing vectors

Michael Tsamparlis[1] **and Andronikos Paliathanasis**[2,3]

[1] Department of Physics, Section of Astronomy, Astrophysics and Mechanics, University of Athens, Panepistemiopolis, Athens 157 83, Greece
[2] Dipartimento di Fisica, Università di Napoli "Federico II" Complesso universitario Monte S. Angelo, Via Cintia - 80126 Napoli, Italy
[3] INFN - Istituto Nazionale di Fisica Nucleare sezione di Napoli Complesso universitario Monte S. Angelo, Via Cintia - 80126 Napoli, Italy

E-mail: mtsampa@phys.uoa.gr, paliathanasis@na.infn.it

Abstract. We study the Lie point symmetries of a general class of partial differential equations (PDE) of second order. An equation from this class naturally defines a second-order symmetric tensor (metric). In the case the PDE is linear on the first derivatives we show that the Lie point symmetries are given by the conformal algebra of the metric modulo a constraint involving the linear part of the PDE. Important elements in this class are the Klein–Gordon equation and the Laplace equation. We apply the general results and determine the Lie point symmetries of these equations in various general classes of Riemannian spaces. Finally we study the type II hidden symmetries of the wave equation in a Riemannian space with a Lorenzian metric.

1. Introduction

In theoretical physics one has two main tools to study the properties of evolution of dynamical systems (a) Symmetries of the equations of motion and (b) Collineations (symmetries) of the background space, where evolution takes place. It is well known that both these tools have the following common characteristics:

1) they form a Lie algebra;

2) they do not fix uniquely either the dynamical system or the space.

The natural question to be to asked is if these two algebras are related and in what way. Equivalently, one may state the question as follows:

To what degree and how the space modulates the evolution of dynamical systems in it? That is, a dynamical system is free to evolve at will in a given space or it is constrained to do so by the very symmetry structure of the space?

This question has been answered many years ago by the Theory of Relativity with the Equivalence Principle, that is, the requirement that free motion in a given gravitational field occurs along the geodesics of the space. However as obvious as this point of view may appear to be it is not easy to comprehend and accept! So let us give a precise formulation now.[1]

[1] This point has been raised during an illuminating discussion with Prof P G Leach in 2000 in Athens while we were driving to the Poseidon Temple in Cape Sounion.

In a Riemannian space the affinely parameterized geodesics are determined uniquely by the metric. The geodesics are a set of homogeneous ordinary differential equations (ODE) linear in the highest order term and quadratically non-linear in the first order terms. A system of such ODEs is characterized (not fully) by its Lie point symmetries. On the other hand a metric is characterized (again not fully) by its collineations. Therefore it is reasonable one to expect that the Lie point symmetries of the system of geodesic equations of a metric will be closely related with the collineations of the metric. That such a relation exists it is easy to see by the following simple example. Consider on the Euclidian plane a family of straight lines parallel to the x-axis. These curves can be considered either as the integral curves of the ODE $\frac{d^2y}{dx^2}=0$ or as the geodesics of the Euclidian metric dx^2+dy^2. Subsequently consider a symmetry operation defined by a reshuffling of these lines without preserving necessarily their parametrization. According to the first interpretation this symmetry operation is a Lie symmetry of the ODE $\frac{d^2y}{dx^2}=0$ and according to the second interpretation it is a (special) projective symmetry of the Euclidian two-dimensional space.

What has been said for a Riemannian space can be generalized to an affine space in which there is only a linear connection. In this case the geodesics are called autoparallels (or paths) and they comprise again a system of ODEs linear in the highest order term and quadratically non-linear in the first order terms. In this case one is looking for relations between the Lie point symmetries of the autoparallels and the projective collineations of the connection.

A Lie point symmetry of an ordinary differential equation (ODE) is a point transformation in the space of variables which preserves the set of solutions of the ODE [1–3]. If we look at these solutions as curves in the space of variables, then we may equivalently consider a Lie point symmetry as a point transformation which preserves the set of the solution curves. Applying this observation to the geodesic curves in a Riemannian (affine) space, we infer that the Lie point symmetries of the geodesic equations in any Riemannian space are the automorphisms which preserve the set of these curves. However we know from Differential Geometry that the point transformations of a Riemannian (affine) space which preserve the set of geodesics are the projective transformations. Therefore it is reasonable to expect a correspondence between the Lie point symmetries of the geodesic equations and the projective algebra of the metric defining the geodesics.

The equation of geodesics in an arbitrary coordinate frame is a second-order ODE of the form

$$\ddot{x}^i+\Gamma^i_{jk}\dot{x}^j\dot{x}^k+F^i(x,\dot{x})=0, \tag{1}$$

where $F^i(x,\dot{x})$ are arbitrary functions of their arguments and the functions Γ^i_{jk} are the connection coefficients of the space. Equivalently equation (1) is also the equation of motion of a dynamical system moving in a Riemannian (affine) space under the action of a velocity dependent force. According to the above argument we expect that the Lie point symmetries of the ODE (1) for given functions $F^i(x,\dot{x})$ will be related with the collineations of the metric. As it will be shown in this case the Lie symmetries of (1) determine a subalgebra of the special projective algebra of the space. The specific subalgebra is selected by means of certain constraint conditions involving geometric quantities of the space and the function $F(x^i,\dot{x}^j)$ [4–9].

The determination of the Lie point symmetries of a given system of ODEs consists of two steps: (a) determination of the conditions which the components of the Lie symmetry vector must satisfy and (b) solution of the system of these conditions. Step (a) is formal and it is outlined, e.g., in [1–3]. The second step is the key one and, for example, in higher dimensions, where one has a large number of equations, the solution can be quite involved and perhaps impossible by algebraic computing. However, if one expresses the system of Lie symmetry conditions of (1) in terms of collineation (i.e. symmetry) conditions of the metric, then the determination of Lie point symmetries is transferred to the geometric problem of determining the special projective group of the metric [9]. In this field there is a significant amount of knowledge from Differential Geometry waiting to be used. Indeed the projective symmetries are already known for many spaces or they can be determined by existing general theorems. For example the projective algebra and all its subalgebras are known for the important case of spaces of constant curvature [10] and in particular for the flat spaces. This implies that, the Lie symmetries of the Newtonian dynamical systems as well as those of Special Relativity can be determined by simple differentiation from the known projective algebra of these spaces!

What has been said for the Lie point symmetries of (1) applies also to Noether point symmetries (provided (1) follows from a Lagrangian). The Noether point symmetries are Lie point symmetries which satisfy the additional constraint

$$X^{[1]}L + L\frac{d\xi}{dt} = \frac{df}{dt}. \tag{2}$$

The Noether point symmetries form a subalgebra of the Lie point symmetry algebra. In accordance with the above this implies that the Noether point symmetries will be related with a subalgebra of the special projection algebra of the space where 'motion' occurs. As it has been shown this subalgebra is the homothetic algebra of the space [9]. It is well known that to each Noether point symmetry it is associated a conserved current (i.e. a Noether first integral). This leads to the important conclusion that the (standard) conserved quantities of a dynamical system depend *on the space it moves and the type of force* $F(x^i)$ *which modulates the motion*. In particular in 'free fall', that is when $F(x^i) = 0$, the orbits are affinely parametrized geodesics and the *geometry* of the space is the sole factor which determines the conserved quantities of motion. This conclusion is by no means trivial and means that the space where motion occurs is not a simple carrier of the motion but it is the major modulator of the evolution of a dynamical system. In other words there is a strong and deep relation between Geometry of the space and Physics (motion) in that space!

The above scenario can be generalized to partial differential equations (PDEs). Obviously, in this case a global answer is not possible. However, it can be shown that for many interesting PDEs the Lie point symmetries are indeed obtained from the collineations of the metric. Pioneering work in this direction is the work of Ibragimov [1]. Recently, Bozhkov et al. [11] studied the Lie and the point Noether symmetries of the Poisson equation and have shown that the Lie symmetries of the Poisson PDE are generated from the conformal algebra of the metric. This result can be generalized and it has been shown [12] that for a general class of PDEs of second order in an n-dimensional Riemannian space, there is a close relation between the Lie point symmetries and the conformal algebra of the space. Examples of such PDEs include some important equations as: the heat equation, the Klein–Gordon equation, the Laplace equation, the Schrödinger equation and others [12–15]. In what follows we discuss in a rather systematic way the aforementioned ideas. The plan of the paper is as follows.

In section 2 we give the basic definitions and properties of the collineations of space times and the Lie symmetries of DEs. In section 3 we study the Lie symmetries of a generic family of second-order partial differential equations and we prove that when the second-order partial differential equation is linear on the derivatives, the corresponding Lie symmetries are generated by the conformal algebra of the underlying geometry. Furthermore, in sections 4 and 5 we apply these general result in order to determine the general form of the Lie symmetry vector of the Poisson equation, of the Klein–Gordon equation and of the Laplace equation. In section 6 we study the application of the conformal Killing vectors in the Laplace equation is some special Riemannian spaces and we study the origin of the type II hidden symmetries. Finally in section 7 we apply the previous results in the case of the Laplace equation in the 1+3 wave equation and in Bianchi I spacetimes.

Journal of Physics: Conference Series **621** (2015) 012014 doi:10.1088/1742-6596/621/1/012014

2. Preliminaries

In this section we give the basic definitions and properties of the collineations of spacetimes and of the point symmetries of differential equations.

2.1. Collineations of Riemannian spaces

A collineation in a Riemannian space of dimension n is a vector field $\mathbf{X}$ which satisfies an equation of the form $\mathcal{L}_X\mathbf{A} = \mathbf{B}$, where $\mathcal{L}_X$ is the Lie derivative with respect to the vector field $\mathbf{X}$, $\mathbf{A}$ is a geometric object (not necessary a tensor field) defined in terms of the metric and its derivatives, and $\mathbf{B}$ is an arbitrary tensor field with the same tensor indices as the geometric object $\mathbf{A}$. The classification of the collineations of Riemannian manifolds can be found in [18]. In the following we are interested in the collineations of the metric tensor, i.e. $\mathbf{A} = g_{ij}$ of the Riemannian space.

A vector field $\mathbf{X}$ is a conformal Killing vector (CKV) of g_{ij} if the following condition holds[2]:

$$\mathcal{L}_{\mathbf{X}} g_{ij} = 2\psi(x^k) g_{ij},$$

where $\psi(x^k) = \frac{1}{n}\mathbf{X}^i_{;i}$. In case $\psi_{;ij} = 0$, $\mathbf{X}$ is called special CKV (sp.CKV), if $\psi(x^k) =$ constant, the vector field X is called homothetic (HV) and if $\psi(x^k) = 0$, the field $\mathbf{X}$ is called a Killing vector (KV).

The CKVs of the metric g_{ij} form a Lie algebra which is called the conformal algebra (CA) of the metric g_{ij} (CKVs). The conformal algebra contains two subalgebras, the Homothetic algebra (HA) and the Killing algebra (KA). These algebras are related as $KA \subseteq HA \subseteq CA$.

Two metrics g_{ij} and $\bar{g}_{ij}$ are conformally related if there exists a function $N^2(x^k)$ such as $\bar{g}_{ij} = N^2(x^k) g_{ij}$. If $\mathbf{X}$ is a CKV of the metric $\bar{g}_{ij}$ so that $L_{\mathbf{X}}\bar{g}_{ij} = 2\bar{\psi}\bar{g}_{ij}$, then $\mathbf{X}$ is also a CKV of the metric g_{ij}, that is $L_{\mathbf{X}} g_{ij} = 2\psi g_{ij}$ with conformal factor $\psi(x^k)$; the two conformal factors are related as follows

$$\psi = \bar{\psi} N^2 - N N^i_{,i} \mathbf{X}^{,i}.$$

The last relation implies that two conformally related metrics have the same conformal algebra, but with different subalgebras; that is, a KV for one may be proper CKV for the other.

A special class of conformally related spaces are the conformally flat spaces. A space V^n is conformally flat if the metric g_{ij} of V^n satisfies the relation $g_{ij} = N^2 s_{ij}$, where s_{ij} is the metric of a flat space which has the same signature as g_{ij}. The maximal dimension of the conformal algebra of a n-dimensional metric ($n > 2$) is $\frac{1}{2}(n+1)(n+2)$ and in that case the space is conformally flat. Moreover, if the conformally flat space V^n admits a $\frac{1}{2}n(n+1)$-dimensional Killing algebra then V^n is a space of constant curvature and admits a proper HV if and only if the space is flat.

Furthermore, if for a CKV $\mathbf{X}$ of the Riemannian space V_G^n the condition $\mathbf{X}_{[i;j]} = 0$ holds, i.e. $\mathbf{X}_{i;j} = \psi(x^k) g_{ij}$, then the CKV will be called gradient CKV. In this case there exists a coordinate system in which the line element of the metric which defines the Riemannian space V_G^n is

$$ds^2 = dx^n + f^2(x^n) h^{AB}(x^A) dx^A dx^B,$$

where $A, B = 1, 2, \ldots, n-1$ [16]. In these coordinates the gradient CKV is $\mathbf{X} = f(x^n)\partial_{x^n}$ with conformal factor $\psi = f_{,x^n}$. In the case when $f(x^n) = x^n$, we have $\psi = 1$; hence, $\mathbf{X}$ becomes gradient HV and if $f(x^n) = f_0$, $\mathbf{X}$ becomes gradient KV.

[2] In the following we use the Einstein notation.

2.2. Point symmetries of differential equations

A partial differential equation (PDE) is defined by a function $H = H(x^i, u^A, u^A_{,i}, u^A_{,ij})$ in the jet space $\bar{B}_{\bar{M}}$, where x^i are the independent variables and u^A are the dependent variables. The infinitesimal point transformation

$$\bar{x}^i = x^i + \varepsilon\xi^i(x^k, u^B), \tag{3}$$

$$\bar{u}^A = \bar{u}^A + \varepsilon\eta^A(x^k, u^B), \tag{4}$$

has the infinitesimal symmetry generator

$$\mathbf{X} = \xi^i(x^k, u^B)\partial_{x^i} + \eta^A(x^k, u^B)\partial_{u^A}. \tag{5}$$

The generator $\mathbf{X}$ of the infinitesimal transformation (3), (4) is called a Lie point symmetry of the PDE H if there exists a function λ such that the following condition holds [1, 2]

$$\mathbf{X}^{[n]}(H) = \lambda H, \quad \mathrm{mod}\, H = 0, \tag{6}$$

where

$$\mathbf{X}^{[n]} = \mathbf{X} + \eta_i^A\partial_{\dot{x}^i} + \eta_{ij}^A\partial_{u^A_{ij...i_n}} + \cdots + \eta^A_{i_1 i_2 ... i_n}\partial_{u^A_{i_1 i_2 ... i_n}} \tag{7}$$

is the n-th prolongation of $\mathbf{X}$ and

$$\eta_i^A = \eta^A_{,i} + u^B_{,i}\eta^A_{,B} - \xi^j_{,i}u^A_{,j} - u^A_{,i}u^B_{,j}\xi^j_{,B} \tag{8}$$

with

$$\begin{aligned}\eta_{ij}^A = {} & \eta^A_{,ij} + 2\eta^A_{,B(i}u^B_{,j)} - \xi^k_{,ij}u^A_{,k} + \eta^A_{,BC}u^B_{,i}u^C_{,j} - 2\xi^k_{,(i|B|}u^B_{j)}u^A_{,k} \\ & - \xi^k_{,BC}u^B_{,i}u^C_{,j}u^A_{,k} + \eta^A_{,B}u^B_{,ij} - 2\xi^k_{,(j}u^A_{,i)k} - \xi^k_{,B}\left(u^A_{,k}u^B_{,ij} + 2u^B_{(,j}u^A_{,i)k}\right).\end{aligned} \tag{9}$$

Lie symmetries of differential equations can be used in order to determine invariant solutions or transform solutions into solutions [3]. From condition (6) one defines the Lagrange system

$$\frac{dx^i}{\xi^i} = \frac{du}{\eta} = \frac{du_i}{\eta_{[i]}} = \cdots = \frac{du_{ij...i_n}}{\eta_{[ij...i_n]}} \tag{10}$$

whose solution provides the characteristic functions

$$W^{[0]}(x^k, u), \quad W^{[1]i}(x^k, u, u_{,i}), \quad \ldots, \quad W^{[n]}(x^k, u, u_{,i}, \ldots, u_{,ij...i_n}).$$

The solution $W^{[n]}$ of the Lagrange system (10) is called the n-th order invariant of the Lie symmetry vector (5) and holds $X^{[n]}W^{[n]} = 0$.

The application of a Lie symmetry to a PDE H leads to a new differential equation $\bar{H}$ which is different from H and it is possible that it admits Lie symmetries which are not Lie symmetries of H. These Lie point symmetries are called Type II hidden symmetries. It has been shown in [17] that if X_1, X_2 are Lie point symmetries of the original PDE with commutator $[X_1, X_2] = cX_1$ where c is a constant, then reduction by X_2 results in X_1 being a point symmetry of the reduced PDE $\bar{H}$, while reduction by X_1 results in a PDE $\bar{H}$ for which X_2 is not a Lie point symmetry.

In the following section we study the Lie point symmetries of a general type of second-order PDEs.

3. Lie symmetries of second-order PDEs and CKVs

It is interesting to examine if the close relation of the Lie and the Noether point symmetries of the second-order ODEs of the form (1) with the collineations of the metric is possible to be carried over to second-order partial differential equations (PDEs). Obviously it will not be possible to give a complete answer, due to the complexity of the study and the great variety of PDEs.

We consider the second-order PDEs of the form

$$A^{ij}u_{ij} - F(x^i, u, u_i) = 0 \tag{11}$$

for which at least one of the A^{ij} is nonzero and derive the point Lie symmetry conditions.

The symmetry condition (6) when applied to (11) gives

$$A^{ij}\eta^{(2)}_{ij} + \left(XA^{ij}\right)u_{ij} - X^{[1]}(F) = \lambda(A^{ij}u_{ij} - F) \tag{12}$$

that leads to

$$\begin{aligned} &A^{ij}\eta_{ij} - \eta_{,i}g^{ij}F_{,u_j} - X(F) + \lambda F + 2A^{ij}\eta_{ui}u_j - A^{ij}\xi^a_{,ij}u_a - u_i\eta_u g^{ij}F_{,u_j} + \xi^k_{,i}u_k g^{ij}F_{,u_j} \\ &+ A^{ij}\eta_{uu}u_iu_j - 2A^{ij}\xi^k_{.,u_j}u_iu_k + u_iu_k\xi^k_{,u}g^{ij}F_{,u_j} + A^{ij}\eta_u u_{ij} - 2A^{ij}\xi^k_{.,i}u_{jk} \\ &+ (\xi^k A^{ij}_{,k} + \eta A^{ij}_{,u})u_{ij} - \lambda A^{ij}u_{ij} - A^{ij}\left(u_{ij}u_a + u_iu_{ja} + u_{ia}u_j\right)\xi^a_{.,u} - u_iu_ju_aA^{ij}\xi^a_{uu} = 0. \end{aligned} \tag{13}$$

We note that we cannot deduce the symmetry conditions before we select a specific form for the function F. However we may determine the conditions which are due to the second derivative of u because these terms do not involve F. This observation significantly reduces the complexity of the remaining symmetry condition. Following this observation we find the condition

$$\begin{aligned} &A^{ij}\eta_u u_{ij} - A^{ij}(\xi^k_{.,i}u_{ja} + \xi^k_{.,j}u_{ik}) + (\xi^k A^{ij}_{,k} + \eta A^{ij}_{,u})u_{ij} - \lambda A^{ij}u_{ij} \\ &- A^{ij}\left(u_{ij}u_a + u_iu_{ja} + u_{ia}u_j\right)\xi^a_{.,u} - u_iu_ju_aA^{ij}\xi^a_{uu} = 0 \end{aligned}$$

which implies the equations

$$\begin{aligned} A^{ij}\left(u_{ij}u_k + u_{jk}u_i + u_{ik}u_j\right)\xi^k_{.,u} &= 0, \\ A^{ij}\eta_u u_{ij} - A^{ij}(\xi^k_{.,i}u_{jk} + \xi^k_{.,j}u_{ik}) + (\xi^k A^{ij}_{,k} + \eta A^{ij}_{,u})u_{ij} - \lambda A^{ij}u_{ij} &= 0, \\ A^{ij}\xi^a_{uu} &= 0. \end{aligned}$$

The first equation is written as

$$A^{ij}\xi^k_{.,u} + A^{kj}\xi^i_{.,u} + A^{ik}\xi^j_{.,u} = 0 \Leftrightarrow A^{(ij}\xi^{k)}_{.,u} = 0. \tag{14}$$

The second equation gives

$$A^{ij}\eta_u + \eta A^{ij}_{,u} + \xi^k A^{ij}_{,k} - A^{kj}\xi^i_{.,k} - A^{ik}\xi^j_{.,k} - \lambda A^{ij} = 0. \tag{15}$$

Therefore the last equation results in

$$\xi^k_{,uu} = 0. \tag{16}$$

It is straightforward to show that condition (14) implies $\xi^k_{.,u} = 0$ which is a well known result.

From the analysis so far we obtain that for all second-order PDEs of the form $A^{ij}u_{ij} - F(x^i, u, u_i) = 0$, for which at least one of the A^{ij} is nonzero, the coefficients $\xi^i_{.,u} = 0$ or $\xi^i = \xi^i(x^j)$.

Furthermore condition (16) is identically satisfied. We consider that A^{ij} is non-degenerate, furthermore the third symmetry condition (15) can be written as follows

$$L_{\xi^i\partial_i}A^{ij} = \lambda A^{ij} - (\eta A^{ij})_{,u}. \tag{17}$$

This condition implies that for all second-order PDEs of the form $A^{ij}u_{ij} - F(x^i, u, u_i) = 0$ for which $A^{ij}{}_{,u} = 0$, i.e. $A^{ij} = A^{ij}(x^i)$, the vector $\xi^i\partial_i$ is a CKV of the metric A^{ij} with conformal factor $(\lambda - \eta_u)(x)$.

Moreover, using that $\xi^i_{,u} = 0$ when at least one of the $A_{ij} \neq 0$, the symmetry condition (13) is simplified as follows

$$\begin{aligned} &A^{ij}\eta_{ij} - \eta_{,i}A^{ij}F_{,u_j} - X(F) + \lambda F + 2A^{ij}\eta_{ui}u_j - A^{ij}\xi^a_{,ij}u_a - u_i\eta_u A^{ij}F_{,u_j} \\ &+ \xi^k_{,i}u_k A^{ij}F_{,u_j} + A^{ij}\eta_{uu}u_iu_j + A^{ij}\eta_u u_{ij} - 2A^{ij}\xi^k_{,i}u_{jk} + \left(\xi^k A^{ij}_{,k} + \eta A^{ij}_{,u}\right)u_{ij} - \lambda A^{ij}u_{ij} = 0, \end{aligned} \tag{18}$$

which together with the condition (17) are the complete set of conditions *for all* second-order PDEs of the form $A^{ij}u_{ij} - F(x^i, u, u_i) = 0$ for which at least one of the $A_{ij} \neq 0$. This class of PDEs is quite general making the above result very useful.

In order to continue we need to assume that the function $F(x, u, u_i)$ is of a special form.

3.1. The Lie point symmetry conditions for a linear function $F(x, u, u_i)$

We consider the function $F(x, u, u_i)$ to be of the form

$$F(x, u, u_i) = B^k(x, u)u_k + f(x, u), \tag{19}$$

where $B^k(x, u)$ and $f(x, u)$ are arbitrary functions of their arguments. In this case the PDE is of the form

$$A^{ij}u_{ij} - B^k(x, u)u_k - f(x, u) = 0. \tag{20}$$

The Lie symmetries of this type of PDEs have been studied previously by Ibragimov [1]. Assuming that at least one of the $A_{ij} \neq 0$ the Lie point symmetry conditions are (17) and (18).

Replacing $F(x, u, u_1)$ in (18) we obtain the following result [12]:

The Lie point symmetry conditions for the second-order PDEs of the form

$$A^{ij}u_{ij} - B^k(x, u)u_k - f(x, u) = 0, \tag{21}$$

where at least one of the $A_{ij} \neq 0$, are

$$A^{ij}(a_{ij}u + b_{ij}) - (a_{,i}u + b_{,i})B^i - \xi^k f_{,k} - auf_{,u} - bf_{,u} + \lambda f = 0, \tag{22}$$

$$A^{ij}\xi^k_{,ij} - 2A^{ik}a_{,i} + aB^k + auB^k_{,u} - \xi^k_{,i}B^i + \xi^i B^k_{,i} - \lambda B^k + bB^k_{,u} = 0, \tag{23}$$

$$L_{\xi^i\partial_i}A^{ij} = (\lambda - a)A^{ij} - \eta A^{ij}{}_{,u}, \tag{24}$$

$$\eta = a(x^i)u + b(x^i), \tag{25}$$

$$\xi^k_{,u} = 0 \Leftrightarrow \xi^k(x^i). \tag{26}$$

From (24) we infer that for all second-order PDEs of the form $A^{ij}u_{ij} - B^k(x, u)u_k - f(x, u) = 0$ for which $A^{ij}{}_{,u} = 0$, the $\xi^i(x^j)$ is a CKV of the metric A^{ij}. Also in this case for the arbitrary function λ holds $\lambda = \lambda(x^i)$. This result establishes a relation between the Lie point symmetries of this type of PDEs with the collineations of the metric defined by the coefficients A_{ij}.

Furthermore in case[3] when $A^{tt} = A^{tx^i} = 0$ and A^{ij} is a non-degenerate metric we obtain that

$$\xi^t_{,i} = 0 \Leftrightarrow \xi^t(t). \tag{27}$$

[3] The coordinates are t, x^i where $i = 1, \ldots, n$.

These symmetry relations coincide with those given in [1]. Finally we note that equation (23) can be written as

$$A^{ij}\xi^k_{,ij} - 2A^{ik}a_{,i} + [\xi, B]^k + (a-\lambda)B^k + (au+b)B^k_{,u} = 0. \tag{28}$$

Let us see now some applications of these results.

4. Symmetries of the Poisson equation in a Riemannian space

The Lie symmetries of the Poisson equation

$$\Delta u - f\left(x^i, u\right) = 0, \tag{29}$$

where $\Delta = \frac{1}{\sqrt{|g|}}\frac{\partial}{\partial x^i}\left(\sqrt{|g|}g^{ij}\frac{\partial}{\partial x^j}\right)$ is the Laplace operator of the metric g_{ij}, for $f = f(u)$ have been given in [1,11]. Here we generalize these results for the case $f = f\left(x^i, u\right)$.

The Lie symmetry conditions (22)–(26) for the Poisson equation (29) are

$$g^{ij}(a_{ij}u + b_{ij}) - (a_{,i}u + b_{,i})\Gamma^i - \xi^k f_{,k} - auf_{,u} - bf_{,u} + \lambda f = 0, \tag{30}$$

$$g^{ij}\xi^k_{,ij} - 2g^{ik}a_{,i} + a\Gamma^k - \xi^k_{,i}\Gamma^i + \xi^i\Gamma^k_{,i} - \lambda\Gamma^k = 0, \tag{31}$$

$$L_{\xi^i\partial_i}g_{ij} = (a-\lambda)g_{ij}, \tag{32}$$

$$\eta = a(x^i)u + b(x^i), \quad \xi^k_{,u} = 0. \tag{33}$$

Equation (31) becomes (see [11])

$$g^{jk}L_\xi\Gamma^i_{.jk} = 2g^{ik}a_{,i}. \tag{34}$$

From (32), ξ^i is a CKV, then equations (34) give

$$\frac{2-n}{2}(a-\lambda)^{,i} = 2a^{,i} \rightarrow (a-\lambda)^i = \frac{4}{2-n}a^{,i}.$$

We define

$$\psi = \frac{2}{2-n}a + a_0, \tag{35}$$

where $\psi = \frac{1}{2}(a-\lambda)$ is the conformal factor of ξ^i, i.e. $L_\xi g_{ij} = 2\psi g_{ij}$. Furthermore, we have

$$\begin{aligned}(2-n)\lambda^i &= (2-n)a^i - 4a^i,\\ (2-n)\lambda^i &= -(n+2)a^i.\end{aligned}$$

Finally, from (30), we have the constraint

$$g^{ij}a_{i;j}u + g^{ij}b_{;ij} - \xi^k f_{,k} - auf_{,u} + \lambda f - bf_{,u} = 0. \tag{36}$$

For $n = 2$, it holds that $g^{jk}L_\xi\Gamma^i_{.jk} = 0$; this means that $a_{,i} = 0 \rightarrow a = a_0$. From (32), ξ^i is a CKV with conformal factor

$$2\psi = (a_0 - \lambda) \tag{37}$$

and $\lambda = a_0 - 2\psi$. Finally, from (30), we have the constraint

$$g^{ij}b_{;ij} - \xi^k f_{,k} - a_0uf_{,u} + (a_0 - 2\psi)f - bf_{,u} = 0. \tag{38}$$

Hence for the Lie symmetries of the Poisson equation in a general Riemannian space we have the following theorem.

Theorem 1 *The Lie symmetries of the Poisson equation (29) are generated from the CKVs of the metric g_{ij} defining the Laplace operator, as follows*

a) for $n>2$, the Lie symmetry vector is

$$X=\xi^{i}(x^{k})\partial_{i}+\left(\frac{2-n}{2}\psi(x^{k})u+a_{0}u+b(x^{k})\right)\partial_{u}, \tag{39}$$

where $\xi^{i}(x^{k})$ is a CKV with conformal factor $\psi(x^{k})$ and the following condition holds

$$\frac{2-n}{2}\Delta\psi u+g^{ij}b_{i;j}-\xi^{k}f_{,k}-\frac{2-n}{2}\psi u f_{,u}-\frac{2+n}{2}\psi f-bf_{,u}=0. \tag{40}$$

b) for $n=2$, the Lie symmetry vector is

$$X=\xi^{i}(x^{k})\partial_{i}+(a_{0}u+b(x^{k}))\partial_{u}, \tag{41}$$

where $\xi^{i}(x^{k})$ is a CKV with conformal factor $\psi(x^{k})$ and the following condition holds

$$g^{ij}b_{;ij}-\xi^{k}f_{,k}-a_{0}uf_{,u}+(a_{0}-2\psi)f-bf_{,u}=0.$$

5. Lie Symmetries of the Klein–Gordon equation and CKVs

In the special case, where $f(x^{i},u)=-V(x^{i})u$, the Poisson equation (29) is reduced to the Klein–Gordon equation

$$\Delta u+V(x^{k})u=0. \tag{42}$$

Therefore from theorem 1 for the Lie symmetries of (42) we have:

Theorem 2 *The Lie point symmetries of the Klein–Gordon equation (42) are generated from the elements of the conformal algebra of the metric g_{ij} defining the Laplace operator Δ. Equation (42) admits the Lie symmetry vector*

$$X=\xi^{i}(x^{k})\partial_{i}+\left(\frac{2-n}{2}\psi(x^{k})u+a_{0}u+b(x^{k})\right)\partial_{u}, \tag{43}$$

where $\xi^{i}(x^{k})$ is a CKV for the metric g_{ij} with conformal factor $\psi(x^{k})$, the potential $V(x^{k})$ satisfies the condition

$$\xi^{k}V_{,k}+2\psi V-\frac{2-n}{2}\Delta\psi=0, \tag{44}$$

and the function $b(x^{k})$ is a solution of (42).

Of interest are the cases where $V(x^{k})=0$ and $V(x^{k})=\frac{n-2}{4(n-1)}R$, where R is the Ricci scalar of the metric g_{ij} which defines the operator Δ. In these cases, the Klein–Gordon equation (42) becomes the Laplace equation

$$\Delta u=0 \tag{45}$$

and the conformal invariant Laplace equation

$$\bar{L}_{g}u=0, \tag{46}$$

where $\bar{L}_{g}=\Delta+\frac{n-2}{4(n-1)}R$. Therefore, from theorem 2 for the Lie symmetries of equations (45) and (46) we have the following result

Theorem 3 *The generic form of the Lie point symmetry of Laplace equation (45) and of the conformal invariant Laplace equation (46) is the vector field (43) where for the Laplace equation (45), in a n-dimensional space with $n>2$, the conformal factor $\psi(x^{k})$ of the CKV $\xi^{i}(x^{k})$ is a solution of (45).*

6. Reduction of Laplace equation in certain Riemannian spaces

From theorem 3 we have that the Lie symmetries of Laplace equation (45) in a Riemannian space are generated from the CKVs (not necessarily proper) whose conformal factor satisfies Laplace equation. This condition is satisfied trivially by the KVs ($\psi = 0$), the HV ($\psi_{;i} = 0$) and the sp.CKVs ($\psi_{;ij} = 0$). Therefore these vectors (which span a subalgebra of the conformal algebra) are among the Lie symmetries of Laplace equation. Concerning the proper CKVs it is not necessary that their conformal factor satisfies the Laplace equation, therefore they may not produce Lie symmetries for Laplace equation.

Furthermore, the special forms of the metric of a Riemannian space which admits a gradient KV/HV or a sp.CKV are well known in the literature. Therefore it is possible to study the application of the Lie symmetries of the Laplace equation in these Riemannian spaces. In the following, we study the reduction of Laplace equation in these general classes of Riemannian spaces and we also study the origin of type II hidden symmetries. We assume that the dimension n of the space is $n > 2$.

6.1. Reduction with a gradient KV/HV

We consider the $(1+n)$-dimensional metric g_{ij} with line element

$$ds^2 = dr^2 + r^{2K} h_{AB} dy^A dy^B, \quad h_{AB} = h_{AB}\left(y^C\right), \tag{47}$$

where h_{AB} is the metric of the n-dimensional space and $A, B, C = 1, \ldots, n$. For a general functional form of h_{AB} and when $K = 0$, the metric (47) admits the gradient KV ∂_r, however when $K = 1$, the latter admits the gradient HV $r\partial_r$ (see [19]).

For the space with line element (47) Laplace equation (45) takes the form

$$u_{,rr} + K\frac{n}{r}u_{,r} + \frac{1}{r^{2K}}\ {}_h\Delta u = 0, \tag{48}$$

where ${}_h\Delta u = h^{AB}\left(y^C\right)u_{,AB} - \Gamma^A\left(y^C\right)u_A$ is the Laplace operator with metric h_{AB}.

Laplace equation (48) admits extra Lie point symmetries when $K = 0, 1$. In particular when $K = 0$, the extra Lie point symmetry is the gradient KV $X_{KV} = \partial_r + \mu u\partial_u$ and when $K = 1$ the extra Lie point symmetry is the gradient HV $X_{HV} = r\partial_r + \mu u\partial_u$. We will study the reduction of equation (48) using the zero-order invariants of the symmetries X_{KV} and X_{HV}.

The zero-order invariants of X_{KV} are $\left\{y^A, e^{-\mu r}u\right\}$ and of X_{HV} are $\left\{y^A, r^{-\mu}u\right\}$. Hence by replacing in equation (48) we find the reduced equation

$${}_h\Delta w + \mu^2 w = 0, \tag{49}$$

where

$$u\left(r, y^A\right) = \begin{Bmatrix} e^{\mu r}w\left(y^A\right) & \text{when } K = 0 \\ r^{\mu}w\left(y^A\right) & \text{when } K = 1 \end{Bmatrix}.$$

Equation (49) is the linear Klein–Gordon equation in the space with metric h_{AB}. According to theorem 2 the Lie point symmetries of the reduced equation (49) follow from the CKVs of the metric h_{AB}.

The relation of the conformal algebra of the n metric h_{AB} and of the $1+n$ metric (47) have been studied in [20]. In particular the KVs of the metrics g_{ij} and h_{AB} are the same. Furthermore, for $K = 0$, the $1+n$ metric g_{ij} admits a HV if and only if the n metric admits one and if ${}_nH^A$ is the HV of the n metric then the HV of the $1+n$ metric is given by the expression

$${}_{1+n}H^{\mu} = z\delta^{\mu}_z +_n H^A\delta^{\mu}_A, \quad \mu = x, 1, \ldots, n. \tag{50}$$

However, for $K = 1$, the HV of the metric g_{ij} is independent from that of h_{AB}.

Finally, the metric (47) admits proper CKVs if and only if the n metric h_{AB} admits gradient CKVs. This is because (47) is conformally related with the decomposable metric

$$ds^{2\prime} = d\bar{r}^2 + h_{AB}\left(y^C\right)dy^A dy^B. \tag{51}$$

which admits CKVs if and only if the h_{AB} metric admits gradient CKVs.

The last implies, that Type II hidden symmetries are generated from the elements of the (proper) conformal algebra of the n-dimensional metric h_{AB} (for $K = 0, 1$) whose conformal factor is a solution of the Klein–Gordon equation (49), according to theorem 2. Furthermore when $K = 1$, the HV is a Type II hidden symmetry.

In the following we will study the origin of type II hidden symmetries in Riemannian spaces which admit a sp.CKV.

6.2. Reduction with a sp.CKV

It is known [21], that if an $n = m + 1$-dimensional ($n > 2$) Riemannian space admits a non null sp.CKVs then also admits a gradient HV and as many gradient KVs as the number of sp.CKVs. In these spaces there exists always a coordinate system in which the metric is written in the form

$$ds^2 = -dz^2 + dR^2 + R^2 f_{AB}\left(y^C\right)dy^A dy^B, \tag{52}$$

where $f_{AB}\left(y^C\right)$, $A, B, C, \ldots = 1, 2, \ldots, m-1$ is an $(m-1)$-dimensional metric. For a general metric f_{AB} the n-dimensional metric (52) admits a three-dimensional conformal algebra with elements

$$K_G = \partial_z, \quad H = z\partial_z + R\partial_R, \quad C_S = \frac{z^2 + R^2}{2}\partial_z + zR\partial_R,$$

where K_G is a gradient KV, H is a gradient HV and C_S is a sp.CKV with conformal factor $\psi_{C_S} = z$. In these coordinates Laplace equation (45) takes the form

$$-u_{zz} + u_{RR} + \frac{(m-1)}{R}u_R + \frac{1}{R^2}\,{}_f\Delta u_{AB} = 0. \tag{53}$$

From theorem 3, we have that the extra Lie point symmetries of (53) are the vectors

$$X^1 = K_G + \mu_G X_u, \quad X^2 = H + \mu_H X_u, \quad X^3 = C_S + 2pzX_u,$$

where $2p = \frac{1-m}{2}$. The nonzero commutators of the extra Lie point symmetries are

$$\left[X^1, X^2\right] = K_G, \quad \left[X^2, X^3\right] = X^3, \quad \left[X^1, X^3\right] = X^2 + 2pX_u.$$

The application of the Lie symmetries which are generated by the gradient KV and the gradient HV have been studied in section 6.1. However we would like to note that if we reduce the Laplace equation (53) by use of the Lie symmetry X^1, the reduced equation admits the inherited symmetry X^2 if and only if $\mu_G = 0$. Furthermore the reduction with the gradient HV leads to a PDE which does not admit inherited symmetries. The resulting type II hidden symmetries follow from the results of section 6.1.

Before we reduce (53) with the symmetry generated by the sp.CKV X^3, it is best to write the metric (52) in the coordinates $\left\{x, R, y^A\right\}$, where the variable x is defined by the relation $z = \sqrt{R\left(R - x^{-1}\right)}$. In the new variables the Lie symmetry X^3 becomes

$$X^3 = \sqrt{R\left(R - x^{-1}\right)}\left(R\partial_R + 2pu\partial_u\right). \tag{54}$$

The zero-order invariants of X^3 in the new coordinates are $\{x, y^A, R^{-2p}u\}$. We choose x, y^A to be the independent variables and $w = w(x, y^A)$ to be the dependent one; that is, the solution of the Laplace equation is in the form $u(x, R, y^A) = R^{2p}w(x, y^A)$. Replacing in (53) we find the reduced equation

$$x^2 w_{xx} + f^{AB} w_{AB} - \Gamma^A w_A - 2p(2p+1)w = 0. \tag{55}$$

For different values of the dimension m of the metric f_{AB}, equation (55) can be written in the following forms

$$x^2 w_{xx} + w_{yy} + \frac{1}{4}w = 0, \text{ when } m = 2, \tag{56}$$

$$_{(m=3)}\bar{\Delta}w = 0, \text{ when } m = 3, \tag{57}$$

$$_{(m\succeq 4)}\bar{\Delta}w - 2p(2p+1)V(\phi)w = 0, \text{ when } m \succeq 4, \tag{58}$$

where $_{(m=3)}\bar{\Delta}$ is the Laplace operator for the metric

$$d\bar{s}^2_{(m=3)} = \frac{1}{x^4}dx^2 + \frac{1}{x^2}f_{AB}dy^A dy^B. \tag{59}$$

The Laplace operator $_{(m\succeq 4)}\bar{\Delta}$ is defined by the metric

$$d\bar{s}^2_{(m\succeq 4)} = d\phi^2 + \frac{1}{V}f_{AB}dy^A dy^B, \tag{60}$$

where $V(\phi) = \frac{(2-m)^2}{\phi^2}$ and $d\phi = \frac{1}{xV}dx$.

By applying the Lie symmetry condition (6) for equation (56) we find that the generic Lie symmetry vector is

$$X = \xi^x(x,y)\partial_x + \xi^y(x,y)\partial_y + (a_0 w + b)\partial_w,$$

where

$$\xi^x(x,y) = c_1 x + i(F_1(y + i\ln x) + F_2(y - i\ln x)),$$
$$\xi^y(x,y) = F_2(y - i\ln x) - F_1(y + i\ln x).$$

$\xi^i = (\xi^x, \xi^y)$ is the generic CKV of the two-dimensional metric $A^{ij} = diag(x^2, 1)$. Therefore all the proper CKVs of the two-dimensional metric A^{ij} generate type II hidden symmetries. Recall that the conformal algebra of a two-dimensional space is infinite-dimensional. The Lie point symmetry $x\partial_x$ is the inherited symmetry H.

Furthermore for equations (57) and (58) from theorems 3 and 2 we have that the type II hidden symmetries are generated by the proper CKVs of the metrics (59) and (60), respectively, with conformal factors such as the condition (43) holds. Finally equations (57) and (58) admit the inherited Lie point symmetry H.

7. Application of the reduction of the Laplace equation in Riemannian spaces

In section 6, we studied the reduction of Laplace equation and the origin of type II hidden symmetries in general Riemannian spaces which admit a gradient KV, a gradient HV and a sp.CKV. In this section we apply these general results in order to study the reduction of Laplace equation and the type II hidden symmetries in the case where the Laplace operator is defined by (a) the n-dimensional Minkowski spacetime M^n (b) the four-dimensional conformally flat Bianchi I spacetime which admits a gradient KV.

7.1. The $1+(n-1)$ wave equation

We consider the $1+(n-1)$ wave equation

$$-u_{,tt}+u_{,zz}+\delta^{AB}u_{,A}u_{,B}=0 \tag{61}$$

which is the Laplace equation in the n-dimensional Minkowski spacetime M^n with $n>3$.

The n-dimensional Minkowski spacetime admits a conformal group G_C of dimension $\dim G_c = \frac{1}{2}(n+1)(n+2)$. Furthermore, the conformal group admits the following subalgebras:

a. T^n, the translation group of n gradient KVs

$$K_G^1=\partial_t,\quad K_G^z=\partial_z,\quad K_G^A=\partial_{y^A};$$

b. $SO(n)$, the Lie group of $\frac{1}{2}n(n-1)$ rotations

$$X_R^{1A}=Y^\alpha\partial_t+t\partial_{Y^\alpha},\quad X_R^{\alpha\beta}=Y^\beta\partial_{Y^\alpha}-Y^\alpha\partial_{Y^\beta},$$

where $Y^\alpha=(z,y^A)$;

c. One gradient HV

$$H=t\partial_t+z\partial_z+y^A\partial_{y^A};$$

d. G_{spC}, the Lie group of n sp. CKVs

$$X_C^1=\frac{1}{2}\left(t^2+\sum_a(Y^\alpha)^2\right)\partial_t+t\sum_a(Y^\alpha\partial_{Y^\alpha}),$$

$$X_C^\alpha=tY^\alpha\partial_t+\frac{1}{2}\left(Y^\alpha+t^2-\sum_{\beta\neq\alpha}\left(Y^\beta\right)^2\right)\partial_{Y^\alpha}+Y^\alpha\sum_{\beta\neq\alpha}\left(Y^\beta\partial_{Y^\beta}\right).$$

Therefore from theorem 3 we have that the Lie symmetries of the wave equation (61) are

$$X_u=u\partial_u,\quad K_G^1,\quad K_G^z,\quad K_G^A,\quad X_R^{1A},\quad X_R^{AB},\quad H,\quad X_C^1-tX_u,\quad X_C^\alpha-Y^\alpha X_u$$

and that the nonzero commutators of the Lie symmetries are

$$\left[K_G^I,X_R^{IJ}\right]=-K_G^J,\quad\left[K_G^I,H\right]=K_G^I,\quad\left[K_G^I,X_C^I\right]=H-X_u,$$
$$\left[K_G^I,X_C^J\right]=X_R^{IJ},\quad\left[H,X_C^I\right]=X_C^I,\quad\left[X_R^{IJ},X_C^I\right]=X_C^J,$$

where $I=1,a$.

In order to apply the results of section 6 we reduce (61) by the use of the Lie point symmetries generated by the gradient KV K_G^z, the gradient HV H and the special CKV X_C^1.

Using the zero-order invariants of the Lie symmetries $K_G^z+\mu X_u$ in (61) the reduced equation is the linear Klein–Gordon equation in the M^{n-1} spacetime

$$-w_{,tt}+\delta^{AB}w_{,A}w_{,B}+\mu^2w=0, \tag{62}$$

where $u\left(t,z,y^A\right)=e^{\mu z}w\left(t,y^A\right)$.

For arbitrary constant μ, equation (62) admits the Lie point symmetries:

$$K_G^1,\quad K_G^A,\quad X_R^{1A},\quad X_R^{AB},\quad H,\quad w\partial_w,$$

which are the KVs and the HV of the M^{n-1}; these symmetries are inherited symmetries. However when $\mu = 0$, equation (62) admits the extra Lie point symmetries

$$\bar{X}_C^1 - \frac{1}{2}tw\partial_w, \quad \bar{X}_C^A - \frac{1}{2}y^A w\partial_w,$$

where $\bar{X}_C^1$, $\bar{X}_C^A$ are the $(n-1)$ sp.CKVs of M^{n-1}; these symmetries are type II hidden symmetries.

We continue with the reduction of equation (61) by using the invariants of the Lie point symmetry $H + \mu X_u$. The wave equation (61) in hyperspherical coordinates (r, θ^A) is in the form of (48) with $K = 1$ and ${}_h\Delta u$ is the Laplacian of the $n-1$ hyperbolic sphere S_h^{n-1} with line element

$$ds_h^2 = d\theta_1^2 + \cosh^2\theta_1 \left(d\theta_2^2 + \cosh^2\theta_2 \left(d\theta_3^2 + \cosh^2\theta_3(\dots)\right)\right). \tag{63}$$

Hence, from section 6 we have that the reduced equation is the linear Klein–Gordon equation in the hyperbolic sphere (63)

$${}_h\Delta w + \mu^2 w = 0, \tag{64}$$

where $u\left(r, \theta^A\right) = r^\mu w\left(\theta^A\right)$. The space S_h^{n-1} is a maximally symmetric space and admits a $\frac{1}{2}n(n+1)$-dimensional CKVs. In particular admits the $\frac{1}{2}n(n-1)$ KVs which form the $SO(n)$ Lie group and n proper gradient CKVs with conformal factor $\psi\left(\theta^A\right)$ such as

$${}_h\Delta\psi = \frac{2-n}{4(n-1)}R_h, \tag{65}$$

where R_h is the Ricci scalar of the S_h^{n-1} sphere which is a constant. Therefore from theorem 3 we have that for an arbitrary constant μ,equation (64) admits as Lie point symmetries the elements of the $SO(n)$ and the linear symmetry $w\partial_w$; these are inherited symmetries. However, when $\mu^2 = \frac{n-2}{4(n-1)}R_h$ the Klein–Gordon equation (64) becomes the conformal invariant Laplace equation and admits $\frac{1}{2}n(n+1)$ Lie point symmetries, which span the conformal group of the space S_h^{n-1}. Therefore reduction with the gradient HV leads to the conformal invariant Laplace equation and the type II symmetries are generated by the proper CKVs of hyperbolic sphere S_h^{n-1}.

In order to reduce the wave equation (61) with the invariants of the Lie point symmetry $X_C^1 - tX_u$ we use the coordinates $\left(t, R, \phi^K\right)$ and equation (61) becomes

$$u_{,tt} - u_{,RR} - \frac{n-2}{R}u_R - \frac{1}{R^2}\,{}_h\bar{\Delta}u = 0, \tag{66}$$

where ${}_h\bar{\Delta}u$ is the Laplacian of the $(n-2)$-dimensional hyperbolic sphere S_h^{n-2}. From section 6 we have that under the second coordinate transformation $t = \sqrt{R\left(R+\tau^{-1}\right)}$ the reduced equation is written in the simpler form

$$_{(m=3)}\bar{\Delta}w = 0, \quad \text{when } n = 4, \tag{67}$$

$$_{(m\succeq 4)}\bar{\Delta}w - 2p\left(2p+1\right)\frac{(2-m)^2}{T^2}w = 0, \quad \text{when } n > 4, \tag{68}$$

where $u\left(\tau, R, \phi^K\right) = R^{2p}w\left(\tau, y^A\right)$ and $_{(m=3)}\bar{\Delta}w$, $_{(m\succeq 4)}\bar{\Delta}w$ are the Laplace operators for the $(n-1)$-dimensional flat space $ds_{(n-1)}^2$ with line element

$$ds_{(n-1)}^2 = dT^2 - T^2 ds_{h(n-2)}^2 \tag{69}$$

for the cases where $n = 4$ and $n > 4$ respectively. $ds^2_{h(n-2)}$ is the line element (63) of dimension $n - 2$.

Therefore equation (67) admits ten extra Lie symmetries which form the conformal algebra of the flat space (69), whereas equation (68) admits as extra Lie point symmetries the gradient HV $T\partial_T$ and the rotation group $SO(n)$ of the metric (69) which is an inherited symmetry. Hence type II hidden symmetries we have only when $n = 4$ and they are generated by KVs the translation group T^3 and the sp.CKVs of the flat space (69).

The reduction of the $1+3$ wave equation has been done previously in [22] and our results coincide with theirs when $n = 4$. However, our approach is geometrical and we are able to recognize the type II hidden symmetries from the form of the metric which defines the Laplace operator without further calculations.

7.2. Bianchi I (diagonal) spacetime

The Bianchi I spacetime is defined as the four-dimensional spacetime which admits as KVs the translation group T^3 of the 3D Euclidian space [23]. The line element of the Bianchi I spacetime in the invariant basis where the KVs are $\{\partial_x, \partial_y, \partial_z\}$ is:

$$ds^2 = -dt^2 + A^2(t)dx^2 + B^2(t)dy^2 + C^2(t)dz^2. \tag{70}$$

For general functions $A(t)$, $B(t)$, and $C(t)$ the KVs are non gradient. We restrict our considerations to the case where $C(t) = 1$ so that the KV ∂_z is a gradient KV. In this case Laplace equation for the Bianchi I spacetime (70) becomes

$$-u_{,tt} + u_{,zz} + A^{-2}u_{,xx} + B^{-2}u_{,yy} - \left(\frac{\dot{A}}{A} + \frac{\dot{B}}{B}\right)u_{,t} = 0. \tag{71}$$

From theorem 3, we have that the Laplace equation (71) admits as extra Lie point symmetries the T^3 Lie group. We consider the reduction of equation (71) with the symmetry

$$X_I = \partial_z + \mu u\partial_u.$$

From the application of the zero order invariance of X_I we have the reduced equation

$$_h\Delta w + \mu^2 w = 0, \tag{72}$$

where $_h\Delta w$ is the Laplace operator for the three-dimensional metric

$$ds^2_{(3)} = -dt^2 + A^2(t)dx^2 + B^2(t)dy^2. \tag{73}$$

The three-dimensional spacetime (73) for general functions $A(t)$ and $B(t)$ admits the two-dimensional conformal algebra with elements the KVs $\{\partial_x, \partial_y\}$ which are inherited Lie point symmetries for the Klein–Gordon equation (72).

Moreover, when $A^2(t) = B^2(t)$, the spacetime (73) admits the KV $X_I^3 = y\partial_x - x\partial_y$ which is a Lie point symmetry of equation (72). However, X_I^3 is an inherited symmetry because it is also a KV of (70) and a Lie point symmetry of Laplace equation (71). Furthermore, when $A^2(t) = B^2(t) = t^2$ the space $ds^2_{(3)}$ admits the HV (73) which is a Lie point symmetry of equation (72) when $\mu = 0$. In this case H is a inherited symmetry because in that case metric (70) admits the HV H, which is also a Lie point symmetry for (71) with commutator $[X_I, H] = \partial_z$.

Finally, when $\left(A^2(t), B^2(t)\right) = \left(\sin^2 t, \cos^2 t\right)$ or $\left(A^2(t), B^2(t)\right) = \left(\sinh^2 t, \cosh^2 t\right)$ the three-dimensional spacetime (73) is a maximally symmetric spacetime, i.e. a space of constant non vanishing curvature, and admits six KVs and four gradient CKVs (for the conformal algebra of the Bianchi I spacetime see [24]). Hence equation (72) for arbitrary μ admits six Lie point symmetries, the KVs. However, when $\mu = -\frac{1}{8}, R_{(3)}$ admits ten Lie point symmetries which are generated by the conformal algebra of (73), where $R_{(3)}$ is the Ricci scalar of the three-dimensional space $ds^2_{(3)}$. The type II hidden symmetries are the symmetries which are generated by the gradient CKVs. The KVs are inherited symmetries because they are also symmetries of equation (71).

8. Conclusion

In this work we have studied the connection between the Lie point symmetries of a general type of second-order PDEs and the collineations of the metric defined by the PDE itself. In particular we wrote the Lie symmetry conditions in a geometric form which include the Lie derivatives of geometric objects. Therefore, we proved that the Lie point symmetries of the PDEs of this family, which are linear in the first derivatives, are related with the elements of the conformal algebra of the metric defined by the PDE modulo a constraint condition depending again on the PDE. Important elements in this family are the Poisson equation, the Klein–Gordon equation and the Laplace equation. To these equations we have derived the Lie point symmetries for various classes of Riemannian spaces. In particular we have studied the type II hidden symmetries of the wave equation in Minkowski space and in Bianchi I spacetimes. The methodology we followed is geometric and can be applied to other types of PDEs as well as to this type of PDEs in other Riemannian spaces.

Acknowledgments

AP acknowledge financial support of INFN.

References

[1] Ibragimov N H 1985 *Transformation Groups Applied to Mathematical Physics* (Dordrecht: D. Reidel Publishing Co)
[2] Stephani H 1989 *Differential Equations: Their Solutions Using Symmetry* (New York: Cambridge University Press)
[3] Bluman G W and Kumei S 1989 *Symmetries of Differential Equations* (New York: Springer-Verlag)
[4] Prince G E and Crampin M 1984 Projective differential geometry and geodesic conservation laws in general relativity. I. Projective actions *Gen. Relativity Gravitation* **16** 921
[5] Aminova A V and Aminov N A-M 2006 Projective geometry of systems of second-order differential equations *Sb. Math.* **197** 951
[6] Aminova A V and Aminov N A-M 2010 Projective geometry theory of systems of second-order differential equations: straightening and symmetry theorems *Sb. Math.* **201** 631
[7] Tsamparlis M and Paliathanasis A 2010 Lie symmetries of geodesic equations and projective collineations *Nonlinear Dynam.* **62** 203
[8] Tsamparlis M and Paliathanasis A 2010 Lie and Noether symmetries of geodesic equations and collineations *Gen. Relativity Gravitation* **42** 2957
[9] Tsamparlis M and Paliathanasis A 2011 The geometric nature of Lie and Noether symmetries *Gen. Relativity Gravitation* **43** 1861
[10] Barnes A 1993 Projective collineations in Einstein spaces *Classical Quantum Gravity* **10** 1139
[11] Bozhkov Y and Freire I L 2010 Special conformal groups of a Riemannian manifold and Lie point symmetries of the nonlinear Poisson equation *J. Differential Equations* **249** 872
[12] Paliathanasis A and Tsamparlis M 2012 Lie point symmetries of a general class of PDEs: the heat equation *J. Geom. Phys.* **62** 2443
[13] Paliathanasis A and Tsamparlis M 2014 The reduction of the Laplace equation in certain Riemannian spaces and the resulting Type II hidden symmetries *J. Geom. Phys.* **76** 107
[14] Paliathanasis A and Tsamparlis M 2014 The geometric origin of Lie point symmetries of the Schrödinger and the Klein–Gordon equations *Int. J. Geom. Methods Mod. Phys.* **11** 1450037
[15] Tsamparlis M, Paliathanasis A and Qadir A 2015 Noether symmetries and isometries of the minimal surface Lagrangian under constant volume in a Riemannian space *Int. J. Geom. Methods Mod. Phys.* **12** 1550003

[16] Daftardar V and Dadhich N 1994 Gradient conformal Killing vectors and exact solutions *Gen. Relativity Gravitation* **26** 859
[17] Govinger K S 2001 Lie subalgebras, reduction of order, and group-invariant solutions *J. Math. Anal. Appl.* **258** 720
[18] Katzin G H, Levine J and Davis W R 1969 Curvature collineations: A fundamental symmetry property of the space-times of general relativity defined by the vanishing Lie derivative of the Riemann curvature tensor *J. Math. Phys.* **10** 617
[19] Colley A A and Tupper B O J 1989 Special conformal Killing vector space-times and symmetry inheritance *J. Math. Phys.* **30** 2616
[20] Tsamparlis M, Nikolopoulos D and Apostolopoulos P S 1998 Computation of the conformal algebra of $1+3$ decomposable spacetimes *Classical Quantum Gravity* **15** 2909
[21] Hall G S and Roy I M 1997 Some remarks on special conformal and special projective symmetries in general relativity *Gen. Relativity Gravitation* **29** 827
[22] Abraham-Shrauner B, Govinder K S and Arrigo D J 2006 Type-II hidden symmetries of the linear 2D and 3D wave equations *J. Phys.* A **39** 5739
[23] Rayan M P Jr and Shepley L C 1975 *Homogeneous Relativistic Cosmologies* (Princeton: Princeton University Press)
[24] Tsamparlis M and Apostolopoulos P S 2000 Symmetries of Bianchi I space-times *J. Math. Phys.* **41** 7573

Journal of Physics: Conference Series **621** (2015) 012015 doi:10.1088/1742-6596/621/1/012015

Laplace type invariants for variable coefficient mKdV equations

Christina Tsaousi[1], Rita Tracinà[2], and Christodoulos Sophocleous[1]

[1] Department of Mathematics and Statistics, University of Cyprus, Nicosia CY 1678, Cyprus
[2] Dipartimento di Matematica e Informatica, Viale A. Doria 6, 95125 Catania, Italy

E-mail: tsaousi.christina@ucy.ac.cy, tracina@dmi.unict.it, christod@ucy.ac.cy

Abstract. We consider a class of variable-coefficient mKdV equations. We derive the equivalence transformations in the infinitesimal form and we employ them to construct differential invariants of the respective equivalence algebra. Operators of invariant differentiation are also constructed. Applications, similar to Laplace invariants, are presented.

1. Introduction

Laplace [1] in his general theory of integration of linear hyperbolic partial differential equations

$$u_{xy} + a(x,y)u_x + b(x,y)u_y + c(x,y)u = 0 \tag{1}$$

derived the quantities

$$h = a_x + ab - c, \quad k = b_y + ab - c$$

known as *Laplace invariants*. The expressions h and k do not change under the linear transformation of the dependent variable,

$$u' = \phi(x,y)u. \tag{2}$$

These invariants are useful in various problems, for example in the group classification of differential equations [2] and the solution of initial value problems for hyperbolic equations by Riemann's method [3].

We recall the following simple but fundamental applications of the Laplace invariants:

1. A hyperbolic equation of the form (1) can be transformed into $u_{xy} = 0$ by means of (2) iff $h = k = 0$.
2. A hyperbolic equation of the form (1) can be transformed into $u_{xy} + c(x,y)u = 0$ by means of (2) iff $h = k$.
3. A hyperbolic equation of the form (1) can be transformed into $u_{xy} + cu = 0$, $c =$const, by means of (2) iff $h = k = f(x)g(y)$.
4. A hyperbolic equation of the form (1) can be factorized iff $h = 0$ or $k = 0$. That is, if $L = \partial_x\partial_y + a(x,y)\partial_x + b(x,y)\partial_y + c(x,y)$ then

$$L = [\partial_x + \alpha(x,y)][\partial_y + \beta(x,y)] \quad \text{iff } h = 0$$

and

$$L = [\partial_y + \beta(x,y)][\partial_x + \alpha(x,y)] \quad \text{iff } k = 0.$$

The proofs of the above statements can be found in [4, 5].

The differential invariants of the Lie groups of continuous transformations play important role in mathematical modelling, non-linear science and differential geometry. First it was noted by S. Lie [6], who showed that every invariant system of differential equations [7], and every variational problem [8], could be directly expressed in terms of differential invariants. Lie also demonstrated [7], how differential invariants can be used to integrate ordinary differential equations, and succeeded in completely classifying all the differential invariants for all possible finite-dimensional Lie groups of point transformations in the case of one independent and one dependent variable. Lie's preliminary results on invariant differentiations and existence of finite bases of differential invariants were generalized by Tresse [9] and Ovsiannikov [11]. The general theory of differential invariants of Lie groups including algorithms of construction of differential invariants can be found in [10, 11].

A simple method for constructing differential invariants of families of linear and nonlinear differential equations admitting infinite equivalence transformation groups was developed by Ibragimov [14] (see also [15]). This method was adopted by various scientists and it was then applied to several linear and nonlinear equations with interesting results. [4,16–32]. For example, Ibragimov [16] gave a solution to the Laplace problem which consists of finding all invariants of the hyperbolic equations (1). Namely, in addition to Ovsiannikov's invariants [2], he found three new invariants together with invariant differentiations and he constructed a basis of all invariants. The Laplace problem was also proved by Mahomed and coauthors [33].

We point out that other approaches also exist. See for example, ref [34, 35]. For instance, Yehorchenko [34] introduced a method where the initial basis operators contains no arbitrary functions. In this method we search for differential operators of any specific finite order and hence, we deal with finite dimensional algebra. The arbitrary functions are expanded into Taylor series,

$$A(t) = \sum_{m=0}^{\infty} a_m t^m, \quad A_t(t) = \sum_{m=1}^{\infty} m a_m t^{m-1}, \quad \text{etc.}$$

More detail of this approach can be found in [34].

Here we apply Ibragimov's method for the general class of variable-coefficient mKdV equations

$$u_t + f(t)u^2 u_x + g(t)u_{xxx} + h(t)u + (p(t) + q(t)x)u_x + k(t)uu_x + l(t) = 0, \tag{3}$$

where all the parameters are smooth functions of t and $f(t)g(t) \neq 0$. We derive the equivalence transformations which are employed to construct differential invariants. Certain applications, similar to Laplace invariants, are presented. Finally, we construct an operator of invariant differentiation.

2. Equivalence transformations

Equivalence transformations play an important part in the theory of invariants. Derivation of equivalence transformations for the class of equations under consideration is the first step towards to the target which is the determination of differential invariants. The set of all equivalence transformations of a given family of differential equations forms a group which is called the equivalence group. There exist two methods for calculation of equivalence transformations, the direct which was used first by Lie [7] and the Lie infinitesimal method which was introduced by Ovsyannikov [11]. Although, the direct method involves considerable computational difficulties, it has the benefit of finding the most general equivalence group and also unfolds all form-preserving [12] (also known as admissible [13]) transformations admitted by this class of equations. For recent applications of the direct method one can refer, for example, to references [36–39]. More detailed description and examples of both methods can be found in [40]. The method that we employ here to determine differential invariants requires the equivalence transformations to be in the infinitesimal form. Hence, we use the infinitesimal method to derive the desired equivalence transformations. We search for the equivalence operator X in the following form:

$$X = \xi^1 \frac{\partial}{\partial t} + \xi^2 \frac{\partial}{\partial x} + \eta \frac{\partial}{\partial u} + \sum_{i=1}^{7} \mu_i \frac{\partial}{\partial \zeta_i},$$

where ζ_i correspond to the functions $f, g, \ldots, l$. The functions ξ^1, ξ^2 and η depend on x, t and u, while μ_i depend on x, t, u, f, g, h, p, q, k and l. Without presenting any detailed analysis, we find that

$$\begin{gathered}
\xi^1 = A(t), \quad \xi^2 = B(t)x + \Gamma(t), \quad \eta = \Theta(t)u + \Psi(t), \\
\mu_1 = (B - 2\Theta - A_t)f, \quad \mu_2 = (3B - A_t)g, \quad \mu_3 = -A_t h - \Theta_t, \\
\mu_4 = (B - A_t)p - \Gamma q - \Psi k + \Gamma_t, \quad \mu_5 = -A_t q + B_t, \\
\mu_6 = -2\Psi f + (B - \Theta - A_t)k, \quad \mu_7 = -\Psi h + (\Theta - A_t)l - \Psi_t.
\end{gathered}$$

That is, the class (3) admits an infinite-dimensional continuous group $\mathcal{E}$ of equivalence transformations generated by Lie algebra $L_{\mathcal{E}}$ spanned by the operators:

$$\begin{aligned}
X_A &= A\frac{\partial}{\partial t} - A_t f\frac{\partial}{\partial f} - A_t g\frac{\partial}{\partial g} - A_t h\frac{\partial}{\partial h} - A_t p\frac{\partial}{\partial p} - A_t q\frac{\partial}{\partial q} - A_t k\frac{\partial}{\partial k} - A_t l\frac{\partial}{\partial l}, \\
X_B &= xB\frac{\partial}{\partial x} + Bf\frac{\partial}{\partial f} + 3Bg\frac{\partial}{\partial g} + Bp\frac{\partial}{\partial p} + B_t\frac{\partial}{\partial q} + Bk\frac{\partial}{\partial k}, \\
X_\Gamma &= \Gamma\frac{\partial}{\partial x} + (\Gamma_t - \Gamma q)\frac{\partial}{\partial p}, \\
X_\Theta &= u\Theta\frac{\partial}{\partial u} - 2\Theta f\frac{\partial}{\partial f} - \Theta_t\frac{\partial}{\partial h} - \Theta k\frac{\partial}{\partial k} + \Theta l\frac{\partial}{\partial l}, \\
X_\Psi &= \Psi\frac{\partial}{\partial u} - \Psi k\frac{\partial}{\partial p} - 2\Psi f\frac{\partial}{\partial k} - (\Psi h + \Psi_t)\frac{\partial}{\partial l}.
\end{aligned}$$

The direct method leads to the following equivalence transformations for the class (3) [41]:

$$\tilde{t} = \alpha(t), \quad \tilde{x} = \beta(t)x + \gamma(t), \quad \tilde{u} = \theta(t)u + \psi(t), \tag{4}$$

where α, β, γ, θ and ψ run through the set of smooth functions of t, $\alpha_t\beta\theta \neq 0$. The arbitrary elements of (3) are transformed by the formulas

$$\begin{gathered}
\tilde{f} = \frac{\beta}{\alpha_t\theta^2}f, \quad \tilde{g} = \frac{\beta^3}{\alpha_t}g, \quad \tilde{h} = \frac{1}{\alpha_t}\left(h - \frac{\theta_t}{\theta}\right), \\
\tilde{p} = \frac{1}{\alpha_t}\left(\beta p - \gamma q + \beta\frac{\psi^2}{\theta^2}f - \beta\frac{\psi}{\theta}k + \gamma_t - \gamma\frac{\beta_t}{\beta}\right), \quad \tilde{q} = \frac{1}{\alpha_t}\left(q + \frac{\beta_t}{\beta}\right), \\
\tilde{k} = \frac{\beta}{\alpha_t\theta}\left(k - 2\frac{\psi}{\theta}f\right), \quad \tilde{l} = \frac{1}{\alpha_t}\left(\theta l - \psi h - \psi_t + \psi\frac{\theta_t}{\theta}\right).
\end{gathered}$$

3. Differential invariants

A function of the form

$$I(x, t, u, \zeta_i(t), \zeta_{it}(t), \zeta_{itt}(t), \dots)$$

which remains invariant under the equivalence group $\mathcal{E}$ is called differential invariant of order s of equation (3), where s denotes the maximal order derivative of $\zeta_i(t)$. If no derivatives appear, then it is called differential invariant of order zero. An equation

$$E(x, t, u, \zeta_i(t), \zeta_{it}(t), \zeta_{itt}(t), \dots) = 0$$

that satisfies the conditions

$$X_k^{(s)}(E)\Big|_{E=0} = 0, \quad k = A,\ B,\ \Gamma,\ \Theta,\ \Psi,$$

is called an invariant equation of order s.

In order to determine the differential invariants of order s, we need to calculate the prolongations of the operator X. The procedure for determining the prolongations can be found in [15]. We do not find invariant of zero order. However we find the following invariant equations of zero order:

$$f = 0, \quad g = 0.$$

Although we have taken the functions $f(t)$ and $g(t)$ to be nonzero, the above equations state that there do not exist point transformations which map an equation of the class (3) into an equation of the same class with either $f(t) = 0$ or $g(t) = 0$.

Next step is to derive differential invariants of first order. We introduce the quantities

$$A_1 = fk_t - kf_t + fhk - 2lf^2, \quad A_2 = 2ghf - 2fqg - gf_t + fg_t.$$

We find that the class of equations (3) admits one differential invariant of first order

$$I^{(1)} = \frac{A_1 g^{7/6}}{{A_2}^{4/3} f^{1/6}}$$

and two invariant equations

$$A_1 = 0, \quad A_2 = 0.$$

For the differential invariants of second order we introduce the quantities

$$\begin{aligned} A_3 &= 8h^2gf^2 - 8hfgf_t - 10hf^2qg + 6hf^2g_t + 2gq^2f^2 + 5gqff_t + 3gf_t^2 \\ &\quad - 3qg_tf^2 - ff_{tt}g + 2f^2gh_t - 2f^2gq_t + f^2g_{tt} - 3f_tg_tf, \\ A_4 &= -3kf_t^2 + 3f_tfk_t + 5f_tfhk - 2f_tf^2l + f_tfkq - qf^2hk + 2qf^3l - qf^2k_t \\ &\quad - 3h^2kf^2 + 6lhf^3 - 4hk_tf^2 + ff_{tt}k - f^2kh_t + 2f^3l_t - f^2k_{tt}. \end{aligned}$$

We find two differential invariants of second order

$$I_1^{(2)} = \frac{gA_3}{{A_2}^2}, \quad I_2^{(2)} = \frac{g^{13/6}A_4}{f^{1/6}{A_2}^{7/3}}$$

and two invariant equations

$$A_3 = 0, \quad A_4 = 0.$$

4. Applications

Similar to the applications of Laplace invariants that stated in the Introduction, we have the following results.

Theorem 1. *Equation* (3) *can be transformed into*

$$u_t + u_{xxx} + c_1 u^2 u_x + c_2 u u_x = 0$$

if and only if $A_1 = A_2 = 0$. *That is, if and only if invariant equations are satisfied.*

Equation (3) *can be transformed into*

$$u_t + u_{xxx} + c_1 u^2 u_x + \phi(t) u u_x = 0$$

if and only if $A_2 = 0$ *and* $A_1 \neq 0$.

Equation (3) *can be transformed into*

$$u_t + \phi(t) u_{xxx} + u^2 u_x + c_1 u u_x = 0$$

if and only if $A_1 = 0$ *and* $A_2 \neq 0$.

Equation (3) *can be transformed into*

$$u_t + u_{xxx} + \phi(t) u^2 u_x + c_1 u u_x = 0$$

if and only if $A_1 - c_1 A_2 = 0$ *and* $h = 2q$.

The first part of the Theorem 1 is also presented in the work [41]. Below we present the mappings that connect equation (3) with each one of the four equations that appear in the Theorem 1.

The equation

$$\tilde{u}_{\tilde{t}} + \tilde{u}_{\tilde{x}\tilde{x}\tilde{x}} + c_1 \tilde{u}^2 \tilde{u}_{\tilde{x}} + c_2 \tilde{u}\tilde{u}_{\tilde{x}} = 0$$

is connected with equation (3) under the mapping

$$\begin{aligned}
\tilde{t} &= \int \sqrt{\frac{f^3}{c_1^3 g}} \mathrm{e}^{-3\int h dt} dt, \\
\tilde{x} &= \sqrt{\frac{f}{c_1 g}} \mathrm{e}^{-\int h dt} x + \int \left\{ \frac{1}{4\sqrt{c_1^5 f g}} \mathrm{e}^{-3\int h dt} \left[c_1^2 (k^2 - 4fp) \mathrm{e}^{2\int h dt} - c_2^2 f^2 \right] \right\} dt, \\
\tilde{u} &= \mathrm{e}^{\int h dt} u + \frac{k}{2f} \mathrm{e}^{\int h dt} - \frac{c_2}{2c_1},
\end{aligned}$$

where the invariant equations $A_1 = 0$, $A_2 = 0$ must hold. We point out that c_2 can be taken equal to zero.

The equation

$$\tilde{u}_{\tilde{t}} + \tilde{u}_{\tilde{x}\tilde{x}\tilde{x}} + c_1 \tilde{u}^2 \tilde{u}_{\tilde{x}} + \phi(\tilde{t}) \tilde{u}\tilde{u}_{\tilde{x}} = 0$$

is connected with equation (3) under the mapping

$$\begin{aligned}
\tilde{t} &= \int \frac{g}{c_1^{3/2}} \mathrm{e}^{-3\int q dt} dt, \\
\tilde{x} &= \frac{1}{\sqrt{c_1}} \mathrm{e}^{-\int q dt} x + \int \frac{g}{\sqrt{c_1} f} \mathrm{e}^{-3\int q dt} \left[k \sqrt{\frac{f}{g}} \mathrm{e}^{\int q dt} \int l \sqrt{\frac{f}{g}} \mathrm{e}^{\int q dt} dt \right. \\
&\quad \left. - \frac{pf}{g} \mathrm{e}^{2\int q dt} - f \left(\int l \sqrt{\frac{f}{g}} \mathrm{e}^{\int q dt} dt \right)^2 \right] dt, \\
\tilde{u} &= \sqrt{\frac{f}{g}} \mathrm{e}^{\int q dt} u + \int l \sqrt{\frac{f}{g}} \mathrm{e}^{\int q dt} dt,
\end{aligned}$$

where the invariant equation $A_2 = 0$ must hold and

$$\phi = \frac{c_1}{f}\left(k\sqrt{\frac{f}{g}}e^{\int q dt} - 2f\int l\sqrt{\frac{f}{g}}e^{\int q dt}dt\right).$$

The equation

$$\tilde{u}_{\tilde{t}} + \phi(\tilde{t})\tilde{u}_{\tilde{x}\tilde{x}\tilde{x}} + \tilde{u}^2\tilde{u}_{\tilde{x}} + c_1\tilde{u}\tilde{u}_{\tilde{x}} = 0$$

is connected with equation (3) under the mapping

$$\begin{aligned}
\tilde{t} &= \int f e^{-\int(q+2h)dt}dt,\\
\tilde{x} &= e^{-\int q dt}x + \int \frac{1}{4f}e^{-\int(q+2h)dt}\left[(k^2 - 4fp)e^{2\int h dt} - c_1^2 f^2\right]dt,\\
\tilde{u} &= e^{\int h dt}u + \frac{k}{2f}e^{\int h dt} - \frac{c_1}{2},
\end{aligned}$$

where the invariant equation $A_1 = 0$ must hold and

$$\phi = \frac{g}{f}e^{2\int(h-q)dt}.$$

The equation

$$\tilde{u}_{\tilde{t}} + \tilde{u}_{\tilde{x}\tilde{x}\tilde{x}} + \phi(\tilde{t})\tilde{u}^2\tilde{u}_{\tilde{x}} + c_1\tilde{u}\tilde{u}_{\tilde{x}} = 0$$

is connected with equation (3) where $h = 2q$ under the mapping

$$\begin{aligned}
\tilde{t} &= \int g e^{-3\int q dt}dt,\\
\tilde{x} &= e^{-\int q dt}x + \int \frac{1}{4f}e^{-\int q dt}\left(k^2 - c_1^2 g^2 - 4fp\right)dt,\\
\tilde{u} &= e^{2\int q dt}u + \frac{k - c_1 g}{2f}e^{2\int q dt},
\end{aligned}$$

provided $A_1 - c_1 A_2 = 0$ and

$$\phi = \frac{f}{g}e^{-2\int q dt}.$$

Here $c_1 \neq 0$, otherwise the result is the same as in the previous case with $c_1 = 0$ and $h = 2q$.

5. Operators of invariant differentiation

Here we find an operator ofinvariant differentiation that transform each invariant of equation (3) into invariants of higher-order of the same equation. Since arbitrary elements are functions of t we look for an operator of invariant differentiation of the form

$$\mathcal{D} = \psi D_t,$$

where $\psi = \psi(t, x, u, f, g, \ldots, l, f_t, \ldots, l_t, \ldots)$ and can be found by solving the differential equations

$$X_A^{(n)}(\psi) = A_t\psi, \quad X_B^{(n)}(\psi) = 0, \quad X_\Gamma^{(n)}(\psi) = 0, \quad X_\Theta^{(n)}(\psi) = 0, \quad X_\Psi^{(n)}(\psi) = 0.$$

For zero order we find $\psi = 0$, for first order

$$\psi = \frac{fg}{A_2} H\big(I^{(1)}\big),$$

and for second order

$$\psi = \frac{fg}{A_2} H\big(I^{(1)}, I_1^{(2)}, I_2^{(2)}\big).$$

Since the function H is arbitrary, we can take it, without of generality, $H = 1$. Hence,

$$\mathcal{D} = \frac{fg}{A_2} D_t.$$

Now, if we apply the invariant differentiation to $I^{(1)}$ we obtain

$$\mathcal{D}(I^{(1)}) = \frac{fg}{A_2} D_t\big(I^{(1)}\big) = I_2^{(2)} + \tfrac{7}{6} I^{(1)} - \tfrac{4}{3} I^{(1)} I_1^{(2)}$$

and therefore,

$$I_2^{(2)} = \mathcal{D}\big(I^{(1)}\big) - \tfrac{7}{6} I^{(1)} + \tfrac{4}{3} I^{(1)} I_1^{(2)}.$$

This means that we have one new differential invariant of second order, while the second can be obtained with the application of the invariant differentiation to the differential invariant of first order. Further calculations showed that the class of equations (3) admits two differential invariants of third order and two of fourth order. It appears that the class (3) has a basis of two differential invariants: $\{I^{(1)}, I_1^{(2)}\}$. Any other differential invariant of higher order can be obtained with the employment of the invariant differentiation. However this result needs to be proved and consequently, it will be a task for the near future.

Acknowledgments
The authors would like to thank both referees for their constructive suggestions that improve the paper.

References

[1] Laplace P S 1773/77 Recherches sur le calcul int égral aux différences partielles *Mémoires de l'Acad émie royale des Sciences de Paris* 341; reprinted in 1893 *Oeuvres Complètes* **9** (Paris: Gauthier-Villars)
[2] Ovsiannikov L V 1960 Group properties of the equation of S.A. Chaplygin, *J. Appl. Mech. Tech. Phys.* **3** 126 [in Russian]; English transl. 2004 *Lie Group Analysis: Classical Heritage*, Ed. N.H. Ibragimov, *ALGA Publications*, p 123
[3] Ibragimov N H 1992 Group analysis of ordinary differential equations and the invariance principle in mathematical physics (for the 150th anniversary of Sophus Lie) *Uspekhi Mat. Nauk* **47** 83 [in Russian]; English transl. *Russian Math. Surveys* **47** 89
[4] Ibragimov N H 2002 Laplace type invariants for parabolic equations *Nonlinear Dynam.* **28** 125
[5] Ibragimov N H 2010 *A Practical Course in Differential Equations and Mathematical Modelling*, 2nd ed. (Benjing: World Scientific Publishing)
[6] Lie S 1884 Über Differentialinvarianten *Math. Ann.* **24** 537
[7] Lie S 1888 Klassifikation und Integration von gewohnlichen Differentialgleichungen zwischen x, y, die eine Gruppe von Transformationen gestatten I, II, *Math. Ann.* **32** 213
[8] Lie S 1897 Über Integralinvarianten und ihre Verwertung für die Theorie der Differentialgleichungen *Leipz. Berichte* **4** 369
[9] Tresse A 1894 Sur les invariant différentiels des groupes continus de transformations *Acta Math.* **18** 1
[10] Olver P J 1995 *Equivalence, Invariants and Symmetry*, (Cambridge: Cambridge University Press)
[11] Ovsiannikov L V 1982 *Group Analysis of Differential Equations* (New York: Academic Press)
[12] Kingston J G and Sophocleous C 1998 On form-preserving point transformations of partial differential equations *J. Phys.* A **31** 1597

[13] Popovych R O, Ivanova N M and Eshraghi H 2004 Group classification of (1 + 1)-dimensional Schrödinger equations with potentials and power nonlinearities *J. Math. Phys.* **45** 3049
[14] Ibragimov N H 1997 Infinitesimal method in the theory of invariants of algebraic and differential equations *Not. S. Afr. Math. Soc.* **29** 61
[15] Ibragimov N H 1999 *Elementary Lie Group Analysis and Ordinary Differential Equations* (New York: Wiley)
[16] Ibragimov N H 2004 Invariants of hyperbolic equations: solution of the Laplace problem *J. Appl. Mech. Tech. Phys.* **45** 158
[17] Ibragimov N H 2002 Invariants of a remarkable family of nonlinear equations *Nonlinear Dynam.* **30** 155
[18] Ibragimov N H, Torrisi M and Valenti A 2004 Differential invariants of nonlinear equations $v_{tt} = f(x, v_x)v_{xx} + g(x, v_x)$ *Commun. Nonlinear. Sci. Numer. Simul.* **9** 69
[19] Ibragimov N H and Meleshko S V 2005 Linearization of third-order ordinary differential equations by point and contact transformations *J. Math. Anal. Appl.* **308** 266
[20] Ibragimov N H and Sophocleous C 2007 Differential invariants of the one-dimensional quasi-linear second-order evolution equation *Commun. Nonlinear Sci. Numer. Simul.* **12** 1133
[21] Ibragimov N H and Meleshko S V 2009 A solution to the problems of invariants for parabolic equations *Commun. Nonlinear Sci. Numer. Simul.* **14** 2551
[22] Johnpillai I K and Mahomed F M 2001 Singular invariant equation for the (1 + 1) FokkerPlank equation *J. Phys. A: Math. Gen.* **28** 11033
[23] Mahomed F M 2008 Complete invariant characterization of scalar linear (1 + 1) parabolic equations *J. Nonlinear Math. Phys.* **15** (suppl. 1) 112
[24] Sophocleous C and Tracinà R 2008 Differential invariants for quasi-linear and semi-linear wave-type equations *Appl. Math. Comput.* **202** 216
[25] Torrisi M, Tracinà R and Valenti A 2004 On the linearization of semilinear wave equations *Nonlinear Dynam.* **36** 97
[26] Torrisi M and Tracinà R 2005 Second-order differential invariants of a family of diffusion equations *J. Phys. A: Math. Gen.* **38** 7519
[27] Tracinà R 2004 Invariants of a family of nonlinear wave equations *Commun. Nonlinear Sci. Numer. Simul.* **9** 127
[28] Tsaousi C and Sophocleous C 2008 On linearization of hyperbolic equations using differential invariants *J. Math. Anal. Appl.* **339** 762
[29] Tsaousi C, Sophocleous C and Tracinà R 2009 Invariants of two and three dimensional hyperbolic equations *J. Math. Anal. Appl.* **349** 516
[30] Tsaousi C and Sophocleous C 2010 Differential invariants for systems of linear hyperbolic equations *J. Math. Anal. Appl.* **363** 238
[31] Tsaousi C, Sophocleous C and Tracinà R 2012 On the invariants of two dimensional linear parabolic equations *Commun. Nonlinear Sci. Numer. Simul.* **17** 3673
[32] Tsaousi C, Tracinà R and Sophocleous C 2015 Differential invariants for third-order evolution equations *Commun. Nonlinear Sci. Numer. Simul.* **20** 352
[33] Johnpillai I K, Mahomed F M and Wafo Soh C 2002 Basis of joint invariants for (1 + 1) linear hyperbolic equations *J. Nonlinear Math. Phys.* **9** 49
[34] Yehorchenko I 2004 Differential invariants for infinite-dimensional algebras in *Symmetry and Perturbation Theory* pp 308–312 (Hackensack, NJ: World Sci. Publ.) (*Preprint* arXiv:math-ph/0607002)
[35] Olver P J and Pohjanpelto J 2008 Pseudo-groups, moving frames, and differential invariants in *Symmetries and Overdetermined Systems of Partial Differential Equations* pp 127–149 (*IMA Vol. Math. Appl.*, Vol. 144, New York: Springer)
[36] Vaneeva O O, Johnpillai A G, Popovych R O and Sophocleous C 2007 Enhanced group analysis and conservation laws of variable coefficient reactiondiffusion equations with power nonlinearities *J. Math. Anal. Appl.* **330** 1363
[37] Vaneeva O O, Popovych R O and Sophocleous C 2009 Enhanced group analysis and exact solutions of variable coefficient semilinear diffusion equations with a power source. *Acta Appl. Math.* **106** 1
[38] Vaneeva O O, Popovych R O and Sophocleous C 2012 Extended group analysis of variable coefficient reaction-diffusion equations with exponential nonlinearities *J. Math. Anal. Appl.* **396** 225
[39] Vaneeva O O, Popovych R O and Sophocleous C 2014 Equivalence transformations in the study of integrability *Phys. Scripta* **89** 038003
[40] Ibragimov N H 2004 Equivalence groups and invariants of linear and non-linear equations *Arch. ALGA* **1** 9
[41] Popovych R O and Vaneeva O O 2010 More common errors in finding exact solutions of nonlinear differential equations: Part I *Commun. Nonlinear Sci. Numer. Simul.* **15** 3887

Journal of Physics: Conference Series **621** (2015) 012016 doi:10.1088/1742-6596/621/1/012016

Group analysis of Benjamin–Bona–Mahony equations with time dependent coefficients

Olena Vaneeva[1], **Roman O Popovych**[1,2] **and Christodoulos Sophocleous**[3]

[1] Institute of Mathematics of NAS of Ukraine, 3 Tereshchenkivska Str., Kyiv 01601, Ukraine
[2] Wolfgang Pauli Institute, Nordbergstraße 15, A-1090 Wien, Austria
[3] Department of Mathematics and Statistics, University of Cyprus, Nicosia CY 1678, Cyprus

E-mail: vaneeva@imath.kiev.ua, rop@imath.kiev.ua, christod@ucy.ac.cy

Abstract. Group classification of a class of Benjamin–Bona–Mahony (BBM) equations with time dependent coefficients is carried out. Two equivalent lists of equations possessing Lie symmetry extensions are presented: up to point equivalence within the class of BBM equations and without the simplification by equivalence transformations. It is shown that the complete results can be achieved using either the gauging of arbitrary elements of the class by the equivalence transformations or the method of mapping between classes. As by-product of the second approach the complete group classification of a class of variable-coefficient BBM equations with forcing term is derived.

1. Introduction

The third-order nonlinear partial differential equation

$$u_t + u_x + uu_x - u_{xxt} = 0,$$

named these days the Benjamin–Bona–Mahony (BBM) equation, appeared in [1] as an alternative to the Korteweg–de Vries equation, $u_t + u_x + uu_x + u_{xxx} = 0$, model for the unidirectional propagation of moderately long waves with small but finite amplitude in systems that manifest nonlinear and dispersive effects. Numerical studies showed that the BBM equation admits soliton solutions whose interaction is inelastic though close to elastic [2,3]. It was proved in [4] for the equivalent form $u_t = uu_x + u_{xxt}$ of the BBM equation that it has no conserved quantity in addition to those found by Benjamin, Bona and Mahony: u (mass), $(u^2 + u_x^2)/2$ (energy), and $u^3/3$ (momentum). Lie symmetries and the corresponding reductions of the BBM equation in the above equivalent form were obtained in [5] (these results were also presented in [6, pp 194–196]). It was found that the maximal Lie symmetry algebra of this equation is a three-dimensional Lie algebra of the type $A_{2.1} \oplus A_1$ spanned by the vector fields ∂_t, $t\partial_t - u\partial_u$ and ∂_x.

There are several recent works (see [7] and references therein) devoted to the study of Lie symmetries of variable-coefficient BBM equations of the general form

$$u_t + f(t)u_x + g(t)uu_x + h(t)u_{xxt} = 0, \tag{1}$$

where f, g and h are arbitrary smooth functions of the variable t with $gh \neq 0$. However none of these works contains exhaustive and completely correct results. We aim to fill up this gap by presenting the exhaustive group classification of equations from class (1) and classifying local conservation laws of these equations.

In order to study transformational properties of class (1), in Section 2 we describe the entire set of its admissible transformations (i.e., the equivalence groupoid of this class) by proving that the class is normalized in the usual sense. The assertions concerning reducibility of variable-coefficient BBM equations to their constant-coefficient counterparts are presented therein. The complete classification of Lie symmetries is obtained in Section 3 under gauging the arbitrary element g to 1 by a parameterized family of point transformations related to the associated equivalence group. The method of mapping between classes is utilized in Section 4 to verify the classification results. As by-product of this approach the complete group classification of a related class of variable-coefficient BBM equations with forcing term, $u_t+uu_x+h(t)u_{xxt}=l(t)$, is derived. In Section 5 we construct local conservation laws of equations from the class (1).

2. Admissible transformations

Two differential equations which are connected by a point transformation are called *similar* [8]. Similar differential equations have similar sets of solutions, symmetries, conservation laws and other related objects. Therefore, it is instructive to study point transformations linking equations from a given class of differential equations. Such transformations are called form-preserving [16] or allowed [18]. The rigorous reformulation of this notion leads to the notions of admissible transformations [9, 10] and the equivalence groupoid of the class [11, 12]. An admissible transformation is a triple that consists of two similar equations from the class and a point transformation between them. The set of admissible transformations of the class possesses the groupoid structure with respect to the natural operation of the composition of admissible transformations and hence it is called the *equivalence groupoid* of the class.

A joint transformation of the unknown functions and the arbitrary elements is called an equivalence transformation of the class if it satisfies the following properties: (i) The transformation becomes point with respect to the independent and the dependent variables when the arbitrary elements are fixed. (ii) It is consistent with the contact structure of the underlying jet space. (iii) It maps any equation from the class to an equation from the same class. Depending on additional restrictions, one have different kinds of equivalence transformations (usual, generalized, extended and generalized extended ones). Equivalence transformations of the same kind constitute a group, which is called the equivalence group of the class (of the respective kind), see [10, 13] and references therein. If the equivalence groupoid of the class is generated by the equivalence group of this class, then this class is called *normalized* in the same sense as the kind of the equivalence group involved [9, 10].

One-parameter subgroups of the equivalence group can be found by the Lie infinitesimal method, whereas the direct method [14–16] allows one to construct the entire equivalence group including discrete equivalence transformations and this technique is preferable. The direct method can also be used for finding the equivalence groupoid and checking whether the class is normalized [10–12, 17]. A very useful feature of normalized classes is that the equivalence groups for their subclasses, singled out by setting additional constraints on arbitrary elements, are subgroups of the equivalence group of the entire class. We use this property to derive the equivalence group of the class (1) from the equivalence group of a normalized superclass.

Consider the class of (1+1)-dimensional partial differential equations of order $p+q>1$ with fixed $p, q \in \mathbb{N}_0$,

$$u_{pq} = H\left(t, x, u_{(p+q-1)}\right), \tag{2}$$

where $u_{pq} = \partial^{p+q}u/\partial t^p \partial x^q$ and $u_{(p+q-1)}$ denotes the set of all derivatives of u up to order $p+q-1$

including u as the zero-order derivative. The function H is an arbitrary smooth function of its arguments. The following assertion was proven in [16].

Theorem 1. *A point transformation* $\tilde t = T(t,x,u)$, $\tilde x = X(t,x,u)$, $\tilde u = U(t,x,u)$, *where* $|\partial(T,X,U)/\partial(t,x,u)| \neq 0$, *maps a fixed equation from class* (2) *to an equation from the same class if and only if it satisfies the following conditions:*

(*i*) $T=T(t),\ X=X(x)$ *for* $pq\neq 0,\ p\neq q$;

(*ii*) $T=T(t),\ X=X(x)$ *or* $T=T(x),\ X=X(t)$ *for* $p=q\neq 0$;

(*iii*) $T=T(t)$ *for* $p=0,\ q\neq 0$;

(*iv*) $X=X(x)$ *for* $p\neq 0,\ q=0$.

Corollary 1. *The usual equivalence group of class* (2) *with any fixed* p *and* q *consists of the transformations whose projections to the space of* (t,x,u) *coincides with that described in Theorem 1. Moreover, this class is normalized in the usual sense.*

Therefore, Theorem 1 provides the complete description of admissible (form-preserving) transformations in class (2). As this class is normalized, we can use transformation constraints from Theorem 1 for describing admissible transformations in subclasses of this class, in particular in class (1).

For class (1) we have $p=1$ and $q=2$, which corresponds to Case (*i*) of Theorem 1. Now we directly seek for all point transformations of the general form

$$\tilde t = T(t),\quad \tilde x = X(x),\quad \tilde u = U(t,x,u) \tag{3}$$

with $T_tX_xU_u\neq 0$ that map a fixed equation of the form (1) to an equation from the same class,

$$\tilde u_{\tilde t} + \tilde f(\tilde t)\tilde u_{\tilde x} + \tilde g(\tilde t)\tilde u\tilde u_{\tilde x} + \tilde h(\tilde t)\tilde u_{\tilde x\tilde x\tilde t} = 0. \tag{4}$$

The components of the prolongation of transformation (3) to the involved partial derivatives are

$$\tilde u_{\tilde t} = \frac{1}{T_t}\left(U_t + U_uu_t\right),\quad \tilde u_{\tilde x} = \frac{1}{X_x}\left(U_x+U_uu_x\right),$$
$$\tilde u_{\tilde x\tilde x\tilde t} = -\frac{X_{xx}}{T_tX_x^3}\left(U_{tx}+U_{xu}u_t+U_{tu}u_x+U_{uu}u_tu_x+U_uu_{tx}\right)+$$
$$+\frac{1}{T_tX_x^2}\big(U_{xxt}+U_{xxu}u_t+2U_{txu}u_x+2U_{xuu}u_tu_x+2U_{xu}u_{tx}+U_{tuu}u_x^2+U_{uuu}u_tu_x^2+$$
$$+2U_{uu}u_xu_{tx}+U_{tu}u_{xx}+U_{uu}u_tu_{xx}+U_uu_{xxt}\big).$$

Using these expressions, we rewrite equation (4) in the old variables, which are without tildes. Then we substitute $u_{xxt} = -(u_t+f(t)u_x+g(t)uu_x)/h(t)$ into the obtained equation in order to confine it to the manifold defined by (1) in the third-order jet space with the independent variables (t,x) and the dependent variable u. Splitting the resulting identity with respect to the derivatives u_t, u_x, u_{tx} and u_{xx} leads to the determining equations on the transformation components T, X, and U,

$$U_{uu}=U_{tu}=0,\quad X_{xx}U_u = 2X_xU_{xu},\quad X_x^3U_u = \tilde h(X_{xx}U_{xu}-X_xU_{xxu}) + \frac{\tilde h}{h}X_xU_u,$$
$$U_t + \frac{T_t}{X_x}(\tilde f+\tilde gU)U_x = \tilde h\frac{X_{xx}}{X_x^3}U_{tx} - \tilde h\frac{1}{X_x^2}U_{xxt},\quad \tilde f+\tilde gU = \frac{\tilde h}{h}\frac{f+gu}{T_tX_x}.$$

The equations $U_{uu}=U_{tu}=0$ imply the representation $U=U^1(x)u+U^0(t,x)$, where U^1 and U^0 are smooth functions of their arguments with $U^1\neq 0$. We substitute this representation for U

into the remaining determining equations and further split with respect to u. As a result, we get the equations $U^1_x = U^0_t = U^1_x = X_{xx} = 0$, $\tilde f = \dfrac{X_x}{T_tU^1}(U^1f - U^0g)$, $\tilde g = \dfrac{X_x}{T_tU^1}g$, and $\tilde h = X_x^2h$. We solve the equations for X, U^1 and U^0, and substitute the obtained expressions into the equations representing transformation components for arbitrary elements, which leads to the following assertion.

Theorem 2. *The usual equivalence group $G_1^\sim$ of class (1) is comprised of the transformations*

$$\tilde t = T(t),\quad \tilde x = \delta_1 x + \delta_2,\quad \tilde u = \delta_3 u + \delta_4,\quad \tilde f = \frac{\delta_1}{T_t\delta_3}(\delta_3 f - \delta_4 g),\quad \tilde g = \frac{\delta_1}{T_t\delta_3}g,\quad \tilde h = \delta_1^2 h,$$

where δ_j, $j = 1,2,3,4$, are arbitrary constants with $\delta_1\delta_3 \neq 0$ and $T = T(t)$ is an arbitrary smooth function with $T_t \neq 0$. Class (1) is normalized in the usual sense.

Thus, each point transformation between equations from class (1) is induced by an element of the group $G_1^\sim$. In order to find which variable-coefficient equations of the form (1) admit constant-coefficient counterparts, we assume that the transformed arbitrary elements $\tilde f$, $\tilde g$ and $\tilde h$ are constants in equivalence transformations. This results in the following assertion:

Proposition 1. *A variable-coefficient equation from class (1) is reduced to a constant-coefficient equation from the same class by a point transformation if and only if the corresponding coefficients f, g and h satisfy the conditions*

$$\left(\frac{f}{g}\right)_t = h_t = 0,$$

i.e., h is a constant and f is proportional to g.

Equivalence transformations allow us to simplify the consideration by reducing the number of arbitrary elements. For example, we can set the gauge $g = 1$ using the family of point transformations

$$\tilde t = \int g(t)\mathrm{d}t,\quad \tilde x = x,\quad \tilde u = u \tag{5}$$

parameterized by the arbitrary element g and related to equivalence transformations from the group $G_1^\sim$. Then the other arbitrary elements are $\tilde f(\tilde t) = f(t)/g(t)$, and $\tilde h(\tilde t) = h(t)$. Here and below an integral with respect to t should be interpreted as a fixed antiderivative.

Therefore, without loss of generality we can restrict ourselves by the study of the class

$$u_t + f(t)u_x + uu_x + h(t)u_{xxt} = 0, \tag{6}$$

as all results on symmetries and exact solutions for equations of the form (1) can be recovered from the corresponding results obtained for equations of the form (6) using the above family of point transformations.

Since class (1) is normalized, the equivalence group of its subclass (6) can be easily found as the subgroup of the group $G_1^\sim$ whose elements preserve the gauge $g = 1$.

Corollary 2. *Class (6) is normalized in the usual sense. Its usual equivalence group $G_2^\sim$ is constituted by the transformations*

$$\tilde t = \frac{\delta_1}{\delta_3}t + \delta_0,\quad \tilde x = \delta_1 x + \delta_2,\quad \tilde u = \delta_3 u + \delta_4,\quad \tilde f = \delta_3 f - \delta_4,\quad \tilde h = \delta_1^2 h,$$

where δ_j, $j = 0,\dots,4$, are arbitrary constants with $\delta_1\delta_3 \neq 0$.

There are no truly variable-coefficient equations in class (6) that are reduced by point transformations to constant-coefficient equations from the same class.

In the next section we use the equivalence transformations in the course of group classification of classes (1) and (6).

3. Lie symmetries

We carry out the group classification of class (6) within the framework of the classical Lie approach [8,21]. We look for vector fields of the form

$$\Gamma = \tau(t,x,u)\partial_t + \xi(t,x,u)\partial_x + \eta(t,x,u)\partial_u$$

which generate one-parameter groups of point-symmetry transformations of an equation from class (6) and hence jointly constitute the maximal Lie invariance algebra $A^{\max}$ of this equation. Any such vector field, Γ, satisfies the infinitesimal invariance criterion, i.e., the action of the third prolongation, $\Gamma^{(3)}$, of the vector field Γ on equation (6) results in the conditions being an identity for all solutions of this equation. That is, the criterion is read as

$$\Gamma^{(3)}\left[u_t + f(t)u_x + uu_x + h(t)u_{xxt}\right]\Big|_{u_{xxt}=-(u_t+f(t)u_x+uu_x)/h(t)} = 0. \tag{7}$$

Theorem 2 allows us to simplify the computation since it implies that $\tau = \tau(t)$, $\xi = \xi(x)$, $\eta = \eta(u)$, and $\xi_{xx} = \eta_{uu} = 0$, Then the left hand side of the condition (7) is a polynomial in the variables x, u, u_t and u_x with coefficients depending on t. Hence these coefficients are equal to zero, which gives the determining equations on remaining arbitrarinesses in the components τ, ξ and η of the vector field Γ. Solving the simplest of these equations we get the general form of the components of a Lie symmetry operator of an equation from class (6),

$$\tau = \left(\tfrac{1}{2}c_2 - c_1\right)t + c_0, \quad \xi = c_2x + c_3, \quad \eta = \left(\tfrac{1}{2}c_2 + c_1\right)u + c_4$$

with c_i, $i = 0,\dots,4$, being arbitrary constants. The residuary determining equations involving arbitrary elements have the form

$$\left(\left(\tfrac{1}{2}c_2 - c_1\right)t + c_0\right)h_t = 2c_2h, \quad \left(\left(\tfrac{1}{2}c_2 - c_1\right)t + c_0\right)f_t = \left(\tfrac{1}{2}c_2 + c_1\right)f - c_4 \tag{8}$$

and are called *classifying equations*. Further analysis is carried out using the *method of furcate splitting* introduced in [22]. For each vector field Γ from the algebra $A^{\max}$ equations (8) gives a system on f and h of the general form

$$(\alpha t + \beta)h_t = \gamma h, \quad (\alpha t + \beta)f_t = \left(\tfrac{1}{2}\gamma - \alpha\right)f + \nu, \tag{9}$$

where α, β, γ and ν are constants. Varying vector fields from $A^{\max}$ we obtain a set of systems of the form (9). The number k of such systems with linearly independent tuples of the coefficients α, β, γ and ν is not greater than two; otherwise the set of these systems would be inconsistent in total. Therefore, possible values of k are $k = 0$, $k = 1$ and $k = 2$.

If $k = 0$, then equations (8) are identities with respect to f and h and hence $c_0 = c_1 = c_2 = c_4 = 0$. Thus, we obtain that the maximal Lie invariance algebras of equations from class (6) for general values of f and h is the one-dimensional algebra $\langle\partial_x\rangle$, which gives Case 0 of Table 1. We get the same result when varying the arbitrary elements f and h an splitting equations (8) with respect to them. In other words, the kernel of the maximal Lie invariance algebras of equations from class (6) (i.e., the common part of the maximal Lie invariance algebras of equations from this class, see [8] for the definition) coincides with the maximal Lie invariance algebra $\langle\partial_x\rangle$ of the general case.

If $k = 1$, equations (9) are not identities but equations on f and h. In order to get $G_2^\sim$-inequivalent solutions of these equations, we study how transformations from $G_2^\sim$ act on the coefficients α, β, γ and ν of system (9). Combined with the multiplication by a nonzero constant χ, each transformation from the equivalence group $G_2^\sim$ is extended to the coefficient quadruple of the system (9) as

$$\tilde\alpha = \alpha\chi, \quad \tilde\beta = \left(\frac{\delta_1}{\delta_3}\beta - \delta_0\alpha\right)\chi, \quad \tilde\gamma = \gamma\chi, \quad \tilde\nu = \left(\delta_3\nu + \delta_4\left(\tfrac{1}{2}\gamma - \alpha\right)\right)\chi.$$

Analysis of the induced transformations for α, β, γ and ν results in the following assertion.

Lemma 1. *Up to $G_2^\sim$-equivalence the parameter quadruple $(\alpha,\beta,\gamma,\nu)$ can be assumed to belong to the set*

$$\{(1,0,\rho,0),\ (1,0,2,1),\ (0,1,1,0),\ (0,1,0,1)\},$$

where ρ is an arbitrary constant.

The group classification for the case $k=1$ is obtained as follows. Sequently solving system (9) for each of the four inequivalent values of $(\alpha,\beta,\gamma,\nu)$ presented in Lemma 1, we get the corresponding cases for the arbitrary elements (h,f),

$$(\varepsilon t^\rho, \lambda t^{\frac{\rho-2}{2}}),\quad (\varepsilon t^2, \ln t+\lambda),\quad (\varepsilon e^t, \lambda e^{\frac{t}{2}}),\quad (\varepsilon, t+\lambda),$$

where ε and λ are constant parameters arising in the course of integration, and $\varepsilon\neq 0$. Then we check whether the remaining parameters can be additionally gauged by transformations from the group $G_2^\sim$. It appears that ε can be scaled to 1 if it is positive and to -1 if it is negative. In the second and the fourth cases the constant λ can be set equal to zero by a translation of u. In the first and the third cases λ cannot be gauged. We substitute the obtained forms of h and f into equations (8) to get constraints on constants c_i, $i=0,\dots,4$. This gives the respective forms of Lie symmetry operators. The results are collected in Cases 1–4 of Table 1.[1]

If $k=2$, then up to $G_2^\sim$-equivalence and linear combining there exists a single pair of the coefficient tuples, namely, $(1,0,0,1)$ and $(0,1,0,0)$, for which the corresponding systems of the form (9) are consistent to each other. Then f and h are constants, $f=0 \bmod G_2^\sim$ and $h=\varepsilon \bmod G_2^\sim$, where $\varepsilon=\pm1$, which gives Case 5 of Table 1.

We sum up the above consideration.

Theorem 3. *The kernel of the maximal Lie invariance algebras of equations from class* (6) *is the one-dimensional algebra $\langle\partial_x\rangle$. All possible $G_2^\sim$-inequivalent cases of Lie symmetry extensions are exhausted by Cases 1–5 of Table 1.*

Corollary 3. *The group classification for class* (1) *up to $G_1^\sim$-equivalence results in the list presented in Table 1, where the arbitrary element g is assumed to be equal* 1.

In order to get the classification list for class (1), where forms of arbitrary elements are not simplified by equivalence transformations, we apply transformation (5) combined with transformations from the equivalence group $G_2^\sim$ to equations of the form (6) with f and h presented in Table 1. Basis elements of the corresponding maximal Lie invariance algebras are pushed forward by the same transformations. Then we re-denote the constants and collect the obtained results in Table 2. The detailed procedure of the *equivalence based approach* for deriving most general forms of arbitrary elements and basis elements of the corresponding maximal Lie invariance algebras can be found in [19].

4. Mapping between classes and group classification of the related class

An alternative way for group classification of class (1) is the method of mapping between classes, which was suggested in [23]. This method has been successfully applied to several classes of nonlinear partial differential equations (see, e.g., [20]).

Class (1) can be mapped to a similar class of third-order partial differential equations of the form

$$u_t+uu_x+h(t)u_{xxt}=l(t),\quad h\neq 0. \tag{10}$$

[1] Arguments of logarithm and bases in powers with noninteger exponents containing ρ should be carefully treated throughout the paper. In fact, one should take their absolute values if they can be negative. Usually the positivity of t well agrees with the physical meaning of t as time variable. So, omitting the notation of absolute value is not too essential in Table 1. At the same time, the expression $\varepsilon T+\kappa$ from Table 2 may be negative and hence we implicitly assume replacing it by its absolute value.

Table 1. The group classification of class (6) up to $G_2^{\sim}$-equivalence.

no.	$h(t)$	$f(t)$	Basis of $A^{\max}$
0	$\forall$	$\forall$	∂_x
1	εt^ρ	$\lambda t^{\frac{\rho-2}{2}}$	$\partial_x,\ 2t\partial_t + \rho x\partial_x + (\rho-2)u\partial_u$
2	εt^2	$\ln t$	$\partial_x,\ t\partial_t + x\partial_x - \partial_u$
3	εe^t	$\lambda e^{\frac{1}{2}t}$	$\partial_x,\ 2\partial_t + x\partial_x + u\partial_u$
4	ε	t	$\partial_x,\ \partial_t - \partial_u$
5	ε	0	$\partial_x,\ \partial_t,\ t\partial_t - u\partial_u$

Here ρ and λ are arbitrary constants, $\varepsilon = \pm 1 \bmod G_2^{\sim}$. In Case 1 $(\rho, \lambda) \neq (0,0)$.

Table 2. The group classification of class (1) using no equivalence.

no.	$h(t)$	$f(t)$	Basis of $A^{\max}$
0	$\forall$	$\forall$	∂_x
1	$\mu_1(\varepsilon T+\kappa)^\rho$	$\mu_2 g(\varepsilon T+\kappa)^{\frac{\rho-2}{2}} + \mu_3 g$	$\partial_x,\ \dfrac{2}{g}(\varepsilon T+\kappa)\partial_t + \varepsilon\rho x\partial_x + \varepsilon(\rho-2)(u+\mu_3)\partial_u$
2	$\mu_1(\varepsilon T+\kappa)^2$	$\mu_2 g\ln(\varepsilon T+\kappa) + \mu_3 g$	$\partial_x,\ \dfrac{1}{g}(\varepsilon T+\kappa)\partial_t + \varepsilon x\partial_x - \varepsilon\mu_2\partial_u$
3	$\mu_1\exp(\sigma T)$	$\mu_2 g\exp(\frac{1}{2}\sigma T) + \mu_3 g$	$\partial_x,\ \dfrac{2}{g}\partial_t + \sigma x\partial_x + \sigma(u+\mu_3)\partial_u$
4	μ_1	$\mu_2 gT + \mu_3 g$	$\partial_x,\ \dfrac{1}{g}\partial_t - \mu_2\partial_u$
5	μ_1	$\mu_3 g$	$\partial_x,\ \dfrac{1}{g}\partial_t,\ \dfrac{T}{g}\partial_t - (u+\mu_3)\partial_u$

Here g is an arbitrary nonvanishing smooth function, $T = \int g(t)\,\mathrm{d}t$; $\varepsilon = \pm 1$; μ_1, μ_2, μ_3, ν and ρ are arbitrary constants satisfying the following constraints: $\mu_1\lambda \neq 0$; in Case 1 $\rho\mu_2 \neq 0$; and in Case 4 $\mu_2 \neq 0$.

The map is realized by the family of point transformations

$$\tilde{t} = \int g(t)\mathrm{d}t, \quad \tilde{x} = x, \quad \tilde{u} = u + \frac{f(t)}{g(t)}, \tag{11}$$

parameterized by two arbitrary elements of class (1). The arbitrary elements in the imaged equations take values (tildes in (10) are omitted)

$$\tilde{h}(\tilde{t}) = h(t), \quad l(\tilde{t}) = \frac{1}{g(t)}\left(\frac{f(t)}{g(t)}\right)_t.$$

Following the method of mapping between classes, we first classify Lie symmetries of the imaged class (10) and then use the family of point transformations (11) to extend the result to the initial class (1).

In order to efficiently solve the group classification problem for class (10), we look for admissible transformations in this class using the direct method. It appears that such transformations are exhausted by transformations from the usual equivalence group of this class.

Theorem 4. *The usual equivalence group $G_3^{\sim}$ of class (10) consists of the transformations*

$$\tilde{t} = \frac{\delta_1}{\delta_3}t + \delta_0, \quad \tilde{x} = \delta_1 x + \delta_2, \quad \tilde{u} = \delta_3 u, \quad \tilde{h} = \delta_1^2 h, \quad \tilde{l} = \frac{\delta_3^2}{\delta_1} l,$$

where δ_j, $j = 0, 1, 2, 3$, are arbitrary constants with $\delta_1\delta_3 \neq 0$. Class (10) is normalized in the usual sense.

Using the classical Lie infinitesimal method in the same way as in the previous section, we get the complete group classification of equations from class (10). The results are summarized in the following assertion.

Theorem 5. *The kernel of the maximal Lie invariance algebras of equations from class (10) is the one-dimensional algebra $\langle\partial_x\rangle$. All possible $G_3^{\sim}$-inequivalent cases of extension of the maximal Lie invariance algebras are exhausted by Cases 1–4 of Table 3.*

Table 3. The group classification of class (10) up to $G_3^{\sim}$-equivalence.

no.	$h(t)$	$l(t)$	Basis of $A^{\max}$
0	$\forall$	$\forall$	∂_x
1	εt^ρ	$\lambda t^{\frac{\rho-4}{2}}$	$\partial_x,\ 2t\partial_t + \rho x\partial_x + (\rho-2)u\partial_u$
2	εe^t	$\lambda e^{\frac{1}{2}t}$	$\partial_x,\ 2\partial_t + x\partial_x + u\partial_u$
3	ε	1	$\partial_x,\ \partial_t$
4	ε	0	$\partial_x,\ \partial_t,\ t\partial_t - u\partial_u$

Here ρ and λ are arbitrary constants, $\varepsilon = \pm 1 \bmod G_3^{\sim}$, in Case 1 $(\rho, \lambda) \neq (0, 0)$.

Remark 1. The most general forms of the functions h and l that correspond to equations from class (10) with Lie symmetry extensions are

1. $h = \lambda_1(\varepsilon t + \kappa)^\rho$, $l = \lambda_2(\varepsilon t + \kappa)^{\frac{\rho-4}{2}}$: $A^{\max} = \langle\partial_x,\ 2(\varepsilon t + \kappa)\partial_t + \varepsilon\rho x\partial_x + \varepsilon(\rho-2)u\partial_u\rangle$;
2. $h = \lambda_1 e^{\sigma t}$, $l = \lambda_2 e^{\frac{1}{2}\sigma t}$: $A^{\max} = \langle\partial_x,\ 2\partial_t + \sigma x\partial_x + \sigma u\partial_u\rangle$;
3. $h = \lambda_1$, $l = \lambda_2$: $A^{\max} = \langle\partial_x,\ \partial_t\rangle$;
4. $h = \lambda_1$, $l = 0$: $A^{\max} = \langle\partial_x,\ \partial_t,\ t\partial_t - u\partial_u\rangle$.

Here λ_1, λ_2, ε, κ and ρ are arbitrary constants with $\lambda_1\sigma\varepsilon \neq 0$. Additionally, in Case 1 $(\rho, \lambda_2) \neq (0, 0)$ and in Case 3 $\lambda_2 \neq 0$. Due to the presence of arbitrary constants λ_1 and λ_2, the constant ε can be assumed to take the values ± 1 only.

The following example shows how to recover the group classification of class (1) using the results obtained for class (10).

Consider Case 1 of Table 3 extended by the equivalence transformations from $G_3^{\sim}$, i.e., the first case presented in Remark 3, were $\tilde{h} = \lambda_1(\varepsilon\tilde{t} + \kappa)^\rho$, $l = \lambda_2(\varepsilon\tilde{t} + \kappa)^{\frac{\rho-4}{2}}$. We denote $\int g(t)\,dt$ by T. As $\tilde{t} = T$ and $l(T) = (f/g)_t/g$, we get $(f/g)_t = \lambda_2 g(t)(\varepsilon T + \kappa)^{\frac{\rho-4}{2}}$. Finally,

$$f(t) = \begin{cases} \lambda_2 g(t)\left(\frac{2}{\varepsilon(\rho-2)}(\varepsilon T + \kappa)^{\frac{\rho-2}{2}} + \lambda_3\right), & \text{if} \quad \rho \neq 2, \\ \lambda_2 g(t)\left(\frac{1}{\varepsilon}\ln(\varepsilon T + \kappa) + \lambda_3\right), & \text{if} \quad \rho = 2. \end{cases}$$

Journal of Physics: Conference Series **621** (2015) 012016 doi:10.1088/1742-6596/621/1/012016

After re-denoting the constants λ_i, $i = 1,2,3$, it is easy to see that we get Cases 1 and 2 of Table 2, respectively. To obtain the corresponding Lie symmetry operators one should make the change of variables $\tilde t = T$, $\tilde x = x$, $\tilde u = u + \mu_2(\varepsilon T+\kappa)^{\frac{\rho-2}{2}} + \mu_3$ (resp. $\tilde u = u + \mu_2\ln(\varepsilon T+\kappa)+\mu_3$ for the second case) in the vector fields $X_1 = \partial_{\tilde x}$ and $X_2 = 2(\varepsilon\tilde t+\kappa)\partial_{\tilde t} + \varepsilon\rho\tilde x\partial_{\tilde x} + \varepsilon(\rho-2)\tilde u\partial_{\tilde u}$.

It is interesting to note that the images of two distinct inequivalent cases of Lie symmetry extensions in class (1) (Cases 1 and 2 of Table 2) belong to the same case of Lie symmetry extensions for class (10) (Case 1 of Table 3).

The other cases are easily treated in the same way.

5. Conservation laws

We classify (local) conservation laws of equations from class (1), applying the most direct method based on the definition of conservation laws.

We briefly present necessary notions, specifying them to the relevant case of a single partial differential equation $\mathcal{L}$: $L = 0$ with the two independent variables t, x and the single dependent variable u, where L is a differential function, i.e., a smooth function of t, x and derivatives of u. A tuple (F, G) of differential functions is a conserved current of $\mathcal{L}$ if $D_tF + D_xG = 0$ for any solution of $\mathcal{L}$, where D_t and D_x are the operators of total derivative with respect to t and x, respectively. Then the functions F and G are called the *density* and the *flux* of the conserved current (F, G). Conserved currents (F, G) and (F', G') are equivalent if there exist such differential functions $\hat F$, $\hat G$ and H that $\hat F$ and $\hat G$ vanish for all solutions of $\mathcal{L}$ and $F' = F + \hat F + D_xH$, $G' = G + \hat G - D_tH$. Elements of the factor-set of conserved currents with respect to this equivalence relation are called *conservation laws* of $\mathcal{L}$. Each conservation law $\mathcal{F}$ of $\mathcal{L}$ admits the representation $D_tF + D_xG = \lambda L$ for some conserved current (F, G) containing in $\mathcal{F}$ and a differential function λ, which is called a *characteristic* of the conservation law $\mathcal{F}$. Characteristics λ and $\tilde\lambda$ are equivalent if their difference vanishes on solutions of $\mathcal{L}$. Under certain natural condition on $\mathcal{L}$, there is one-to-one correspondence between conservation laws and equivalence classes of characteristics [21].

Computing conservation laws for equations from class (1), we assume that components of conserved currents are of the general form

$$F = F(t, x, u, u_t, u_x), \qquad G = G(t, x, u, u_t, u_x, u_{tt}, u_{tx}, u_{xx}).$$

Following [4], it is possible to show that in fact any conservation law of each equation from class (1) possesses a characteristic of order not greater than two and, therefore, contains a conserved current whose density and flux orders are not greater than one and two, respectively. The proof of this is quite long and will be a subject of another paper.

The classification of local conservation laws of equations from the class (1) is as follows.

Case 0. Each equation from the class (1) admits the "natural" conservation law with the constant characteristic $\lambda^1 = 1$. The corresponding density and flux are

$$F^1 = u, \quad G^1 = f(t)u + \frac{1}{2}g(t)u^2 + h(t)u_{tx}.$$

For general admitted values of the arbitrary elements f, g and h the associated space of conservation laws is one-dimensional.

Case 1. If the arbitrary elements satisfy the equation $((1/h)_t/g)_t = 0$, and, therefore,

$$h(t) = \left(\rho_1\int g(t)\mathrm{d}t + \rho_2\right)^{-1},$$

where ρ_1 and ρ_2 are constants with $(\rho_1, \rho_2) \neq (0, 0)$, then the space of conservation laws of the corresponding equation of the form (1) is at least two-dimensional. The second basis conservation law can be chosen to have the following characteristic, density and flux:

$$\lambda^2 = \frac{u}{h} - \rho_1\left(x - \int f(t)\mathrm{d}t\right), \quad F^2 = \frac{u^2}{2h(t)} - \frac{u_x^2}{2} - \rho_1\left(x - \int f(t)\mathrm{d}t\right)u,$$
$$G^2 = \frac{g(t)}{3h(t)}u^3 + \frac{f(t)}{2h(t)}u^2 + uu_{tx} + \rho_1 h(t)u_t - \rho_1\left(x - \int f(t)\mathrm{d}t\right)\left(f(t)u + \frac{1}{2}g(t)u^2 + h(t)u_{tx}\right),$$

where $(1/h)_t/g = \rho_1 = \text{const}$.

Using the family of point transformations $\tilde t = \rho_1\int g(t)\mathrm{d}t + \rho_2$, $\tilde x = x$, $\tilde u = u/\rho_1$ related to the group $G_1^\sim$, we can reduce any equation of this case with $\rho_1 \neq 0$ to the form $u_t + f(t)u_x + uu_x + t^{-1}u_{txx} = 0$ (tildes are omitted in the latter equation).

Case 2. One more case with at least two-dimensional spaces of conservation laws is given by the arbitrary elements satisfying the condition $((f/g)_t/g)_t = 0$, i.e., if

$$f(t) = \left(\sigma_1\int g(t)\mathrm{d}t + \sigma_2\right)g(t),$$

where σ_1 and σ_2 are arbitrary constants. The second basis conservation law can be chosen to have the following characteristic, density and flux:

$$\lambda^3 = W^2 - 2\sigma_1 x + 2\frac{h}{g}u_{tx}, \quad F^3 = \frac{1}{3}W^3 - 2\sigma_1 xu - \frac{1}{3}\frac{f(t)^3}{g(t)^3},$$
$$G^3 = \frac{h(t)}{g(t)}W_t^2 + \frac{h(t)^2}{g(t)}u_{tx}^2 + h(t)W^2u_{tx} - 2\sigma_1 h(t)xu_{tx} - 2\sigma_1 f(t)xu - \sigma_1 g(t)xu^2 + \frac{g(t)}{4}W^4,$$

where $W = u + f(t)/g(t)$, $(f/g)_t/g = \sigma_1 = \text{const}$.

Using the family of point transformations $\tilde t = \sigma_1\int g(t)dt + \sigma_2$, $\tilde x = \sigma_1 x$, $\tilde u = u$ related to the group $G_1^\sim$, we can reduce any equation of this case with $\sigma_1 \neq 0$ to the form $u_t + tu_x + uu_x + h(t)u_{txx} = 0$ (tildes are omitted in the latter equation).

Case 3. The maximal dimension of the spaces of conservation laws for equations from class (1) equals three and is reached for the intersection of Cases 1 and 2, where arbitrary elements satisfy the both constraints, $((f/g)_t/g)_t = 0$ and $((1/h)_t/g)_t = 0$. Then for each of the spaces, a basis consists of conservation laws with the characteristics λ^1, λ^2 and λ^3 and the conserved currents (F^1, G^1), (F^2, G^2) and (F^3, G^3), respectively. The corresponding equation can be reduced to the form $u_t + (\sigma_1 t + \sigma_2)u_x + uu_x + (\rho_1 t + \rho_2)^{-1}u_{txx} = 0$ by transformation (5). The further simplification is possible by transformations from the group $G_2^\sim$. For example, we can set one of the linear combinations $\sigma_1 t + \sigma_2$ or $\rho_1 t + \rho_2$ to t if $\sigma_1 \neq 0$ or $\rho_1 \neq 0$, respectively.

A well-studied subcase of Case 3 is constituted by constant-coefficient equations, for which $\rho_1 = \sigma_1 = 0$ [1,4,24]. Up to $G_1^\sim$-equivalence any constant-coefficient equation from class (1) can be mapped to the equation $u_t = uu_x + \varepsilon u_{xxt}$, where $\varepsilon = \operatorname{sgn} h = \pm 1$. Then the characteristics and the components of the conserved currents of the above basis conservation laws take the form (cf. [6, p 195])

$$\lambda^1 = 1, \quad F^1 = u, \quad G^1 = -\frac{1}{2}u^2 - \varepsilon u_{tx},$$
$$\lambda^2 = \varepsilon u, \quad F^2 = \frac{u^2}{2} + \varepsilon\frac{u_x^2}{2}, \quad G^2 = -\frac{1}{3}u^3 - \varepsilon uu_{tx},$$
$$\lambda^3 = u^2 + 2\varepsilon u_{tx}, \quad F^3 = \frac{1}{3}u^3, \quad G^3 = \varepsilon u_t^2 - u_{tx}^2 - \varepsilon u^2u_{tx} - \frac{1}{4}u^4.$$

6. Conclusion

The aim of the present paper is to enhance and generalize existing results on Lie symmetries and conservation laws of variable-coefficient BBM equations of the form (1). Comparing the results of [7] with those collected in Table 2, we conclude that Lie symmetry extensions presented in [7] are particular specifications of Cases 1–5 from Table 2 for certain fixed values of the arbitrary element g. For example, there are two cases in the classification list derived in [7] with the three-dimensional maximal Lie symmetry algebras (Cases 4 and 10 of Table 1 therein). These cases are particular subcases of Case 5 of Table 2 for $g = g_0 = \text{const}$ and $g = g_0 e^{kt}$.

The results of Section 5 on conservation laws are quite expectable and, at the same time, are not trivial. They naturally generalize well-known results of constant-coefficient BBM equations and need the completion of the most significant and tricky part of the proof, which is deriving an upper bound for order of conservation laws similarly to [4].

Acknowledgments

This work and the participation of OV in the Workshop was partially supported by a grant from the Niels Henrik Abel Board. OV would like to thank the University of Cyprus for hosting her during this project. The research of ROP was supported by the Austrian Science Fund (FWF), project P25064. ROP is also grateful for the hospitality and financial support provided by the University of Cyprus.

References

[1] Benjamin T B, Bona J L and Mahony J J 1972 Model equations for long waves in nonlinear dispersive systems *Philos. Trans. Roy. Soc. London Ser.* A **272** 47
[2] Abdulloev Kh O, Bogolubsky I L and Makhankov V G 1976 One more example of inelastic soliton interaction *Phys. Lett.* A **56** 427
[3] Bona J L, Pritchard W G and Scott L R 1983 A comparison of solutions of two model equations for long waves *Lectures in Appl. Math.* **20** 235
[4] Duzhin S V and Tsujishita T 1984 Conservation laws of the BBM equation *J. Phys.* A: *Math. Gen.* **17** 3267
[5] Kostin V M 1969 Certain invariant solutions of equations of the Korteweg–de Vries type *Zh. Prikl. Mekh. Tekh. Fiz.* **4** 69
[6] Ibragimov N H (ed) 1994 *CRC Handbook of Lie Group Analysis of Differential Equations* (Symmetries, Exact Solutions and Conservation Laws vol 1) (Boca Raton: Chemical Rubber Company)
[7] Molati M and Khalique C M 2013 Symmetry classification and invariant solutions of the variable coefficient BBM equation *Appl. Math. Comp.* **219** 7917
[8] Ovsiannikov L V 1982 *Group Analysis of Differential Equations* (New York: Academic Press)
[9] Popovych R O 2006 Classification of admissible transformations of differential equations *Collection of Works of Institute of Mathematics* vol 3 (Kyiv: Institute of Mathematics) p 239
[10] Popovych R O, Kunzinger M and Eshraghi H 2010 Admissible transformations and normalized classes of nonlinear Schrödinger equations *Acta Appl. Math.* **109** 315
[11] Popovych R O and Bihlo A 2012 Symmetry preserving parameterization schemes *J. Math. Phys.* **53** 073102
[12] Bihlo A, Dos Santos Cardoso-Bihlo E M and Popovych R O 2012 Complete group classification of a class of nonlinear wave equations *J. Math. Phys.* **53** 123515
[13] Kuriksha O, Pošta S and Vaneeva O 2014 Group classification of variable coefficient generalized Kawahara equations *J. Phys.* A: *Math. Theor.* **47** 045201
[14] Kingston J G and Sophocleous C 1991 On point transformations of a generalised Burgers equation *Phys. Lett.* A **155** 15–19.
[15] Kingston J G 1991 On point transformations of evolution equations *J. Phys.* A **24** L769
[16] Kingston J G and Sophocleous C 1998 On form-preserving point transformations of partial differential equations *J. Phys.* A **31** 1597
[17] Pocheketa O A, Popovych R O and Vaneeva O O 2014 Group classification and exact solutions of variable-coefficient generalized Burgers equations with linear damping *Appl. Math. Comput.* **243** 232
[18] Winternitz P and Gazeau J P 1992 Allowed transformations and symmetry classes of variable coefficient Korteweg–de Vries equations *Phys. Lett.* A **167** 246
[19] Vaneeva O 2012 Lie symmetries and exact solutions of variable coefficient mKdV equations: an equivalence based approach *Commun. Nonlinear Sci. Numer. Simul.* **17** 611

[20] Vaneeva O, Kuriksha O and Sophocleous C 2015 Enhanced group classification of Gardner equations with time-dependent coefficients *Commun. Nonlinear Sci. Numer. Simul.* **22** 1243

[21] Olver P 1993 *Applications of Lie Groups to Differential Equations* (New York: Springer-Verlag)

[22] Nikitin A G and Popovych R O 2001 Group classification of nonlinear Schrödinger equations *Ukr. Math. J.* **53** 1255 (*Preprint* arXiv:math-ph/0301009)

[23] Vaneeva O O, Popovych R O and Sophocleous C 2009 Enhanced group analysis and exact solutions of variable coefficient semilinear diffusion equations with a power source *Acta Appl. Math* **106** 1

[24] Olver P J 1979 Euler operators and conservation laws of the BBM equation *Math. Proc. Cambridge Philos. Soc.* **85** 143

Gauge equivalent integrable equations on $\mathfrak{sl}(3)$ with reductions

Alexandar B Yanovski

Department of Mathematics and Applied Mathematics
University of Cape Town, 7700 Rondebosch, Cape Town, South Africa

E-mail: `Alexandar.Ianovsky@uct.ac.za`

Abstract. We consider an auxiliary linear problem on the algebra $\mathfrak{sl}(3)$ with reductions of $\mathbb{Z}_2 \times \mathbb{Z}_2$ type introduced recently by Gerdjikov, Mikhailov and Valchev (GMV), and discuss its relation with a Generalized Zakharov–Shabat system having the same type of reduction symmetry. We describe the gauge-equivalent hierarchies and find explicitly the analog of the first nontrivial soliton equation in the hierarchy of equations related to GMV linear problem.

1. Introduction

The concept of gauge-equivalent soliton equations originates from the paper of Mikhailov and Zakharov [1], where it was applied to reveal relations between several different models of classical field theory, such as the n-field, the sine-Gordon equation and the Thirring model. As far as we know the next example was the discovery of the gauge equivalence between the nonlinear Schrödinger equation (NLS),

$$\mathrm{i}\varphi_t + \varphi_{xx} + 2\varphi|\varphi|^2 = 0, \tag{1}$$

(here φ is a complex–valued function going to zero when $|x|$ goes to infinity) and the Heisenberg Ferromagnet equation,

$$S_t = -\frac{1}{2\mathrm{i}}[S, S_{xx}], \tag{2}$$

with $S(x)$ being a $\mathfrak{sl}(2,\mathbb{C})$-valued function such that

$$S^\dagger = -S, \quad \lim_{x\to\pm\infty} S(x) = \sigma_3 = \begin{pmatrix} 1 & 0 \\ 0 & -1 \end{pmatrix}, \quad S^2(x) = \mathbf{1},$$

("$\dagger$" means Hermitian conjugation). As it is well known the NLS equations is represented in a Lax form $[L, A] = 0$, where the operator L defines the auxiliary problem on $\mathfrak{sl}(2,\mathbb{C})$,

$$L\psi = (\mathrm{i}\partial_x + q - \lambda\sigma_3)\psi = 0, \tag{3}$$

known as classical Zakharov–Shabat system. Here the 'potential' $q(x)$ is a smooth function of the type

$$\begin{pmatrix} 0 & q_+(x) \\ q_-(x) & 0 \end{pmatrix},$$

where $q_\pm(x)$ are complex-valued functions that rapidly decay when $|x|$ tends to infinity. In order to get the NLS one must have $q_+(x) = \varphi(x)$ and $q_-(x) = \varphi^*(x)$, where * means complex conjugation. From its side HF equation has a Lax representation $[\tilde{L}, \tilde{A}] = 0$ with $\tilde{L}$

$$(\mathrm{i}\partial_x - \lambda S(x))\tilde{\psi} = \tilde{L}\tilde{\psi} = 0.$$

Zakharov and Takhtadjan [2] showed that L and $\tilde{L}$ are gauge-equivalent, that is:

$$\tilde{L} = \psi_0^{-1} L \psi_0,$$

where ψ_0 is the Jost solution of (3) for $\lambda = 0$ having the properties:

$$\lim_{x\to\infty} \psi_0(x) = \mathbf{1}, \qquad \lim_{x\to-\infty} \psi_0(x) = T^{-1}(0),$$

and $T(\lambda)$ is the transition matrix.

Remark 1.1 *It should be mentioned that $T(0)$ should also satisfy the relation $T^{-1}(0)\sigma_3 T(0) = \sigma_3$, we shall make some comments about that later.*

As a result NLS and HF systems are equivalent because one has Lax representation $[L, A] = 0$ and the other has Lax representation $[\tilde{L}, \tilde{A}] = 0$ with $\tilde{A} = \psi_0^{-1} A \psi_0$.

We have been able to extend the above result to the hierarchies of soliton systems associated with L and $\tilde{L}$, as well known the NLS and HF are the first nontrivial equations in these hierarchies, see [3]. This was achieved through the theory of Recursion Operators Λ ($\tilde{\Lambda}$) related to L and $\tilde{L}$ respectively. Since Recursion Operators give not only the hierarchies of equations but also the hierarchies of their conservation laws, Hamiltonian structures etc., etc., we in fact established a gauge-covariant theory of the Recursion Operators for these systems.

Later we generalized the theory, see [4–6] for a system of the type

$$(\mathrm{i}\partial_x + q(x) - \lambda J)\,\psi = L\psi = 0 \tag{4}$$

called generalized Zakharov–Shabat system (GZS) when J is real and Caudrey–Beals–Coifman system (CBC system) when J is complex. In the above $q(x)$ and J belong to some fixed simple Lie algebra $\mathfrak{g}$ in some finite dimensional irreducible representation; (3) is obtained for $\mathfrak{g} = \mathfrak{sl}(2, \mathbb{C})$. The element J is regular, that is, the kernel of ad_J ($\mathrm{ad}_J X \equiv [J, X]$, $X \in \mathfrak{g}$) is a Cartan subalgebra $\mathfrak{h} \subset \mathfrak{g}$. The potential $q(x)$ belongs to the orthogonal completion $\mathfrak{h}^\perp$ of $\mathfrak{h}$ with respect to the Killing form:

$$\langle X, Y \rangle = \mathrm{tr}\,(\mathrm{ad}_X\, \mathrm{ad}_Y), \qquad X, Y \in \mathfrak{g},$$

$q(x)$ is smooth and tends to zero as $x \to \pm\infty$. We shall restrict ourselves below to the case when $q(x)$ is of Schwartz type. This is enough for our purposes. The theory for (4) was first developed for the case $\mathfrak{g} = \mathfrak{sl}(n, \mathbb{C})$ and it turned out the case when J is not real presents more difficulties which is the reason for the different names. It is not possible here to give a historic review, for such a review and an extended bibliography see [7]. Let us only mention that the case when we have CBC system in arbitrary representation of the algebra $\mathfrak{g}$ has been studied in [8]. The corresponding gauge-equivalent system is of the form

$$(\mathrm{i}\partial_x - \lambda S(x))\,\tilde{\psi} = \tilde{L}\tilde{\psi} = 0 \tag{5}$$

and is referred as GZS system (or CBC) system in pole gauge in contrast to (4) which is referred as GZS system (or CBC) system in canonical gauge. Here $S(x)$ takes values in the orbit of J in the coadjoint representation of $\mathfrak{g}$ and satisfies $\lim_{x\to\pm\infty} S(x) = J$. Let us mention that as before the first nontrivial systems in the hierarchies of soliton equations corresponding to (4) and (5) are called NLS and HF type equations, respectively.

The correspondence between the soliton equations associated with the GZS (CBC) system in pole and canonical gauge goes even further, to the geometric interpretation of the soliton equations as fundamental fields of a certain Poisson–Nijenhuis structure, see [7], but in this short note we just mention this issue.

2. Systems on $\mathfrak{sl}(3)$ and the Gerdjikov–Mikhailov–Valchev (GMV) system

The GZS in pole gauge on $\mathfrak{sl}(3)$ has been studied as an application of the general results about GZS and CBC system in pole gauge. The Generating Operator has been calculated and some systems of HF type with possible physical applications have been found [9]. However, the interest in it was renewed after the system that we refer below as GMV has been introduced [10, 11]. At the beginning the GMV system study started independently, spectral properties were studied and Generating Operators were calculated. Later it was pointed out that GMV could be treated as $\mathfrak{sl}(3)$ GZS system in pole gauge with additional reductions of Mikhailov type, so that the Generating Operators found for the GMV system could be obtained from the Generating Operator for the general $\mathfrak{sl}(3)$ system and geometric interpretation has been clarified [12, 13]. We want to take this further and describe the first nontrivial soliton system related to GMV linear problem as gauge-equivalent to some NLS type system. Now we introduce the GMV system. By this name we shall call the auxiliary linear problem

$$\tilde{L}_0\tilde{\psi} = (\mathrm{i}\partial_x + \lambda S_1)\tilde{\psi} = 0, \quad S_1 = \begin{pmatrix} 0 & u & v \\ u^* & 0 & 0 \\ v^* & 0 & 0 \end{pmatrix}.$$

In the above u, v (the potentials) are smooth complex valued functions on x belonging to the real line and by * we denote the complex conjugation. In addition, the functions u and v satisfy the relation

$$|u|^2 + |v|^2 = 1.$$

Natural asymptotic conditions for u and v are

$$\lim_{x\to\pm\infty} u(x) = 0, \qquad \lim_{x\to\pm\infty} v(x) = v_\pm = e^{\mathrm{i}\eta_\pm},$$

or

$$\lim_{x\to\pm\infty} u(x) = u_\pm = e^{\mathrm{i}\mu_\pm}, \qquad \lim_{x\to\pm\infty} v(x) = 0,$$

where $\eta_\pm$, $\mu_\pm$ are real constants.

As described in [10, 11] the GMV system arises when one looks for integrable system having a Lax representation $[\tilde{L}_0, \tilde{A}] = 0$ with $\tilde{L}_0$ and $\tilde{A}$ subject to Mikhailov-type reduction requirements, see [14]. In this particular case the Mikhailov reduction group G_R acting on the fundamental solutions of the system (6) is generated by the two elements g_1 and g_2 acting in the following way:

$$g_1(\tilde{\psi})(x,\lambda) = \left[\tilde{\psi}(x,\lambda^*)^\dagger\right]^{-1},$$
$$g_2(\tilde{\psi})(x,\lambda) = H\tilde{\psi}(x,-\lambda)H, \quad H = \mathrm{diag}(-1,1,1),$$

where $\dagger$ denotes Hermitian conjugation. Since $g_1g_2 = g_2g_1$ and $g_1^2 = g_2^2 = \mathrm{id}$, $G_R = \mathbb{Z}_2 \times \mathbb{Z}_2$. The map $\mathcal{H}\colon X \mapsto HXH = HXH^{-1}$ is involutive automorphism of $\mathfrak{sl}(3,\mathbb{C})$ which commutes with the complex conjugation σ defining the real form $\mathfrak{su}(3)$ of $\mathfrak{sl}(3,\mathbb{C})$, $(\sigma(X) = -X^\dagger)$. Then introducing the spaces

$$\mathfrak{g}^{[j]} = \{X\colon \mathcal{H}(X) = (-1)^j X\}, \quad j = 0, 1,$$

we get the splittings

$$\mathfrak{sl}(3,\mathbb{C}) = \mathfrak{g}^{[0]} \oplus \mathfrak{g}^{[1]}, \quad \mathfrak{su}(3) = \left(\mathfrak{g}^{0} \cap \mathfrak{su}(3)\right) \oplus \left(\mathfrak{g}^{[1]} \cap \mathfrak{su}(3)\right).$$

The spaces $\mathfrak{g}^{[0]}, \mathfrak{g}^{[1]}$ consist: $\mathfrak{g}^{[1]}$ of all off-diagonal matrices $X = (x_{ij})$ for which $x_{23} = x_{32} = 0$ and $\mathfrak{g}^{[0]}$ of all traceless matrices $X = (x_{ij})$ for which $x_{12} = x_{21} = x_{13} = x_{31} = 0$. Of course, $S \in \mathfrak{g}^{[1]}$. Also, since $\mathcal{H}$ is automorphism, the spaces $\mathfrak{g}^{[0]}$ and $\mathfrak{g}^{[1]}$ are orthogonal with respect to the Killing form.

The invariance under the group generated by g_1, g_2 means that if $\tilde{\psi}$ is the common G_R-invariant fundamental solution of (6) and a linear problem of the type

$$\tilde{A}\psi = \mathrm{i}\partial_t\psi + \left(\sum_{k=0}^{n} \lambda^k \tilde{A}_k\right)\psi = 0, \quad \tilde{A}_k \in \mathfrak{sl}(3,\mathbb{C})$$

we must have

$$\tilde{A}_{2k+1} \in \mathfrak{g}^{[1]} \cap \mathrm{i}\mathfrak{su}(3), \quad A_{2k} \in \mathfrak{g}^{[0]} \cap \mathrm{i}\mathfrak{su}(3), \quad k = 0, 1, 2, \ldots. \tag{6}$$

In the same way $S_1 \in \mathfrak{g}^{[1]} \cap \mathrm{i}\mathfrak{su}(3)$ which forces S_1 to be in the form we have written.

In the papers [10, 11] the study of the spectral properties of (6) has been initiated and the generating operators for the system have been calculated. We shall show now that the GMV problem is gauge-equivalent to a generalized Zakharov–Shabat problem (GZS problem) also with $\mathbb{Z}_2 \times \mathbb{Z}_2$ reduction.

3. GZS type system gauge-equivalent to GMV system

It can be checked the matrix S has constant eigenvalues. As pointed in [10], we have $g^{-1}Sg = J_1$, where g is unitary matrix ($g^\dagger = g^{-1}$) of the form

$$g = \frac{1}{\sqrt{2}} \begin{pmatrix} 1 & 0 & -1 \\ u^* & \sqrt{2}v & u^* \\ v^* & -\sqrt{2}u & v^* \end{pmatrix}, \quad J_1 = \mathrm{diag}(1, 0-1).$$

Since $g(x) \in \mathrm{SU}(3)$ the values of $S_1(x)$ will be in the orbit $\mathcal{O}_{J_1}(\mathrm{SU}(3))$ of J_1 with respect to SU(3) (it is a submanifold of $\mathrm{i}\mathfrak{su}(3)$ of course). Thus $S_1(x) \in \mathcal{O}_{J_1}(\mathrm{SU}(3)) \cap \mathfrak{g}^{[1]}$. Let us note also that conversely, if we assume that $S_1(x) \in \mathcal{O}_{J_1}(\mathrm{SU}(3)) \cap \mathfrak{g}^{[1]}$ then first S_1 has the form as in (6). Next, as easily checked, the eigenvalues of the matrix S_1 are $\mu_1 = 0, \mu_2 = -\mu_3 = \sqrt{|u|^2 + |v|^2}$ but since they coincide with $0, \pm 1$ we must have $|u|^2 + |v|^2 = 1$.

Thus we speak about a problem of GZS type in pole gauge with reductions. The above suggests that maybe we can use g in order to find the gauge-equivalent GZS system in canonical gauge. So let us make a gauge transformation $L' = g^{-1}Lg$. Then the GMV system is transformed into a system of the type

$$L'\psi' = (\mathrm{i}\partial_x + q' + \lambda J_1)\psi' = 0,$$

where $q' = \mathrm{i}g^{-1}g_x$, or after expressing g with u and v,

$$q' = \frac{\mathrm{i}}{2} \begin{pmatrix} uu_x^* + vv_x^* & \sqrt{2}(uv_x - vu_x) & uu_x^* + v^*v_x \\ \sqrt{2}(v^*u_x^* - u^*v_x^*) & 2(v^*v_x + uu_x^*) & \sqrt{2}(v^*u_x^* - u^*v_x^*) \\ uu_x^* + vv_x^* & \sqrt{2}(uv_x - vu_x) & uu_x^* + vv_x^* \end{pmatrix}.$$

One can see that the condition $|u|^2 + |v|^2 = 1$ ensures that $q'^{\dagger} = q'$ and that $\operatorname{tr} q' = 0$ but q' is not off-diagonal as GZS requires. We must address that but let us make first an important observation. q' has a symmetry that is the counterpart to the symmetry defined by the automorphism $\mathcal{H}$. Indeed, a simple calculation shows that

$$g^{-1}Hg = K = -\begin{pmatrix} 0 & 0 & 1 \\ 0 & 1 & 0 \\ 1 & 0 & 0 \end{pmatrix} \tag{7}$$

and as a consequence

$$\mathrm{Ad}(g^{-1}) \circ \mathcal{H} = \mathcal{K} \circ \mathrm{Ad}(g^{-1}),$$

where $\mathcal{K}$ is the involutive automorphism $X \mapsto KXK$ (of course $K^{-1} = K$). One checks that

$$\mathcal{K}q' = q', \quad \mathcal{K}J_1 = -J_1, \quad \mathcal{K}J_2 = J_2,$$

where

$$J_2 = J_1^2 - \frac{1}{3}\operatorname{tr}(J_1^2)\mathbf{1} = \frac{1}{3}\operatorname{diag}(1, -2, 1).$$

Now let us consider the diagonal part of q'. We see that it is equal to $\frac{3\mathrm{i}}{2}(uu_x^* + vv_x^*)J_2$ and the function $\frac{3\mathrm{i}}{2}(uu_x^* + vv_x^*)$ is real, let us denote it by $a(x)$. We shall make now corrections to g in order to obtain a gauge transformation which will give a potential having the same symmetries as q' but having all the entries on its diagonal equal to zero. Let us try to find $\tilde{g} = g(x)(\exp \mathrm{i}\xi(x)J_2)$, where $\xi(x)$ a real function, in such a way that the potential constructed through $\tilde{g}$ will have the symmetries of q' but will be with zero diagonal elements. First we see that $\tilde{g}^{-1}S\tilde{g} = J_1$ since J_1 and J_2 are both diagonal, hence commute. Next, the new potential $q = \mathrm{i}\tilde{g}^{-1}\tilde{g}_x$ equals

$$q = \exp(-\mathrm{i}\xi J_2)q' \exp(\mathrm{i}\xi J_2) - \xi_x J_2.$$

The diagonal part of the first term in the right-hand side is clearly the same as that of q' so it will suffice to take $\xi(x)$ satisfying $\xi_x = a(x)$ in order to ensure that q will have diagonal elements equal to zero. This of course can be done, one can set for example

$$\xi(x) = \int_{-\infty}^{x} a(y)\mathrm{d}y.$$

Now the important thing is that q also satisfies $\mathcal{K}q = q$. This follows from the fact that since $\mathcal{K}J_2 = J_2$ then K commutes with J_2. Also, since ξ is real we have $q^{\dagger} = q$. Finally, from our construction follows that $q(x) \to 0$ when $x \to \pm\infty$. How fast $q(x)$ goes to zero depends of course how fast $u(x)$ and $v(x)$ go to their constant values. Thus we obtained that our initial linear problem should be gauge-equivalent to GZS type linear problem

$$L_0\psi = (\mathrm{i}\partial_x + q + \lambda J_1)\psi = 0, \tag{8}$$

where we have the symmetry requirements

$$q^{\dagger} = q, \quad J_1^{\dagger} = J_1, \quad \mathcal{K}q = q, \quad \mathcal{K}J_1 = -J_1. \tag{9}$$

It is readily checked that if q satisfies the restrictions (9) then it must have the form

$$q(x) = \begin{pmatrix} 0 & w(x) & \beta(x) \\ w^*(x) & 0 & w^*(x) \\ \beta(x) & w(x) & 0 \end{pmatrix},$$

where $\beta(x)$ is real function and $w(x)$ is complex function. Thus q is parametrized by 3 real scalar functions (as is $S(x)$). The linear problem (8) is naturally a GZS type linear problem with $\mathbb{Z}_2 \times \mathbb{Z}_2$ reduction given by the elements h_1, h_2:

$$h_1(\psi)(x,\lambda) = \left[\psi(x,\lambda^*)^\dagger\right]^{-1}, \tag{10}$$
$$h_2(\psi)(x,\lambda) = \mathcal{K}\psi(x,-\lambda) = K\psi(x,-\lambda)K. \tag{11}$$

It is instructive to show how one can obtain from L_0 the GMV problem $\tilde{L}_0$. Let ψ_0 be the Jost solution of L_0 for $\lambda = 0$, that is

$$\mathrm{i}\partial_x\psi_0 + q\psi_0 = 0, \qquad \lim_{x\to-\infty}\psi_0 = \mathbf{1}.$$

Since in the future using upper indexes will become complicated we shall use hat to denote the inverse element. With this notation we have that $Kq\hat{K} = q$ immediately gives $K\psi_0\hat{K} = \psi_0$, one also sees that ψ_0 is unitary matrix. Let us take a constant orthogonal matrix R such that $H = \hat{R}KR$. (A glance at (7) shows that such R exists). Then let us take the function $\psi^H = \psi_0 R$. One readily checks that $\psi^H H = K\psi^H$. Of course, ψ^H satisfies the same equation as ψ_0 but have different asymptotic when $x \to -\infty$. It is also a unitary matrix. If we perform a gauge transformation $L \mapsto \hat{\psi}^H L \psi^H$ we shall get the linear operator

$$\mathrm{i}\partial_x + \lambda S, \quad S = \hat{\psi}^H J_1 \psi^H. \tag{12}$$

Then obviously $S^\dagger = S$ and as it can be checked also $HS\hat{H} = -S$ because $KJ_1\hat{K} = -J_1$. Thus, L_0 defined in (8) subject to reductions (10) is gauge-equivalent to the GMV system.

Of course, there remains the question about the asymptotic of ψ_H that ensure the proper asymptotic of S in the GMV problem.

Remark 3.1 *The same issue arises in the gauge-equivalence of the Nonlinear Schrödinger equation and the Heisenberg Ferromagnet equation, [2], we remind that in order to ensure that* $\lim_{x\to\pm\infty} S = \sigma_3$ *one needs to assume that* $T(0)\sigma_3\hat{T}(0) = \sigma_3$, *where* $T(0)$ *is the transition matrix for the classical Zakharov–Shabat system.*

Let us define $g_\pm = \lim_{x\to\infty} g(u(x), v(x))$. When $x \to -\infty$ we have that S in (12) tends to $\hat{R}J_1R$ so one simply can choose $\hat{R} = g_-$ to ensure the proper limit. When $x \to +\infty$ in order to have the proper limit, just as in the $\mathfrak{sl}(2)$ case, one needs to impose the restriction:

$$\hat{R}T(0)R = g_+, \quad \text{or equivalently,} \quad g_-T(0)\hat{g}_- = g_+.$$

Now assume that χ is a fundamental analytic solution (FAS) of $L_0\psi = 0$ such that $\chi \to \exp \mathrm{i}J_1x$ when $x \to -\infty$. Then $\tilde{\chi} = \hat{\psi}^H\chi$ is a fundamental solution of $L\psi = 0$ and when $x \to -\infty$ one has $\tilde{\chi} \to g_- \exp \mathrm{i}J_1x$. Thus knowing FAS for one of the systems one can find the FAS for the other. From the other hand the FAS are essential for the theory of the Recursion Operators so between the theories in canonical and in pole gauge there is a close relation, a fact reported as gauge-covariance of the theory.

We shall give a brief sketch about this in the next sections and shall outline what happens with the theory when we have reductions. We shall omit most of the detail, for them see [8] for the general case and [15] for the system with $\mathbb{Z}_p$ reductions. From the other hand we shall consider the case of a general $\mathbb{Z}_p$ reduction of the GZS systems on simple Lie algebras or more precisely Caudrey–Beals–Coifman (CBC) system, [16] since the element J in front of λ is complex, because for GZS on $\mathfrak{sl}(3)$ there are some specifics that make the things not so transparent as we would wish. The full description of the theory for the GMV system and its gauge-equivalent will be given in a separate article.

Now we are going to give some examples of gauge-equivalent equations. As it is known for the GZS (CBC) system the evolution defined by the NLEEs is linearized in terms of the transition matrix $T(\lambda)$ and has the form $\mathrm{i}\dot{T} = [F(\lambda), T]$, where $F(\lambda)$ is a polynomial in λ with values in the Cartan subalgebra $\mathfrak{h}$ of the diagonal matrices called dispersion. The equations corresponding to linear dependence on λ are linear, quadratic dependence (no linear term) corresponds to Schrödinger type equations in canonical gauge and Heisengerg Ferromagnet type equation for the pole gauge etc.

The first two equations in the hierarchy related to GMV system correspond to the choices of the dispersion

(a) $F(\lambda) = \lambda J_1$,
(b) $F(\lambda) = -\lambda^2 J_2 + c\lambda J_1$ where $c = \text{const}$.

The system corresponding to the case (a) is

$$u_t - u_x = 0, \quad v_x - v_t = 0.$$

The system corresponding to the case (b) is (see [10])

$$\begin{aligned} &\mathrm{i}u_t + u_{xx} + (uu_x + vv_x^* - \mathrm{i}c)u_x + (uu_x^* + vv_x^*)u = 0, \\ &\mathrm{i}v_t + v_{xx} + (uu_x + vv_x^* - \mathrm{i}c)v_x + (uu_x^* + vv_x^*)v = 0. \end{aligned}$$

In particular, the analog of the HF-type equation is the system obtained for $c = 0$

$$\begin{aligned} &\mathrm{i}u_t + u_{xx} + (uu_x + vv_x^*)u_x + (uu_x^* + vv_x^*)u = 0, \\ &\mathrm{i}v_t + v_{xx} + (uu_x + vv_x^*)v_x + (uu_x^* + vv_x^*)v = 0. \end{aligned} \tag{13}$$

Now consider the gauge-equivalent hierarchy related to GZS system with reductions. As usual, the system corresponding to the case (a) is linear, it is

$$w_x - w_t = 0, \quad \beta_x - \beta_t = 0.$$

The system corresponding to the second choice of the dispersion law is

$$\begin{aligned} &\mathrm{i}w_t + w_{xx} - \mathrm{i}cw_x + (\beta^2 + \mathrm{i}\beta_x + 2|w|^2)w = 0, \\ &\beta_t + (2|w|^2 - c\beta)_x = 0, \end{aligned}$$

which is the gauge-equivalent to the second equation in the GMV hierarchy. In particular, for $c = 0$ we get the gauge-equivalence of (13) and

$$\begin{aligned} &\mathrm{i}w_t + w_{xx} + (\beta^2 + \mathrm{i}\beta_x + 2|w|^2)w = 0, \\ &\beta_t + 2(|w|^2)_x = 0, \end{aligned}$$

which is the NLS type equation gauge-equivalent to the HF-type equation (13). It is a Gross–Pitaevskii type of NLS for w with potential defined by another field $\beta(x)$.

4. Completeness relations in canonical gauge and pole gauge for the CBC system
Consider the CBC system $L\psi = (\mathrm{i}\partial_x - \lambda J + q)\psi = 0$ on some simple Lie algebra $\mathfrak{g}$ in a typical representation. As mentioned, J is such that $\mathfrak{h} = \ker \mathrm{ad}_J$ defines a Cartan subalgebra (we shall call it 'the Cartan subalgebra') and $q(x)$ takes values in the orthogonal complement $\mathfrak{h}^{\perp} = \bar{\mathfrak{g}}$ with respect to the Killing form. Let Δ be the root system of $\mathfrak{g}$ (defined by $\mathfrak{h}$) and let $\Sigma = \bigcup_{\alpha\in\Delta} \zeta_\alpha$ be the bunch of straight lines ζ_α where $\zeta_\alpha = \{\lambda : \mathrm{Im}(\lambda\alpha(J)) = 0\}$. The connected components of the set $\mathbb{C}\setminus\Sigma$ are open sectors in the λ-plain. We shall denote these sectors by Ω_ν ordering them anti-clockwise. Clearly ν takes values from 1 to some even number $2M$. Then the boundary of the sector Ω_ν consists of two rays: l_ν and $l_{\nu+1}$ (l_ν comes before $l_{\nu+1}$ when we turn anti-clockwise) so that $\overline{\Omega}_\nu \cap \overline{\Omega}_{\nu-1} = l_\nu$. (The 'bar' here denotes the closure). We shall understand the number ν modulo $2M$. Naturally,

$$\mathbb{C}\setminus\Sigma = \bigcup_{\nu=1}^{2M} \Omega_\nu, \quad \Omega_\nu \bigcap \Omega_\mu = \emptyset, \quad \nu \neq \mu.$$

If $\alpha, \beta \in \Delta$, $\alpha \neq \beta$ and ν is fixed, then for $\lambda \in \Omega_\nu$ the value $\mathrm{Im}[\lambda(\alpha-\beta)(J)]$ does not change sign. This permits to define ordering of the roots relative to the sector Ω_ν (ν-ordering), we write $\alpha >_\nu \beta$ iff $\mathrm{Im}\,\lambda(\alpha-\beta)(J) > 0$. Consequently, in each Ω_ν we have splitting into positive and negative roots $\Delta = \Delta_\nu^+ \cup \Delta_\nu^-$.

Limiting ourselves to the typical representation of $\mathfrak{g}$, for a large class of potentials $q(x)$ it can be shown that in each of the sectors Ω_ν there exists fundamental solution (FAS) $\chi_\nu(x,\lambda)$ with the properties:

(a) $\chi_\nu(x,\lambda)$ is meromorphic in Ω_ν and has only finite number of poles. The poles define the discrete spectrum of the problem and for simplicity below we shall assume that $\chi_\nu(x,\lambda)$ is analytic in Ω_ν.
(b) $\chi_\nu(x,\lambda)$ allows extension by continuity to the boundary of Ω_ν (to the rays l_ν and $l_{\nu+1}$).
(c) For $\lambda \in \Omega_\nu$ the function $\chi_\nu(x,\lambda)e^{i\lambda Jx}$ is bounded and we have $\lim_{x\to-\infty} \chi_\nu(x,\lambda)e^{i\lambda Jx} = \mathbf{1}$ and $\lim_{\lambda\to+\infty} \chi_\nu(x,\lambda)e^{i\lambda Jx} = \mathbf{1}$.

Below we shall have a situation when some set of functions $f_\nu(\lambda)$ is such that each $f_\nu(\lambda)$ is analytic in Ω_ν and allows extension to the boundary of this sector $l_\nu \cup l_{\nu+1}$. Then say on l_ν we have the extensions from the left and from the right. We shall denote the extension from the left by $f_\nu^+(\lambda)$ and from the right by $f_\nu^-(\lambda)$ (of course $\lambda \in l_\nu$). Thus for example on each l_ν we have the solutions $\chi_\nu^{\pm}(\lambda, x)$ of the CBC problem.

The fundamental solutions $\chi_\nu(x,\lambda)$ permit to construct the resolvent of the operator $\Psi \mapsto [L,\Psi]$ which naturally must be strongly related to the NLEEs having Lax representation $[L,A] = 0$. This is indeed, so, the resolvent kernel of the above operator gives rise to completeness relations. Let π_0 be the orthogonal projector on $\bar{\mathfrak{g}}$, let E_α, $\alpha\in\Delta$ are the root vectors and let us define the following functions, called Generalized Exponents or adjoint solutions.

$$e_\alpha^\nu(x,\lambda) = \pi_0(\chi_\nu(x,\lambda)E_\alpha\chi_\nu^{-1}(x,\lambda)), \quad \lambda \in \bar{\Omega}_\nu.$$

As we agreed, for $\lambda \in l_\nu$ we shall write $e_\alpha^{(+;\nu)}(x,\lambda)$ if the solution is extended from the sector Ω_ν and $e_\alpha^{(-;\nu)}(x,\lambda)$ if the solution is extended from the sector $\Omega_{\nu-1}$. In other words, for $\lambda \in l_\nu$

$$e_\alpha^{(\nu;-)}(x,\lambda) = \pi_0(\chi_{\nu-1}(x,\lambda)E_\alpha\chi_{\nu-1}^{-1}(x,\lambda)),$$
$$e_\alpha^{(\nu;+)}(x,\lambda) = \pi_0(\chi_\nu(x,\lambda)E_\alpha\chi_\nu^{-1}(x,\lambda)).$$

Journal of Physics: Conference Series **621** (2015) 012017 doi:10.1088/1742-6596/621/1/012017

Taking all this into account, the completeness relation (we remind that there is no discrete spectrum) can be cast into the following form:

$$\Pi_0\, \delta(x-y) = \frac{1}{2\pi}\sum_{\nu=1}^{M}\int_{l_\nu}\Big\{\sum_{\alpha\in\delta_\nu^+}\big(e_\alpha^{(+;\nu)}(x)\otimes e_{-\alpha}^{(+;\nu)}(y) - e_{-\alpha}^{(-;\nu)}(x)\otimes e_\alpha^{(-;\nu)}(y)\big)\Big\}\mathrm{d}\lambda,$$

where

$$\Pi_0 = \sum_{\gamma\in\Delta}\frac{E_\gamma\otimes E_{-\gamma}}{\gamma(J)},\quad \delta_\nu^\pm = \Delta_\nu^\pm\cap\delta_\nu,\quad \delta_\nu = \{\alpha\in\Delta\colon\ \mathrm{Im}(\lambda\alpha(J)) = 0 \text{ for } \lambda\in l_\nu\}.$$

In the above we also assumed that the rays are oriented from 0 to ∞ and for shortness we have omitted the dependence on λ. The formula itself must be understood in the following way. First, it is assumed that $\mathfrak{g}^*$ is identified with $\mathfrak{g}$, the pairing between them being given by the Killing form. So for example, for $X, Y, Z\in\mathfrak{g}$ making a contraction of $X\otimes Y$ with Z on the right we obtain $X\langle Y, Z\rangle$ and making contraction from the left we get $\langle Z, X\rangle Y$. Next, the formula for Π_0 implies that making a contraction with Π_0 from the right we get $\Pi_0 X = \mathrm{ad}_J^{-1}\pi_0 X$ and similarly from the left $X\Pi_0 = -\mathrm{ad}_J^{-1}\pi_0 X$. (On the space $\bar{\mathfrak{g}}$ the operator ad_J is invertible).

Suppose that we have a L^1-integrable function $g:\mathbb{R}\mapsto\bar{\mathfrak{g}}$. Then we have the expansions (for $\epsilon = +1$ and $\epsilon = -1$ respectively)

$$g(x) = \frac{1}{2\pi}\sum_{\nu=1}^{2M}\int_{l_\nu}\Big\{\sum_{\alpha\in\delta_\nu^+}\big(e_{\epsilon\alpha}^{(+;\nu)}(x)\langle\langle e_{-\epsilon\alpha}^{(+;\nu)}, [J, g]\rangle\rangle - e_{-\epsilon\alpha}^{(-;\nu)}(x)\langle\langle e_{\epsilon\alpha}^{(-;\nu)}, [J, g]\rangle\rangle\big)\Big\}\mathrm{d}\lambda. \tag{14}$$

In the above we used the following notation: for two functions $f_1(x)$, $f_2(x)$ with values in $\mathfrak{g}$ we put

$$\langle\langle f_1, f_2\rangle\rangle = \int_{-\infty}^{+\infty}\langle f_1(x), f_2(x)\rangle\mathrm{d}x.$$

It can be shown that for either of the choices for ϵ the expansion (14) converges in the same sense as the Fourier expansions for $g(x)$. Finally, if one introduces the operators

$$\Lambda_\pm(X(x)) = \mathrm{ad}_J^{-1}\left(\mathrm{i}\partial_x X + \pi_0[q, X] + \mathrm{i}\,\mathrm{ad}_q\int_{\pm\infty}^{x}(\mathrm{id}-\pi_0)[q(y), X(y)]\mathrm{d}y\right).$$

One can see that

$$(\Lambda_- - \lambda)e_\alpha^{(+;\nu)} = 0,\qquad (\Lambda_- - \lambda)e_{-\alpha}^{(-;\nu)} = 0,\quad \alpha\in\delta_\nu^+,$$

$$(\Lambda_+ - \lambda)e_{-\alpha}^{(+;\nu)} = 0,\qquad (\Lambda_+ - \lambda)e_{\alpha}^{(-;\nu)} = 0,\quad \alpha\in\delta_\nu^+.$$

The operators $\Lambda_\pm$ are the famous Generating Operators, for the expansions (14) they play the role that $\mathrm{i}\partial_x$ plays for the Fourier expansion. The importance of the above expansions for the theory of NLEEs (they appeared first in the seminal work of [17] for the ZS system and later were generalized) is based on the fact that when one expands over the Generalized Exponents the potential $q(x)$ and $\delta q(x)$ one gets as coefficients a set of minimal scattering data and its variation. A crucial fact from the theory is that the coefficients in these expansions have linear evolution for the NLEEs related to L. Through the operators $\Lambda_\pm$ could be obtained the NLEEs, their conservation laws, the hierarchy of symplectic structures, etc, see [7].

The system, gauge-equivalent to the CBC system L, is defined as $\tilde{L}\tilde{\psi} = (\mathrm{i}\partial_x - \lambda S)\tilde{\psi}$ where $S(x)$ takes values in the orbit $\mathcal{O}_J$ of the element J in the adjoint representation of the algebra $\mathfrak{g}$. One must write here also the limits of $S(x)$ when $x \to \pm\infty$ but we do not want to specify them now. The system on $\mathcal{O}_J$ could be considered in its own right but in case there is a gauge transformation from L to $\tilde{L} = \hat{\psi}_0 L \psi_0$ with ψ_0 which is a solution of CBC for $\lambda = 0$ and such that it gives the proper values of $\lim_{x\to\pm\infty} S(x)$ the expansions for $\tilde{L}$ could be obtained immediately from those we know for L using a transformation $\mathrm{Ad}(\hat{\psi}_0(x)) \otimes \mathrm{Ad}(\hat{\psi}_0(y))$. Indeed, in this way we obtain

$$\tilde{\Pi}_0\, \delta(x-y) = \frac{1}{2\pi}\sum_{\nu=1}^{2M}\int_{l_\nu} \mathrm{d}\lambda \Big\{ \sum_{\alpha\in\delta_\nu^+} \Big(\tilde{e}_\alpha^{(+;\nu)}(x) \otimes \tilde{e}_{-\alpha}^{(+;\nu)}(y) - \tilde{e}_{-\alpha}^{(-;\nu)}(x) \otimes \tilde{e}_\alpha^{(-;\nu)}(y) \Big) \Big\},$$

where

$$\begin{aligned} \tilde{\Pi}_0 &= \mathrm{Ad}(\hat{\psi}_0(x)) \otimes \mathrm{Ad}(\hat{\psi}_0(y))\, \Pi_0, \\ \tilde{e}_\alpha^\nu(x) &= \mathrm{Ad}(\hat{\psi}_0) e_\alpha^\nu(x) = \tilde{\pi}_0(\tilde{\chi}_\nu E_\alpha \hat{\tilde{\chi}}_\nu), \\ \tilde{\pi}_0 &= \mathrm{Ad}(\hat{\psi}_0(x))\pi_0 \mathrm{Ad}(\psi_0(x)), \end{aligned}$$

and the functions $\tilde{g}(x)$ are such that for each x we have $\tilde{g}(x) \in \tilde{\mathfrak{h}}^\perp(x)$ where $\tilde{\mathfrak{h}} = \ker \mathrm{ad}_S$. The Generating Operators for $\tilde{L}$ are

$$\tilde{\Lambda}_\pm = \mathrm{Ad}(\hat{\psi}_0(x))\Lambda_\pm \mathrm{Ad}(\psi_0(x)).$$

Of course, there is the question how to calculate $\tilde{\Lambda}_\pm$ in terms of $S(x)$ and its derivatives. This is not very easy, but there is a clear procedure how to do that. The above theory is referred as a gauge-covariant theory of the Recursion Operators since it permits to reformulate immediately results from canonical to pole gauge [3].

The existence of reductions complicates the picture. Let us first see what happens in canonical gauge if there are $\mathbb{Z}_p$ reductions.

5. Completeness relations in the presence of reductions defined by automorphisms

Assume that for the CBC system on the algebra $\mathfrak{g}$ we have a reduction group G_M generated by one element h acting as

$$h(\psi(x,\lambda)) = \mathcal{K}(\psi(x,\omega^{-1}\lambda)), \quad \omega = \exp\frac{2\pi\mathrm{i}}{p},$$

where $\mathcal{K}$ is automorphism of order p of the Lie group G corresponding to the algebra $\mathfrak{g}$. It generates an automorphism of $\mathfrak{g}$ which we shall denote by the same letter $\mathcal{K}$. Since this immediately leads to $\mathcal{K}J = \omega J$ and $\mathcal{K}q = q$ the automorphism $\mathcal{K}$ preserves the Cartan subalgebra. Next, since $h^p = \mathrm{id}$, G_M is isomorphic to $\mathbb{Z}_p$. The automorphism $\mathcal{K}$ acts also on the set of roots, that action we shall denote by the same letter again, and then for the root vectors we have $\mathcal{K}(E_\alpha) = \zeta(\alpha)E_{\mathcal{K}\alpha}$, where $\zeta(\alpha)$ are numbers, such that $\zeta(\alpha)\zeta(-\alpha) = 1$ and $\zeta(\alpha)\zeta(\beta) = \zeta(\alpha+\beta)$ if $\alpha+\beta \in \Delta$ (see [18] for the details).

Since the automorphisms of $\mathfrak{g}$ leave the Killing form invariant, the algebra $\mathfrak{g}$ splits into a direct sum of eigenspaces of $\mathcal{K}$ (upper indexes are understood modulo p), that is:

$$\mathfrak{g} = \oplus_{s=0}^{p-1}\mathfrak{g}^{[s]}, \quad [\mathfrak{g}^{[k]}, \mathfrak{g}^{[l]}] \subset \mathfrak{g}^{[k+l]}.$$

For each $X \in \mathfrak{g}^{[s]}$ we have $\mathcal{K}X = \omega^s X$ and the spaces $\mathfrak{g}^{[s]}$, $\mathfrak{g}^{[k]}$ are orthogonal with respect to the Killing form if $k+s \neq 0 \pmod p$. Because $\mathcal{K}$ is an automorphism of $\mathfrak{g}$ leaving $\mathfrak{h}$ invariant, it leaves invariant also the orthogonal complement $\bar{\mathfrak{g}} = \mathfrak{h}^{\perp}$ of $\mathfrak{h}$. Thus each $\mathfrak{g}^{[s]}$ splits into $\bar{\mathfrak{g}}^{[s]} \oplus \mathfrak{h}^{[s]}$ and

$$\bar{\mathfrak{g}} = \oplus_{s=0}^{p-1} \bar{\mathfrak{g}}^{[s]}, \qquad \mathfrak{h} = \oplus_{s=0}^{p-1} \mathfrak{h}^{[s]}.$$

The spaces $\bar{\mathfrak{g}}^{[k]}$ and $\mathfrak{h}^{[s]}$ are orthogonal with respect to the Killing form for arbitrary k and s.

The invariance of the set of the fundamental solutions can be additionally specified if we take the FAS $\chi_\nu(x,\lambda)$ defined in Ω_ν, $\nu = 1,2,\dots,2M$. Then one easily sees that $\mathcal{K}(\chi_\nu(x,\lambda))$ must be analytic solution in the sector containing $\omega\lambda$, one has an action of $\mathcal{K}$ (multiplication by ω on the sectors Ω_ν and on the rays l_ν. One can see that if $\omega\Omega_\nu = \Omega_{\nu+a}$ then $\omega l_\nu = l_{\nu+a}$, $\delta^{\pm}_{\nu+a} = \delta^{\pm}_{\nu}$. Using that one can cast the completeness relations into the form:

$$\begin{aligned}\Pi_0\, \delta(x-y) = \frac{1}{2\pi p} \sum_{\nu=1}^{2M} \sum_{k=1}^{p} \int_{l_\nu} \Bigg\{ \Bigg[\sum_{\alpha\in\delta_\nu^+} \omega^k \mathcal{K}^k \otimes \mathcal{K}^k (e_\alpha^{(+;\nu)}(x) \otimes e_{-\alpha}^{(+;\nu)}(y)) \\ -\omega^k \mathcal{K}^k \otimes \mathcal{K}^k (e_{-\alpha}^{(-;\nu)}(x) \otimes e_{\alpha}^{(-;\nu)}(y)) \Bigg] \Bigg\} d\lambda,\end{aligned}$$

where

$$(\mathcal{K}\otimes\mathcal{K})(X\otimes Y) = \mathcal{K}(X)\otimes\mathcal{K}(Y).$$

Note that the numbers $\zeta(\alpha)$ don't appear any more, this occurs because we apply $\mathcal{K}$ always on products of the type $E_\alpha \otimes E_{-\alpha}$. The expansions of a function $g(x)$ over the adjoint solutions can be simplified further, if for arbitrary x the value $g(x) \in \mathfrak{g}^{[s]}$. In that case

$$\begin{aligned}g(x) = \frac{\epsilon}{2\pi p} \sum_{\nu=1}^{2M} \int_{l_\nu} \Bigg\{ \sum_{\alpha\in\delta_\nu^+} \Bigg[\sum_{k=1}^{p} \omega^{-ks} \mathcal{K}^k (e_{\epsilon\alpha}^{(+;\nu)}(x,\lambda)) \langle\langle (e_{-\epsilon\alpha}^{(+;\nu)}, [J,h] \rangle\rangle \\ - \sum_{k=1}^{p} \omega^{-ks} \mathcal{K}^k (e_{-\epsilon\alpha}^{(-;\nu)}(x,\lambda)) \langle\langle (e_{\epsilon\alpha}^{(-;\nu)}, [J,g] \rangle\rangle \Bigg] \Bigg\} d\lambda.\end{aligned}$$

In the above are written two expansions, one for $\epsilon = +1$ and the other for $\epsilon = -1$. Making a contraction one can see that $g(x)$ is actually expanded over the functions:

$$e_\alpha^{(\pm;\nu;s)}(x,\lambda) = \sum_{k=1}^{p} \omega^{-ks} \mathcal{K}^k (e_\alpha^{(\pm;\nu)}(x,\lambda)), \quad \nu = 1,2,\dots,a,$$

which are up to a multiplier p the orthogonal projections of $e_\alpha^{(\pm;\nu)}(x,\lambda)$ on $\bar{\mathfrak{g}}^{[s]}$.

The operators $\Lambda_\pm$ map functions with values in $\bar{\mathfrak{g}}^{[s]}$ into functions with values in $\bar{\mathfrak{g}}^{[s-1]}$. In particular, for the new Generalized Exponents we have

$$\begin{aligned}\Lambda_- e_\alpha^{(+;\nu;s)} &= \lambda e_\alpha^{(+;\nu,s-1)}, & \Lambda_- e_{-\alpha}^{(-;\nu,s)} &= \lambda e_{-\alpha}^{(-;\nu,s-1)}, & \alpha\in\delta_\nu^+,\\ \Lambda_+ e_{-\alpha}^{(+;\nu,s)} &= \lambda e_{-\alpha}^{(+;\nu,s-1)}, & \Lambda_- e_{\alpha}^{(-;\nu,s)} &= \lambda e_{\alpha}^{(-;\nu,s-1)}, & \alpha\in\delta_\nu^+.\end{aligned}$$

Therefore $e_\alpha^{(\pm;\nu,s)}$ are not eigenfunctions of $\Lambda_\pm$. However, we obviously have:

$$\begin{aligned}\Lambda_-^p e_\alpha^{(+;\nu;s)} &= \lambda^p e_\alpha^{(+;\nu,s)}, & \Lambda_-^p e_{-\alpha}^{(-;\nu,s)} &= \lambda^p e_{-\alpha}^{(-;\nu,s)}, & \alpha\in\delta_\nu^+,\\ \Lambda_+^p e_{-\alpha}^{(+;\nu,s)} &= \lambda^p e_{-\alpha}^{(+;\nu,s)}, & \Lambda_+^p e_{\alpha}^{(-;\nu,s)} &= \lambda^p e_{\alpha}^{(-;\nu,s)}, & \alpha\in\delta_\nu^+.\end{aligned}$$

This is important, so let us formulate it separately:

Theorem 5.1 *For the expansions* (15) *the role of the Recursion Operators are played by the p-th powers of the operators* $\Lambda_\pm$.

All this can be reformulated for the gauge-equivalent system, one has in fact all these relations but with tilde and the role of $\mathcal{K}$ is played by $\mathcal{H} = \mathrm{Ad}(\hat{\psi}_0)\mathcal{K}\,\mathrm{Ad}(\psi_0)$. The new Generating Operators will be $\tilde{\Lambda}^p_\pm$ since the operators $\tilde{\Lambda}_\pm$ map functions with values in $\tilde{\bar{\mathfrak{g}}}^{[s]}$ into functions with values in $\tilde{\bar{\mathfrak{g}}}^{[s-1]}$, where $\tilde{\bar{\mathfrak{g}}}^{[s]} = \mathrm{Ad}(\hat{\psi}_0)\bar{\mathfrak{g}}^{[s]}$.

All the above is applicable to the GMV system. Here $p = 2$, $\omega = -1$, we have two sectors – the upper and the lower half-plane, the two rays are the the positive and the negative semi-axes. For that reason the integrals over them are written as integral on the whole axis, etc. We have the automorphism given by H which is treated according to the general theory above. Of course must remember that there is one more $\mathbb{Z}_2$ reduction, given by g_1, see (6). However, its effect is reduced to the requirement that the coefficients belong to $i\mathfrak{su}(3)$, see (6). In fact our pairs can be written on the real form $\mathfrak{su}(3)$, that is 'algebraically' nothing serious happens.

Let us end with some comments. As one sees, the automorphism $\mathcal{H}$ depends ψ_0. Therefore, in principle it is x-dependent. The situation that we had for the GMV system and its gauge-equivalent when $\mathcal{H}$ is constant is very pleasant, since it permits to parametrize the potential S in a very nice way. This may seem quite unique, but it is not so. In fact, returning back to Section 3 we see that when $\mathcal{K}$ is inner automorphism given by $X \mapsto KX\hat{K}$ we can make a gauge transformation to a system having constant H. Finally, we want to underline that the Mikhailov-type reductions have the nice property that they are not destroyed by the gauge transformations and the theory still remains gauge-covariant.

6. Conclusion

In this article we considered the problem of finding gauge-equivalent hierarchies of soliton equations related to Generalized Zakharov–Shabat system (GZS system) on $\mathfrak{sl}(3)$ in canonical and in pole gauge when reductions are present. This question is of theoretical importance for the soliton equations (one could recall the impact caused by the discovery that the Nonlinear Schrödinger equation and the Heisenberg ferromagnet (HF) equation are gauge-equivalent). In particular, we found the gauge-equivalent to the system introduced recently by Gerdjikov, Mikhailov and Valchev (GMV system) and we have found the gauge-equivalent counterpart of the first two equations in the hierarchy related to the GMV system. It will be interesting to find some other low rank examples of CBC systems with reductions in pole and in canonical gauge thus increasing the pool of gauge-equivalent NLEEs.

There are hopes that the theory of the systems in general position which is put already in gauge-covariant form will be generalized when there are reductions of $\mathbb{Z}_p$ type and we made some steps in that direction. It should be underlined that since the Generating Operator theory for the GMV system has been developed independently (and specially for it), see [10, 11], it is still not in the gauge-covariant form we sketched in the previous sections and this is a task still to be done.

There is one more aspect about our results we would like to mention. Though one can suggest that the HF-type equation related tom GMV system should have similar physical applications as HF equation (of course, related to models of interacting spin 1 particles rather than spin 1/2 particles as HF) we are not aware of such application yet. From the other side, the gauge-equivalent to this system turned out to be of the Gross–Pitaevskii type and there are strong hopes that it has physical applications since Gross–Pitaevskii type equations are used in number of models describing Bose–Einstein condensate.

Acknowledgments

The author is grateful to NRF South Africa incentive grant 2014 for the financial support.

References

[1] Zakharov V and Mikhailov A 1978 Relativistically invariant two-dimensional models of field theory which are integrable by means of the inverse scattering problem method *Soviet Phys. JETP* **74**:6 (6) 1953

[2] Zakharov V and Takhtadzhyan L 1979 Equivalence of the nonlinear Schrödinger equation and the equation of a Heisenberg ferromagnet *Theoret. and Math. Phys.* **38**:1 17

[3] Gerdjikov V and Yanovski A 1986 Gauge-covariant theory of the generating operator. I *Comm. Math. Phys.* **103**:4 549

[4] Gerdjikov V and Yanovski A 1985 Gauge-covariant formulation of the generating operator. II. Systems on homogeneous spaces *Phys. Lett.* A **110**:2 53

[5] Gerdjikov V 1986 Generalized Fourier transforms for the soliton equations. Gauge-covariant formulation *Inverse Problems* **2**:1 51

[6] Yanovski A 1987 PhD thesis 5-87-222 *Gauge-Covariant Approach to the Theory of the Generating Operators for Soliton Equations* (Dubna: Joint Institute for Nuclear Research)

[7] Gerdjikov V, Vilasi G and Yanovski A 2008 *Integrable Hamiltonian Hierarchies. Spectral and Geometric Methods* (Berlin: Springer-Verlag)

[8] Gerdjikov V and Yanovski A 1994 Completeness of the eigenfunctions for the Caudrey–Beals–Coifman system *J. Math. Phys.* **35**:7 3687

[9] Yanovski A 1993 Generating operators for the generalized Zakharov–Shabat system and its gauge equivalent system in $\mathrm{sl}(3,\mathbb{C})$ case *Preprint* Universität Leipzig, Naturwissenchaftlich Theoretisches Zentrum Report N 20 http://cdsweb.cern.ch/record/256804/files/P00019754.pdf

[10] Gerdjikov V, Mikhailov A and Valchev T 2010 Reductions of integrable equations on **A.III**-symmetric spaces *J. Phys.* A **43** 434015

[11] Gerdjikov V, Mikhailov A and Valchev T 2010 Recursion operators and reductions of integrable equations on symmetric spaces *J. Geom. Symmetry Phys.* **20** 1

[12] Yanovski A 2011 On the recursion operators for the Gerdjikov, Mikhailov and Valchev system *J. Math. Phys.* **52**:8 082703

[13] Yanovski A 2012 Geometry of the recursion operators for Caudrey–Beals–Coifman system in the presence of Mikhailov $\mathbb{Z}_p$ reductions *J. Geom. Symmetry Phys.* **25** 77

[14] Mikhailov A 1981 The reduction problem and the inverse scattering method *Phys.* D **3** 73

[15] Yanovski A 2013 Recursion operators and expansions over adjoint solutions for the Caudrey–Beals–Coifman system with $\mathbb{Z}_p$ reductions of Mikhailov type *J. Geom. Symmetry Phys.* **30** 105

[16] Beals R and Coifman R 1984 Scattering and inverse scattering for first order systems *Comm. Pure Asppl. Math.* **37**:1 39

[17] Ablowitz M, Kaup D, Newell A and Segur H 1974 The inverse scattering method – Fourier analysis for nonlinear problems *Studies in Appl. Math.* **53**:4 249

[18] Goto M and Grosshans F 1978 *Semisimple Lie algebras* (Lecture Notes in Pure and Applied Mathematics vol 38) (New-York – Basel: Marcel Dekker, Inc.)

Solutions of the system of d'Alembert and eikonal equations, and classification of reductions of PDEs

Irina Yehorchenko

Institute of Mathematics of NAS of Ukraine, 3 Tereshchenkivs'ka Str., 01601 Kyiv, Ukraine

E-mail: iyegorch@imath.kiev.ua

Abstract. We present an approach to systematic description and classification of solutions of partial differential equations that are obtained by means of reduction of these equations to other equations with smaller number of independent variables. We propose to classify such reductions by means of classification of reduction conditions. The approach is illustrated by an example of the system of d'Alembert and eikonal equations. Solutions of this system were used to outline classification of reductions for the general nonlinear d'Alembert equation, with generalisation to arbitrary Poincaré invariant equations.

1. Introduction

In this paper we propose a new view to a problem that seemed to be solved long ago – how to classify solutions of partial differential equations (PDEs), obtained by means of reduction to new equations with smaller number of independent variables. We will discuss from this point some well-known methods, such as Lie algorithm of symmetry reduction and various non-classical approaches.

This paper continues development of ideas presented in the talks at the previous GADEIS conferences [1–3]. In the papers [1,2] we found and studied reduction conditions of the nonlinear d'Alembert equations and possible reduced equations. In [3] we considered classification of equations with respect to their possible reductions.

Here we consider a new idea – classification of reductions of PDEs to new PDEs with smaller number of independent variables or to ODEs. This idea stems from observation that many methods for searching of such reductions produce redundant results. A generally accepted method for classification of symmetry-induced reductions provides for using of subalgebras inequivalent up to conjugacy (see [4, 5]). First, it is necessary to find the invariance algebra of the equation under consideration. Subalgebras of Lie algebras can be classified with respect to conjugacy (see, e.g., [6] for classification of subalgebras of the Poincaré algebra). Then the original equation can be reduced with respect to each of the obtained inequivalent subalgebras, and solutions of the reduced equations are constructed. It is usually assumed that this method gives reductions that are inequivalent.

However, this method may produce the same reduction conditions and whence the same reductions for different non-conjugate subalgebras, and it does not work well for the general reductions of PDEs (then we have an arbitrary number of independent variables).

We propose to classify reductions of PDEs by means of symmetry classification of the reduction conditions. Thus, inequivalent solutions of the reduction conditions will provide inequivalent reductions. This way we may obtain fewer reductions than by means of the standard Lie procedure, cutting some reductions obtainable from subalgebras of the invariance algebra, and more reductions due to nonclassical reductions if they exist.

The only problem with such equivalence and further classification of reductions – it is very difficult to solve equations of the reduction conditions. However, in many interesting cases, e.g.. for the general nonlinear d'Alembert equation, it appears possible.

It is worth noting that the direct method presented by Clarkson and Kruskal [7] actually gives classification of reductions even if it gives no new solutions, though this particular advantage of their method was not pointed out by its authors.

Previously [8] we presented application of such direct reduction method by means of the ansatz

$$u = \exp\{if(t,\vec{x})\}\,\varphi(\omega)$$

to the nonlinear Schrödinger equation with an arbitrary function on $|u|$ in the nonlinear part

$$2iu_t + \triangle u - uF(|u|) = 0,$$

where u is a complex valued function, $u = u(t,\vec{x})$, is an n-dimensional vector of space variables $\vec{x} = (x_1, x_2, x_3)$, $|u| = \sqrt{uu^*}$, an asterisk designates complex conjugation,

$$\triangle u = \sum_{a=1}^{3} \frac{\partial^2 u}{\partial x_a^2}, \quad a = 1,2,3.$$

Though no new reductions were obtained as compared to the Lie algorithm, we may see that the listed reductions are inequivalent with respect to equivalence transformations of the reduction conditions.

In this paper we present an approach to classification of reductions of the nonlinear d'Alembert equation with arbitrary number of independent variables

$$u_{\mu\mu} = F(u). \tag{1}$$

Here and below the indices mean differentiation with respect to the relevant independent variables, $u_\mu = \frac{\partial u}{\partial x_\mu}$, $u_{\mu\nu} = \frac{\partial u}{\partial x_\mu \partial x_\nu}$, Greek indices run from 0 to n, Latin indices run from 1 to n and summation over repeated indices is assumed if nothing is indicated overwise. Here we have n space variables and $n+1$ independent variables.

We consider classification of reductions of the equation (1) to ODEs by means of the ansatz with one new independent variable

$$u = \varphi(v), \tag{2}$$

where v is a new variable.

Further we give some basic definitions.

2. What are reduction and classification of reductions

The equation

$$\Phi(x, u, \underset{1}{u}, \ldots, \underset{l}{u}) = 0, \tag{3}$$

where $\underset{k}{u}$ is the set of all kth-order partial derivatives of the function $u = (u^1, u^2, \ldots, u^m)$, is called reducible by means of an ansatz $u = \varphi(\omega)$, if substitution of this ansatz into the equation gives an equation on ω. Here $x = (x_1, x_2, \ldots, x_n)$, and in this section we present definitions for arbitrary equations with n independent and m dependent variables.

There could be various formats and generalisations of the above ansatz, but for presentation of the general idea this simplest form is sufficient. The resulting equations are called reduced equations and are used to find exact solutions. Ansatzes that reduce equations are connected with the sets of operators of conditional invariance, as explained in [9].

We have the equation (3) and k first-order operators in involution $\langle Q_1, Q_2, \ldots, Q_k \rangle$ (in the general case these operators do not form an algebra). These operators reduce the equation $\Phi = 0$, if the system

$$\Phi = 0, \quad Q_i u = 0 \tag{4}$$

is compatible. Technically, we take the general solution of the system $Q_i u = 0$ – the ansatz, and substitute it into the equation $\Phi = 0$, obtaining an equation with smaller number of independent variables.

The concept of "reduction" includes both the reduced equation and the set of new variables, as well as the ansatz. We can consider as classification of reductions either symmetry classification of the system (4) or classification of solutions of its compatibility conditions (reduction conditions).

3. Review of some methods for PDEs reduction

Let us list some symmetry-related methods used to find exact solutions of PDEs by means of their reduction to equations with smaller number of independent variables or to ordinary differential equations:

1. Symmetry reduction or Lie algorithm (for the detailed description of this method see, e.g., the books by Ovsyannikov [4] or Olver [5]). This approach is quite algorithmic and thus the most popular for the systematic search for families of exact solutions of PDEs.

2. The direct method (often giving wider classes of solutions than the symmetry reduction) was proposed by P. Clarkson and M. Kruskal [7]. This method for majority of equations results in considerable difficulties as it requires investigation of compatibility and solution of cumbersome reduction conditions of the initial equation. These reduction conditions are much more difficult for investigation and solution in the case of equations containing second and/or higher derivatives for all independent variables, and for multidimensional equations – e.g., in the situation of the nonlinear wave equations.

This approach is equivalent to finding of Q-conditional symmetries of PDEs (for the proof of this statement see [9]) and methods for finding of conditional and nonclassical symmetries (the same reductions may be found by both methods).

3. The direct method for finding of classical symmetry solutions. It is possible to find reductions using not arbitrary operators of Q-conditional symmetry, but only operators from the invariance algebra. This way we also will obtain inequivalent reductions.

This method (without identification as a distinct method and without pointing out classification of reductions) was actually used in the paper by Fushchych and Serov [10].

4. Other methods (including ad-hoc methods and guessing) that produce specific reductions or exact solutions, e.g., methods named after specific ansatzes being used, or methods similar to the method of differential constraints. Many famous exact solutions of various famous PDEs were found by just these methods. Ad-hoc methods include many methods that work only for some equations such as substitution of some prescribed ansatzes and looking for equations that they reduce. More solutions and reductions are suggested every day in multiple papers.

However, we need to determine whether these reductions (and thus solutions) are not equivalent to known ones; to find interesting solutions and reductions within equivalence classes; to systematise known solutions.

4. Example of reductions of the noninear d'Alembert equation

Here we consider conditions of reduction of the multidimensional d'Alembert equation (1) by means of the ansatz (2) with one new independent variable, $u_{\mu\mu} = u_{00} - u_{11} - \cdots - u_{nn}$.

The reduction conditions are a system of the d'Alembert and eikonal equations (sometimes they are also called Hamilton equations) [11, 12] and have the following form:

$$v_{\mu\mu} = R(v), \quad v_\mu v_\mu = r(v). \tag{5}$$

Sufficient conditions of reduction of the wave equation to an ODE and the general solution of the system (5) in the case of three spatial dimensions were found by Fushchych, Zhdanov, Revenko in [13].

It is evident that the d'Alembert–eikonal system (5) may be reduced by local transformations to the form

$$v_{\mu\mu} = f(v), \quad v_\mu v_\mu = \lambda, \quad \lambda = 0, 1. \tag{6}$$

Let us note that the case of $\lambda = 0$ leads only to a degenerated reduction equation $F(v) = 0$ that may give solutions of the original equation $u = \varphi(v)$, $\varphi(v)$ being an arbitrary function, that we will not consider here. Only the case of $\lambda = 1$ in the system (6) provides reductions of equation (1) to ODEs.

It is possible to generalise statements on compatibility of the d'Alembert–eikonal system (6) formulated in [13] for three space variables to an arbitrary number of space variables.

Statement. *For the system* (6) ($v = v(x_0, x_1 \ldots, x_n)$) *to be compatible it is necessary and sufficient that the function f have the following form:*

$$f = \frac{\lambda N}{v + C}, \quad N = 0, 1, \ldots, n.$$

Proof of this statement can be obtained as a generalisation of the proof adduced in [13].

5. General solution of the d'Alembert–eikonal system

We will consider in more detail solutions of the system

$$v_{\mu\mu} = \frac{N}{v + C}, \quad N = 0, 1, \ldots, n, \quad v_\mu v_\mu = 1. \tag{7}$$

These systems for different N are inequivalent with respect to local transformations.

Our results on compatibility of the reduction conditions actually mean that any multidimensional (in n space variables) nonlinear wave equation of the form (1) can be reduced by ansatzes (2) only to ODEs equivalent up to local transformations of dependent and independent variables to equations of the following form:

$$\varphi'' + \varphi' \frac{N}{v} = F(\varphi),$$

where N is an integer, $N = 0, 1, \ldots, n$.

We presented classification of the reduction conditions (5) up to the local transformations. For the case of n independent variables we receive $n + 1$ inequivalent systems of reduction conditions of the form (7). Thus, we can receive only $n + 1$ inequivalent reduced equations by means of the ansatz (2). However, classification of reductions requires also to classify possible new variables in the ansatz (2).

The solutions of the system (7) can be classified by their ranks. The rank of the solution is the rank of the $(n + 1) \times (n + 1)$ matrix $(v_{\mu\nu})$ of the second derivatives of the function v. It follows from the condition $v_\mu v_\mu = 1$ that $\det(v_{\mu\nu}) = 0$, so we cannot have solutions of the rank $n+1$. It is obvious that solutions of different ranks cannot be equivalent. Examples of solutions of the rank N are

$$v = \left(x_0^2 - \left(x_1^2 + x_2^2 + \cdots + x_N^2\right)\right)^{1/2}.$$

Examples of the zero-rank solutions are linear functions.

Statement. *Rank of solutions of the system (7) is equal to N. These solutions can be written in parametric form similarly to [13] as follows:*

$$v = x_0\left(1 + \tau_k^2 + B_l^2(\tau_k)\right)^{1/2} + x_k\tau_k + x_l B_l + \Phi(\tau_k),$$
$$l = N+1, \ldots, n, \quad \tau_k = \tau_k(x), \quad k = 1, \ldots, N,$$
$$x_k + x_0\left(\tau_k + B_l\frac{\partial B_l}{\partial \tau_k}\right)\left(1 + \tau_k^2 + B_l(\tau_k)\right)^{-1/2} + x_l\frac{\partial B_l}{\partial \tau_k} + \frac{\partial \Phi}{\partial \tau_k} = 0,$$

with functions Φ having certain special forms (examples are adduced below). $B_l(\tau_k)$ are arbitrary properly differentiable functions of the parameter functions τ_k.

Further classification of solutions of the reduction conditions will involve the substitution

$$z_k = v_a\left(1 + \tau_m^2\right)^{-1/2}, \quad p(z_k) = \left(1 + \tau_m^2\right)^{-1/2}\Phi(\tau_k).$$

Conditions for the functions Φ are transformed into conditions for p, and further classification of solutions of the d'Alembert–eikonal system (7) is done in accordance to the rank of the matrix $(p_{z_k z_m})$.

Examples of the form of the function $\Phi(\tau)$:

$$\Phi(\tau) = C_k\tau_k + C_0\left(1 + \tau_k^2\right)^{1/2},$$
$$\Phi(\tau) = B_m(\rho)\tau_m + \rho\tau_N - \left(1 + \tau_k^2\right)^{1/2}Q(\rho), \quad m = 1, \ldots, N-1,$$
$$\dot{B}_m(\rho)\tau_m + \tau_N - \left(1 + \tau_k^2\right)^{1/2}\dot{Q}(\rho) = 0,$$

where $B_m(\rho)$, $Q(\rho)$ are arbitrary properly differentiable functions of the parameter function ρ for the ranks of the matrix $(p_{z_k z_m})$ greater than 1.

Thus, we showed classification of solutions of the system (7) by their ranks and by the ranks of parameter functions $p(z_k)$ in the presentation of the general solution.

6. Conclusions and further work

The idea of classification of reductions was presented with the objective of giving hints and directions for further research – classification of reductions for equations of mathematical physics, and studying of different reduction methods with consideration of equivalence analysis of the resulting reductions.

Another direction of further work may be based upon the fact that ansatzes and methods used for reduction of the d'Alembert (n-dimensional wave) equation can be also used for arbitrary Poincaré-invariant equations. Arbitrary ansatzes of the form (2) that reduce the equation (1) will also reduce all Poincaré-invariant scalar equations. It can be easily proved based upon the fact that all Poincaré-invariant scalar equations can be represented as functions of differential invariants of the Poincaré algebra. If we substitute the ansatz (2) into any invariant from the basis of differential invariants of the form $u_\mu u_\nu u_{\mu\nu}$, $u_{\mu\nu}u_{\mu\nu}$ etc. we will receive the product of derivatives of the function $\varphi(v)$ and the same forms of differential invariant of the function v.

From the reduction conditions (7) on the function v we get

$$v_\mu v_\nu v_{\mu\nu} = 0, \quad v_{\mu\nu}v_{\mu\nu} = -\frac{N}{v^2}, \quad v_\mu v_\nu v_{\mu\alpha}v_{\nu\alpha} = 0, \quad v_{\mu\nu}v_{\mu\alpha}v_{\nu\alpha} = \frac{2N}{v^3}$$

and so on, thus, the original equation will be reduced to an ODE including only new variable v and derivatives of the new function $\varphi(v)$.

Another idea for further research – it may be interesting to look at connections between "old" view to equivalence of symmetry-induced solutions and classification of solutions of the reduction conditions. One such link is obvious – the reduced equations obtained from subalgebras that include the operator $P_0 + P_3$ would correspond to solutions of the reduction conditions

$$v_{\mu\mu} = 0, \quad v_\mu v_\mu = 1.$$

It will be also extremely interesting to study equivalence of reductions of PDEs that were obtained by various ad-hoc and other methods involving reduction of PDEs.

Acknowledgments
The author would like to thank all the organisers of the Workshop on Group Analysis of Differential Equations and Integrable Systems for the wonderful conference. I also would like to thank the University of Cyprus for support of my participation. The author also greatly appreciates comments of the reviewers that very considerably contributed to improvement of this paper.

References
[1] Yehorchenko I A 2009 Reduction of non-linear d'Alembert equations to two-dimensional equations *Proc. 4th Int. Workshop on Group Analysis of Differential Equations and Integrable Systems (Protaras, 2008)* (Nicosia: University of Cyprus) p 243
[2] Yehorchenko I A 2011 Reduction of multidimensional wave equations to two-dimensional equations: investigation of possible reduced equations *Proc. 5th Int. Workshop on Group Analysis of Differential Equations and Integrable Systems (Protaras, 2010)* (Nicosia: University of Cyprus) p 225
[3] Yehorchenko I A 2013 Hidden and conditional symmetry reductions of second-order PDEs and classification with respect to reductions *Proc. 6th Int. Workshop on Group Analysis of Differential Equations and Integrable Systems (Protaras, 2012)* (Nicosia: University of Cyprus) p 237
[4] Ovsyannikov L V 1982 *Group Analysis of Differential Equations* (New York: Academic Press)
[5] Olver P 1989 *Applications of Lie Groups to Differential Equations* (New York: Springer)
[6] Patera J, Sharp R T, Winternitz P and Zassenhaus H Subgroups of the Poincaré group and their invariants 1976 *J. Math. Phys.* **17** 977
[7] Clarkson P A and Kruskal M D 1989 New similarity reductions of the Boussinesq equation *J. Math. Phys.* **30** 2201
[8] Yehorchenko I A 2014 Ansatzes and exact solutions for nonlinear Schrödinger equations *Preprint* arXiv:1412.1889
[9] Zhdanov R Z, Tsyfra I M and Popovych R O 1999 A precise definition of reduction of partial differential equations *J. Math. Anal. Appl.* **238** 101
[10] Fushchych W I and Serov N I 1983 The symmetry and some exact solutions of the nonlinear many-dimensional Liouville, d'Alembert and eikonal equations *J. Phys.* A **22** 3645
[11] Cieciura G, Grundland A 1984 A certain class of solutions of the nonlinear wave equation *J. Math. Phys.* **25** 3460
[12] Fushchych W I, Zhdanov R Z and Yegorchenko I A 1991 On the reduction of the nonlinear multi-dimensional wave equations and compatibility of the d'Alembert–Hamilton system *J. Math. Anal. Appl.* **160** 352
[13] Fushchych W I, Zhdanov R Z and Revenko I V 1990 Compatibility and solutions of nonlinear d'Alembert and Hamilton equations (*Preprint* no. 90-39, Kiev: Institute of Mathematics) (in Russian)

Quantum integrable systems in three-dimensional magnetic fields: the Cartesian case

Alexander Zhalij

Institute of Mathematics of NAS of Ukraine, 3 Tereshchenkivska Str., 01601 Kyiv, Ukraine

E-mail: zhaliy@imath.kiev.ua

Abstract. In this paper we construct integrable three-dimensional quantum-mechanical systems with magnetic fields, admitting pairs of commuting second-order integrals of motion. The case of Cartesian coordinates is considered. Most of the systems obtained are new and not related to the separation of variables in the corresponding Schrödinger equation.

1. Introduction

We consider the stationary Schrödinger equation for particle moving in external electromagnetic field in three-dimensional Euclidean space

$$H\psi = E\psi, \quad H = \frac{1}{2}\vec{p}^{\,2} + V(x,y,z) + A_i(x,y,z)p_i + p_iA_i(x,y,z), \tag{1}$$

where $V(x,y,z)$ and $\vec{A} = (A_1(x,y,z), A_2(x,y,z), A_3(x,y,z))$ are scalar and vector potentials of electromagnetic field, respectively. Here and below we use the notation $\vec{p} = -ih\vec{\nabla}$ and the summation from 1 to 3 over the repeated indices is understood.

By analogy with classical Hamiltonian mechanics this system is called integrable if there exists a pair of quantum-mechanical operators P and Q which commute with each other as well as with the Hamiltonian H, i.e., the following relations hold

$$[H,Q] = [H,P] = [P,Q] = 0.$$

Moreover, all three operators H, Q and P are algebraically independent, i.e., any of them cannot be represented as polynomial of two others [1].

In this paper we restrict ourselves by the case of Q and P being quadratic polynomials of $\vec{p}$

$$\begin{aligned} Q &= \alpha_{ik}(x,y,z)p_ip_k + f_i(x,y,z)p_i + \gamma_1(x,y,z), \\ P &= \beta_{ik}(x,y,z)p_ip_k + g_i(x,y,z)p_i + \gamma_2(x,y,z). \end{aligned} \tag{2}$$

In classical mechanics integrable systems are interesting since their motion in a phase space is more ordered, namely, it is restricted to a torus. In quantum mechanics integrability of an n-dimensional quantum-mechanical system, i.e., existing of n quantum integrals of motion (operators, which commute with each other as well as with the operator of equation (1)), simplifies the problem of determining of energy spectrum and wave functions even when integrability does not lead to the variables separation, but only to the so-called "quasi-separation of variables" [2].

Therefore, the classification problem of all potentials V and $\vec{A}$ of electromagnetic field, for which quantum-mechanical problem is integrable in the sense described above, is of current importance.

In the three-dimensional case of scalar potential, when there is no any magnetic field, this problem was solved as early as in 1967 by Yakov A. Smorodinsky with coauthors [3, 4]. They have proved that if there exist two quantum integrals of motion of first or second order with respect to $\vec{p}$ (in the sense described above) then there exists a possibility of variable separation in the corresponding Schrödinger equation [4]. The inverse statement is also true [3]. Thus, eleven classes of integrable potentials V obtained by them coincide with the results of classic paper of Eisenhart [5], where he as early as in 1948 has described all scalar potentials for which corresponding Schrödinger equation (or Hamilton–Jacobi equation in classical mechanics) admits variable separation at least in one of eleven coordinate systems.

The next step was done in 1972 by V.N. Shapovalov with coauthors [6]. They have obtained complete classification list of vector-potentials with nonzero magnetic field, for which corresponding Schrödinger equation (1) admits variable separation at least in one of eleven coordinate systems, and list of corresponding pairs of operators which commute with each other and with operator of the equation.

Posterior results in this direction are connected with works of P. Winternitz and his coauthors [7–9]. For two-dimensional case they have discovered that existing of second-order integrals of motion in magnetic field does not guarantee possibility of variable separation. Nevertheless, even in this case, integrals of motion are classified on equivalence classes under action of Euclidian group and second-order terms with respect to p_i in these integrals have the same form as in the case of pure scalar potential. It was also shown that in magnetic field quantum case [9] does not obviously coincide with classical one [7, 8], namely, constructed vector-potentials can depend on Plank constant $\hbar$ in a nontrivial way.

In this paper we make the next step in classification of potentials of electromagnetic field V and $\vec{A}$ in three-dimensional Euclidian space, for which corresponding quantum-mechanical system described by equation (1) is integrable in the sense explained above, i.e., for which there exists a pair of operators (2) that commute with each other as well as with the operator of the equation. As a result, we obtain a number of vector-potentials for which corresponding Schrödinger equation is integrable but does not admit variable separation. Therefore, these potentials did not appear in the Shapovalov's classification [6] and are new.

2. Classification scheme

At first we consider single operator Q of the form (2) which commutes with the operator H of Schrödinger equation (1). The commutator $[Q, H]$ contains terms of zero, first, second and third orders with respect to p_i, which coefficients have to vanish. Coefficients of the third power of p_i give the following system:

$$\frac{\partial \alpha_{ik}}{\partial x_m} p_m p_i p_k = 0.$$

After solving this system we obtain that the operator Q can be presented as symmetric bilinear polynomial of the infinitesimal generators of a group of motions of three-dimensional Euclidean space E_3, i.e., symmetry group of the three-dimensional Schrödinger equation for free particle (the Helmholtz equation):

$$Q = a_{ik} M_i M_k + b_{ik}(p_i M_k + M_k p_i) + c_{ij} p_i p_k + f_i(x, y, z) p_i + \gamma_1(x, y, z), \tag{3}$$

where a_{ik}, b_{ik} and c_{ij} are constants, M_i is operator of rotation, namely, $M_i = \varepsilon_{ikl} x_k p_l$, where ε_{ikl} is completely antisymmetric tensor.

Comparing constructed form (3) of operator with similar form of operator in the case of pure scalar potential [4], we conclude that second-order terms with respect to p_i remain the same after appearing of nonzero magnetic field (as well as in two-dimensional case [9]).

Therefore, analogously to [4], a pair of commuting operators P and Q having the form (3) can be reduced by rotations and translations of coordinate system and by the transformations

$$Q' = \mu P + \nu Q + \lambda H$$

to one of eleven classes corresponding to eleven classical coordinate systems which provide separation of variable in the three-dimensional Schrödinger equation for free particle.

That is why in our case of nonzero magnetic field we obtain that the second-order terms with respect to p_i in these eleven commuting pairs of the operators P and Q are of the same form as ones presented in [4]. However, in contrast to the pure scalar case, some or all of the coefficients of the first power of p_i are nonzero functions. The forms of these functions determine the form of magnetic field (in scalar case they equal zero).

In this paper we completely solve the simplest "Cartesian" case, i.e. the case

$$Q = p_1^2 + \vec{f}(x,y,z)\vec{p} + \gamma_1(x,y,z), \tag{4}$$

$$P = p_2^2 + \vec{g}(x,y,z)\vec{p} + \gamma_2(x,y,z). \tag{5}$$

Splitting the equations $[H,Q] = [H,P] = [P,Q] = 0$ with respect to different powers of p_i, we obtain an overdetermining system of PDEs on unknown functions f_1, f_2, f_3, g_1, g_2, g_3, γ_1, γ_2 and V, A_1, A_2, A_3.

The coefficients of the highest powers of p_i give the following system

$$\begin{gathered}
f_2 = f_2(x), \quad f_3 = f_3(x), \quad g_1 = g_1(y), \quad g_3 = g_3(y), \quad f_{1y} = g_{2x}, \\
f_2'(x) + f_{1y} = 4A_{2x}, \quad g_3'(y) + g_{2z} = 4A_{3y}, \quad 4A_{1x} = f_{1x}, \\
f_3'(x) + f_{1z} = 4A_{3x}, \quad g_1'(y) + g_{2x} = 4A_{1y}, \quad 4A_{2y} = g_{2y}.
\end{gathered}$$

Its general solution for $\vec{A}$ is

$$\begin{aligned}
4A_1 &= s_x + k_{1x} + g_1(y) + r_1(z), \\
4A_2 &= s_y + k_{2y} + f_2(x) + r_2(z), \\
4A_3 &= s_z + k_{1z} + k_{2z} + f_3(x) + g_3(y) + r_3'(z),
\end{aligned}$$

where $s = s(x,y,z)$, $k_1 = k_1(x,z)$ and $k_2 = k_2(y,z)$.

The gauge transformation

$$\vec{A} \to \vec{A} + \vec{\nabla}F, \quad F = s(x,y,z) + k_1(x,z) + k_2(y,z) + r_3(z),$$

simplifies the obtained expression for $\vec{A}$ in the following way

$$A_1 = \frac{1}{4}(g_1(y) + r_1(z)), \quad A_2 = \frac{1}{4}(f_2(x) + r_2(z)), \quad A_3 = \frac{1}{4}(f_3(x) + g_3(y)).$$

With this expression for $\vec{A}$ in hand, we obtain from the coefficients of the lowest powers of p_i the system of ODEs on the functions $g_1(y)$, $r_1(z)$ $f_2(x)$, $r_2(z)$, $f_3(x)$ and $g_3(y)$:

$$\begin{aligned}
f_2(x)g_3'(y) &= g_1(y)f_3'(x), \\
r_1(z)f_2'(x) &= f_3(x)r_2'(z), \\
r_2(z)g_1'(y) &= g_3(y)r_1'(z).
\end{aligned} \tag{6}$$

It is obvious that the Schrödinger equation with vector-potential (1) is invariant with respect to permutations of A_1, A_2 and A_3, which are done simultaneously with permutations of the variables x_1, x_2 and x_3. Equations (6) are invariant with respect to permutations of the functions $g_1(y)$, $r_1(z)$, $f_2(x)$, $r_2(z)$, $f_3(x)$ and $g_3(y)$. These equivalence transformations can be represented in such a way

$$\begin{pmatrix} f_2 & g_3 & r_1 \\ f_3 & g_1 & r_2 \\ A_1 & A_2 & A_3 \\ x & y & z \end{pmatrix} \sim \begin{pmatrix} r_1 & f_2 & g_3 \\ r_2 & f_3 & g_1 \\ A_3 & A_1 & A_2 \\ z & x & y \end{pmatrix} \sim \begin{pmatrix} g_3 & r_1 & f_2 \\ g_1 & r_2 & f_3 \\ A_2 & A_3 & A_1 \\ y & z & x \end{pmatrix} \sim$$

$$\begin{pmatrix} f_3 & r_2 & g_1 \\ f_2 & r_1 & g_3 \\ A_1 & A_3 & A_2 \\ x & z & y \end{pmatrix} \sim \begin{pmatrix} g_1 & f_3 & r_2 \\ g_3 & f_2 & r_1 \\ A_2 & A_1 & A_3 \\ y & x & z \end{pmatrix} \sim \begin{pmatrix} r_2 & g_1 & f_3 \\ r_1 & g_3 & f_2 \\ A_3 & A_2 & A_1 \\ z & y & x \end{pmatrix}.$$

Usage of these equivalence transformations allows us to describe exhaustively all solutions of equations (6). This gives the complete description of all possible forms of vector-potentials $\vec{A}$. The residual coefficients of powers of p_i serve for determining of scalar component V of vector-potential and put additional constraints on the functions $g_1(y)$, $r_1(z)$, $f_2(x)$, $r_2(z)$, $f_3(x)$ and $g_3(y)$.

3. List of results

Below we adduce the final results of our calculations. Here and below $\vec{\Omega}$ denotes magnetic field, namely, $\vec{\Omega} = \text{rot}\,\vec{A}$.

Case 1.

$$\vec{A} = 0, \quad \vec{\Omega} = 0, \quad V = u_1(x) + u_2(y) + u_3(z), \quad Q = p_1^2 + 2u_1(x), \quad P = p_2^2 + 2u_2(y).$$

This case corresponds to zero magnetic field and is contained in the Eisenhart's classification [5]. According to his results all scalar potentials for which corresponding Schrödinger equations admit variable separation in Cartesian coordinates are exhausted by ones having the form

$$V = u_1(x) + u_2(y) + u_3(z).$$

Case 2.

$$\vec{A} = \begin{pmatrix} v_1(z) \\ v_2(z) \\ 0 \end{pmatrix}, \quad \vec{\Omega} = \begin{pmatrix} -v_2'(z) \\ v_1'(z) \\ 0 \end{pmatrix}, \quad V = v_3(z), \quad Q = p_1^2, \quad P = p_2^2.$$

Case 3.

$$\vec{A} = \begin{pmatrix} 0 \\ 0 \\ f(x) + g(y) \end{pmatrix}, \quad \vec{\Omega} = \begin{pmatrix} g'(y) \\ -f'(x) \\ 0 \end{pmatrix}, \quad V = u_1(x) + u_2(y),$$

$$Q = p_1^2 + 4f(x)p_3 + 2u_1(x), \quad P = p_2^2 + 4g(y)p_3 + 2u_2(y).$$

The cases 2 and 3 where obtained by Shapovalov et al. [6]. According to their results these two cases exhaust all vector-potentials with nonzero magnetic field for which corresponding Schrödinger equations admit variable separation in Cartesian coordinates.

Next three cases are not connected with variable separation and, therefore, these potentials did not appear in the Shapovalov's classification [6] and they are new.

Case 4.

$$\vec{A} = \begin{pmatrix} g'(y) \\ f'(x) \\ 0 \end{pmatrix}, \quad \vec{\Omega} = \begin{pmatrix} 0 \\ 0 \\ f''(x) - g''(y) \end{pmatrix},$$

$$V = -(C_3 f(x) + C_3 g(y) + 2C_2 f^2(x) + 2C_1 g^2(y) + r(z) + 4g(y)f''(x) + 4f(x)g''(y)),$$

$$Q = p_1^2 + 4f'(x)p_2 - 2(4g(y)f''(x) + 2C_2 f(x)^2 + C_3 f(x)),$$

$$P = p_2^2 + 4g'(y)p_1 - 2(4f(x)g''(y) + 2C_1 g(y)^2 + C_3 g(y)),$$

where the functions $f(x)$ and $g(y)$ are solutions of the ODEs

$$f''(x) = Cf^2(x) + C_1 f(x) + C_4, \tag{7}$$

$$g''(y) = Cg^2(y) + C_2 g(y) + C_5. \tag{8}$$

If $C = 0$ these are linear second-order ODEs. The case $C \neq 0$ is more interesting. If additionally $C_1 \neq 0$ (resp. $C_2 \neq 0$), then solution of equation (7) (resp. (8)) is first Painlevé transcendent. If $C_1 = 0$ (resp. $C_2 = 0$), then (7) (resp. (8)) is the Weierstrass equation which solutions are expressed either via Weierstrass functions or via elementary ones depending on values of the parameter C_4 (resp. C_5) and integration constants. See, e.g. [10] for details.

Case 5.

$$\vec{A} = \begin{pmatrix} g'(y) \\ f'(x) \\ Cf(x) + Cg(y) \end{pmatrix}, \quad \vec{\Omega} = \begin{pmatrix} Cg'(y) \\ -Cf'(x) \\ f''(x) - g''(y) \end{pmatrix},$$

$$V = -(C_3 f(x) + C_3 g(y) + 2C_2 f^2(x) + 2C_1 g^2(y) + 4g(y)f''(x) + 4f(x)g''(y)),$$

$$Q = p_1^2 + 4(f'(x)p_2 + Cf(x)p_3) - 2(4g(y)f''(x) + 2C_2 f(x)^2 + C_3 f(x)),$$

$$P = p_2^2 + 4(g'(y)p_1 + Cg(y)p_3) - 2(4f(x)g''(y) + 2C_1 g(y)^2 + C_3 g(y)),$$

where the functions $f(x)$ and $g(y)$ are solutions of the ordinary differential equations

$$f''(x) = C_6 f^2(x) + C_1 f(x) + C_4,$$

$$g''(y) = C_6 g^2(y) + C_2 g(y) + C_5,$$

which are integrated analogously to the previous case.

Case 6.

$$\vec{A} = \frac{1}{4}\begin{pmatrix} w_1'(y) + v_1'(z) \\ u_2'(x) + v_2'(z) \\ u_3'(x) + w_3'(y) \end{pmatrix}, \quad \vec{\Omega} = \frac{1}{4}\begin{pmatrix} w_3''(y) - v_2''(z) \\ v_1''(z) - u_3''(x) \\ u_2''(x) - w_1''(y) \end{pmatrix},$$

$$V = -\frac{1}{4}(u_1(x) + w_2(y) + v_3(z) + w_1(y)u_2''(x) + v_1(z)u_3''(x) + u_2(x)w_1''(y) + v_2(z)w_3''(y) + u_3(x)v_1''(z) + w_3(y)v_2''(z)),$$

$$Q = p_1^2 + u_2'(x)p_2 + u_3'(x)p_3 - \frac{1}{2}(w_1(y)u_2''(x) + v_1(z)u_3''(x) + u_1(x)),$$

$$P = p_2^2 + w_1'(y)p_1 + w_3'(y)p_3 - \frac{1}{2}(u_2(x)w_1''(y) + v_2(z)w_3''(y) + w_2(y)),$$

Journal of Physics: Conference Series **621** (2015) 012019 doi:10.1088/1742-6596/621/1/012019

where the functions $u_2(x)$, $u_3(x)$, $w_1(y)$, $w_3(y)$, $v_1(z)$, $v_2(z)$, $u_1(x)$, $w_2(y)$ and $v_3(z)$ are defined in a special way and described by the following four cases:

Case 6.1.

$$u_2(x) = a_3(r_1\cosh(a_1x) + k_1\sinh(a_1x)), \quad u_3(x) = a_2(r_1\sinh(a_1x) + k_1\cosh(a_1x)),$$
$$w_1(y) = a_3(r_2\cosh(a_2y) + k_2\sinh(a_2y)), \quad w_3(y) = a_1(r_2\sinh(a_2y) + k_2\cosh(a_2y)),$$
$$v_1(z) = a_2(r_3\cosh(a_3z) + k_3\sinh(a_3z)), \quad v_2(z) = a_1(r_3\sinh(a_3z) + k_3\cosh(a_3z)),$$
$$u_1(x) = \frac{a_2^2a_3^2}{4}\left((r_1^2+k_1^2)\cosh(2a_1x) + 2r_1k_1\sinh(2a_1x)\right) + C(r_1\cosh(a_1x) + k_1\sinh(a_1x)),$$
$$w_2(y) = \frac{a_1^2a_3^2}{4}\left((r_2^2+k_2^2)\cosh(2a_2y) + 2r_2k_2\sinh(2a_2y)\right) + C(r_2\cosh(a_2y) + k_2\sinh(a_2y)),$$
$$v_3(z) = \frac{a_1^2a_2^2}{4}\left((r_3^2+k_3^2)\cosh(2a_3z) + 2r_3k_3\sinh(2a_3z)\right) + C_1(r_3\cosh(a_3z) + k_3\sinh(a_3z)),$$

with 5 possible subcases:

a) $C = 0, \quad C_1 = 0;$
b) $r_1 = k_1, \quad r_2 = k_2, \quad r_3 = k_3, \quad C_1 = C;$
c) $r_1 = k_1, \quad r_2 = -k_2, \quad r_3 = -k_3, \quad C_1 = C;$
d) $r_1 = -k_1, \quad r_2 = k_2, \quad r_3 = -k_3, \quad C_1 = -C;$
e) $r_1 = -k_1, \quad r_2 = -k_2, \quad r_3 = k_3, \quad C_1 = -C.$

Case 6.2.

$$u_2(x) = a_3(r_1\sin(a_1x) - k_1\cos(a_1x)), \quad u_3(x) = a_2(r_1\cos(a_1x) + k_1\sin(a_1x)),$$
$$w_1(y) = a_3(r_2\sin(a_2y) - k_2\cos(a_2y)), \quad w_3(y) = a_1(r_2\cos(a_2y) + k_2\sin(a_2y)),$$
$$v_1(z) = a_2(r_3\cosh(a_3z) + k_3\sinh(a_3z)), \quad v_2(z) = a_1(r_3\sinh(a_3z) + k_3\cosh(a_3z)),$$
$$u_1(x) = \frac{a_2^2a_3^2}{4}\left((r_1^2-k_1^2)\cos(2a_1x) + 2r_1k_1\sin(2a_1x)\right) + C(r_1\sin(a_1x) - k_1\cos(a_1x)),$$
$$w_2(y) = \frac{a_1^2a_3^2}{4}\left((r_2^2-k_2^2)\cos(2a_2y) + 2r_2k_2\sin(2a_2y)\right) + C(r_2\sin(a_2y) - k_2\cos(a_2y)),$$
$$v_3(z) = -\frac{a_1^2a_2^2}{4}\left((r_3^2+k_3^2)\cosh(2a_3z) + 2r_3k_3\sinh(2a_3z)\right) + C_1(r_3\cosh(a_3z) + k_3\sinh(a_3z)),$$

with 5 possible subcases:

a) $C = 0, \quad C_1 = 0;$
b) $r_1 = ik_1, \quad r_2 = -ik_2, \quad r_3 = -k_3, \quad C_1 = iC;$
c) $r_1 = ik_1, \quad r_2 = ik_2, \quad r_3 = k_3, \quad C_1 = iC;$
d) $r_1 = -ik_1, \quad r_2 = -ik_2, \quad r_3 = k_3, \quad C_1 = -iC;$
e) $r_1 = -ik_1, \quad r_2 = ik_2, \quad r_3 = -k_3, \quad C_1 = -iC.$

Case 6.3.

$$u_2(x) = a_3(r_1\cos(a_1x) + k_1\sin(a_1x)), \quad u_3(x) = ia_2(r_1\sin(a_1x) - k_1\cos(a_1x)),$$
$$w_1(y) = a_3(r_2\cos(a_2y) + k_2\sin(a_2y)), \quad w_3(y) = ia_1(r_2\sin(a_2y) - k_2\cos(a_2y)),$$
$$v_1(z) = a_2(r_3\cos(a_3z) + k_3\sin(a_3z)), \quad v_2(z) = ia_1(r_3\sin(a_3z) - k_3\cos(a_3z)),$$
$$u_1(x) = -\frac{a_2^2a_3^2}{4}\left((r_1^2 - k_1^2)\cos(2a_1x) + 2r_1k_1\sin(2a_1x)\right) + C(r_1\cos(a_1x) + k_1\sin(a_1x)),$$
$$w_2(y) = -\frac{a_1^2a_3^2}{4}\left((r_2^2 - k_2^2)\cos(2a_2y) + 2r_2k_2\sin(2a_2y)\right) + C(r_2\cos(a_2y) + k_2\sin(a_2y)),$$
$$v_3(z) = -\frac{a_1^2a_2^2}{4}\left((r_3^2 - k_3^2)\cos(2a_3z) + 2r_3k_3\sin(2a_3z)\right) + C_1(r_3\sin(a_3z) - k_3\cos(a_3z)),$$

with 5 possible subcases:

a) $C = 0, \quad C_1 = 0;$
b) $r_1 = ik_1, \quad r_2 = -ik_2, \quad r_3 = ik_3, \quad C_1 = iC;$
c) $r_1 = -ik_1, \quad r_2 = -ik_2, \quad r_3 = -ik_3, \quad C_1 = iC;$
d) $r_1 = -ik_1, \quad r_2 = ik_2, \quad r_3 = ik_3, \quad C_1 = -iC;$
e) $r_1 = ik_1, \quad r_2 = ik_2, \quad r_3 = -ik_3, \quad C_1 = -iC.$

Case 6.4.

$$u_2(x) = a_3(r_1\cosh(a_1x) + k_1\sinh(a_1x)), \quad u_3(x) = -ia_2(r_1\sinh(a_1x) + k_1\cosh(a_1x)),$$
$$w_1(y) = a_3(r_2\cosh(a_2y) + k_2\sinh(a_2y)), \quad w_3(y) = -ia_1(r_2\sinh(a_2y) + k_2\cosh(a_2y)),$$
$$v_1(z) = a_2(r_3\cos(a_3z) + k_3\sin(a_3z)), \quad v_2(z) = ia_1(r_3\sin(a_3z) - k_3\cos(a_3z)),$$
$$u_1(x) = \frac{a_2^2a_3^2}{4}\left((r_1^2 + k_1^2)\cosh(2a_1x) + 2r_1k_1\sinh(2a_1x)\right) + C(r_1\cosh(a_1x) + k_1\sinh(a_1x)),$$
$$w_2(y) = \frac{a_1^2a_3^2}{4}\left((r_2^2 + k_2^2)\cosh(2a_2y) + 2r_2k_2\sinh(2a_2y)\right) + C(r_2\cosh(a_2y) + k_2\sinh(a_2y)),$$
$$v_3(z) = \frac{a_1^2a_2^2}{4}\left((r_3^2 - k_3^2)\cos(2a_3z) + 2r_3k_3\sin(2a_3z)\right) + C_1(r_3\sin(a_3z) - k_3\cos(a_3z)),$$

with 5 possible subcases:

a) $C = 0, \quad C_1 = 0;$
b) $r_1 = -k_1, \quad r_2 = -k_2, \quad r_3 = -ik_3, \quad C_1 = C;$
c) $r_1 = k_1, \quad r_2 = -k_2, \quad r_3 = ik_3, \quad C_1 = C;$
d) $r_1 = k_1, \quad r_2 = k_2, \quad r_3 = -ik_3, \quad C_1 = -C;$
e) $r_1 = -k_1, \quad r_2 = k_2, \quad r_3 = ik_3, \quad C_1 = -C.$

Therefore, we obtain a number of new vector-potentials, for which the corresponding Schrödinger equation (1) is integrable in the sense described above.

Acknowledgments
The author is grateful to Prof. P. Winternitz for useful discussions. This research was partially supported by a grant from NATO. The author would also like to acknowledge partial support of his participation in GADEIS VII provided by the University of Cyprus.

References

[1] Hietarinta J 1998 Pure quantum integrability *Phys. Lett.* A **246** 97 (*Preprint* arXiv:solv-int/9708010)
[2] Charest F, Hudon C and Winternitz P 2007 Quasiseparation of variables in the Schrödinger equation with a magnetic field *J. Math. Phys.* **48** 012105 (*Preprint* arXiv:math-ph/0502046)
[3] Smorodinskii Ya A and Tugov I I 1966 On complete sets of observables *J. Exp. Theor. Phys.* **50** 653 [in Russian]; English transl. in *Soviet Physics JETP* **23** 434
[4] Makarov A A, Smorodinsky J A, Valiev Kh and Winternitz P 1967 A systematic search for nonrelativistic systems with dynamical symmetries *Nuovo Cimento* A **52** 1061
[5] Eisenhart L P 1948 Enumeration of potentials for which one-particle Schrödinger equations are separable *Phys. Rev.* **74** 87
[6] Shapovalov V N, Bagrov V G and Meshkov A G 1972 Separation of variables in the stationary Schrödinger equation *Izv. Vyssh. Uchebn. Zaved. Fizika* **8** 45 [in Russian]; English transl. in *Russ. Phys. J.* **15** 1115
[7] Dorizzi B, Grammaticos B, Ramani A and Winternitz P 1985 Integrable Hamiltonian systems with velocity-dependent potentials *J. Math. Phys.* **26** 3070
[8] McSween E and Winternitz P 2000 Integrable and superintegrable Hamiltonian systems in magnetic fields *J. Math. Phys.* **41** 2957
[9] Berube J and Winternitz P 2004 Integrable and superintegrable quantum systems in a magnetic field *J. Math. Phys.* **45** 1959 (*Preprint* arXiv:math-ph/0311051)
[10] Kamke E 1977 *Differentialgleichungen Lösungsmethoden und Lösungen* (Stuttgart: Teubner)

AUTHOR INDEX

Institute of Physics
Dirac House, Temple Back
Bristol BS1 6BE UK

CURRAN ASSOCIATES INC
proceedings.com